D. S. L. CARDWELL

From Watt to Clausius

THE RISE OF THERMODYNAMICS IN THE EARLY INDUSTRIAL AGE

CORNELL UNIVERSITY PRESS
ITHACA, NEW YORK

International Standard Book Number 0-8014-0678-1
Library of Congress Catalog Card Number 72-163129

PRINTED IN GREAT BRITAIN

Contents

LIST OF ILLUSTRATIONS   vii

PREFACE   xi

Chapter One   The Origins of the Science of Heat   1

'Fire' in Seventeenth-Century Philosophy and Arts;
The Fire-Engine; Physics and Chemistry: Some Basic
Concepts

Chapter Two   The Eighteenth-Century Contribution   32

The Spread of the Fire-Engine; The Scottish School;
The Establishment of the Science of Heat

Chapter Three   The Rival Technologies: Steam and Water   67

Power Technology at the End of the Eighteenth
Century; Steam-Power and Water-Power

Chapter Four   The Beginnings of a New Cosmology   89

A Retrospect; Geophysics and Meteorology; Rumford; the False Dawn of the Kinetic Theory; Radiant
Heat

Chapter Five   The Science of Heat becomes Autonomous   121

Theories of Heat and the Physics of Gases; The Laws
of Expansion

Chapter Six   The Developments of the Power Technologies   150

The Technology of Heat; Power Technology in the
Early Nineteenth Century; Later Studies of the Heat-
Engine; Water-Power Technology

Chapter Seven   The Convergence of Technology and Science   186

Sadi Carnot and the Beginnings of Thermodynamics;
Carnot and the Physics of Gases; After Carnot; The
Consequences: an Outline of the Later History of
Thermodynamics; Emile Clapeyron; J. R. Mayer
and J. P. Joule

Chapter Eight   The New Science   239
Kelvin, Rankine and Clausius; The Concept of
Entropy; Some Loose Ends; Science and Modern
Culture

NOTES AND REFERENCES   297
INDEX   330

List of Illustrations

Between pages 64 and 65

Plate I   Halley's map of the wind systems of the world.

Plate II   The mill at Rossett, near Chester, with a simple undershot waterwheel.

Plate III   'Fairbottom Bobs'. The Newcomen engine which was used to pump out a mine at Fairbottom, between Oldham and Ashton-under-Lyne.

Plate IV   An early eighteenth-century machine intended to develop the power of falls of water too high for a single water-wheel.

Plate V   The column-of-water engine erected by William Westgarth at the Allenheads lead mine in Northumberland.

Plate VI   Design for an early column-of-water engine by Gensanne.

Between pages 128 and 129

Plate VII   Masonry arch dam on the river Bollin built by Peter Ewart to provide power for Quarry Bank Mill.

Plate VIII   Mill at Congleton, Cheshire, showing weir and breast-wheel.

Plate IX   Mills at Congleton, Cheshire, showing weirs for breast-wheels.

Plate X   Water-turbine installed about seventy years ago in the old water-wheel house at Quarry Bank Mill.

Plate XI   Diagram of John Smeaton's model for the experimental study of the efficiencies of water-wheels.

Plate XII   Column-of-water engine at Schemnitz in about 1770.

Plate XIII   A water-powered cotton mill of the industrial revolution.

Between pages 192 and 193

Plate XIV   Early nineteenth-century industrial water-wheels designed to achieve the maximum possible efficiency.

Plate XV   Diagram of a double-acting column-of-water engine.

Plate XVI   A German double-acting column-of-water engine fitted with Watt's parallel motion.

Plate XVII   George Manwaring's column-of-water engine of 1812.

Plate XVIII   Masterman's rotative engine. A development of the engine invented by Amontons.

Plate XIX   Hornblower's engine.

Plate XX   Woolf's engine.

Between pages 256 and 257

Plate XXI   James Prescott Joule as a young man.

Plate XXII   Peter Ewart.

Plate XXIII   Apparatus used by Joule in his demonstration that when a gas expands without doing work its 'internal energy' remains unchanged.

Plate XXIV   Joule's idea of the transformation of mechanical energy into heat energy. From one of his laboratory notebooks.

Plate XXV   A rotative engine installed in a cotton mill near Darwen, Lancs., *c* 1880.

Plate XXVI   A steam engine, built by Galloway's, which was one of the last to be installed in a cotton mill.

FOR OLIVE

Preface

The sources of this book are twofold. In the first place I have long been interested in the relationships, historical and contemporary, between science and technology, or the industrial arts. The rise of the science of thermodynamics, and its dependence on the development of the technology of heat power, is a fascinating instance of this sort of relationship. In the second place there is the more general problem of identifying the factors that govern scientific advance in any age, and the question that this poses concerning the scope of the history of science.

These two problems overlap when we consider the period during which thermodynamics was being founded. According to what is still perhaps the predominant view among historians, science is a philosophical activity, unconcerned with the practical arts, and advancing from generation to generation along the lines prescribed by each separate 'discipline' and by virtue of the practice of 'scientific method'. Thus it is assumed that chemistry is and always has been advanced by chemists, physics by physicists and so on. Now this is, I think, a partial truth which can be dangerously misleading when applied to the history of science. In fact it can amount to a mere projection into the past of our present arrangements for, and ideas about, science.

It is in one way unfortunate that the expression 'the scientific revolution', which is now often used to sum up the extraordinary progress made in the sixteenth century between the times of Galileo and Newton, should have been given so much *general* currency. It seems to imply that science was established in the seventeenth century, and that everything that has followed has amounted to an almost automatic, almost inevitable progress, thanks to the application of scientific method within the framework laid down by Newton and his contemporaries (regarded, of course, as mathematicians, chemists, 'physicists' etc.).

It is precisely this attitude that I wish to challenge. The progress of science has, I believe, at all times involved intense creative effort on the part of a small number of extremely able men, supported, or at least not hindered in their efforts, by the societies of western Europe. It is hardly an exaggeration to say that in almost every generation science has had to be recast and re-interpreted, sometimes very fundamentally indeed. Thus even the least historically minded know something of the changes wrought in physics by relativity and the quantum theory, which were initiated at the beginning of this century. It has been said that these were the first radical breaks in Newtonian science since its establishment in the seventeenth century. But this was not the case. Relativity and quantum theory were major transpositions in a nineteenth-century physics that had long outgrown Newtonianism. The major topics of nineteenth-century physics were energy, thermo-dynamics and field theory. To the extent that these can be traced back to the sixteenth century, Huygens and Leibnitz, and not Newton, were among the ancestors of the first; Des-cartes, Huygens, and perhaps Hooke, of the last. I shall argue that the second of these three, thermodynamics, originated substantially (but not of course wholly) in the power tech-nologies of the eighteenth and early nineteenth centuries—and these had virtually nothing to do with Newtonianism.

Now this is in no sense an attack on Newton: his status is inviolate. But certain Newtonians of the late eighteenth century and to some extent the nineteenth century are not beyond reproach. It seems to be a law of human nature, easily verifiable by reference to many examples, that very great men like Newton have faithful disciples to whom the lightest pronounce-ments of the master are established certainties, and who violently oppose anyone who ventures to question them. These people apart, all men of science habitually paid tribute to Newton as the man who, more than anyone else, had established the prestige and authority of natural science. However, the cumulative effect of these attitudes, rooted in the past, does not make it easy for the historian to show that scientific ideas change in character, often fundamentally, from generation to genera-tion; and that they are not merely deductions from one great system established in the sixteenth century.

In fact the sources of science are many and diverse. In any age scientific progress depends on a wide variety of ideas and data, which may come from any source—astronomical and geographical discovery, meteorology, industry and technology, philosophical speculations—as well as from the laboratories and studies of conventional scientists. We may, with our highly organised science, tend to forget this necessary diversity and to expect that all advances must be made along approved channels: like welfare and charity, matters for the appropriate departmental authority. If this is the case then it is indeed a pity, for in the last resort a science whose form and content is decided by organisations and not by men can hardly be a prosperous one. But this speculation has taken us away from the main discussion, which is the relationship between technology and an important branch of science, thermodynamics, during its all-important formative stage.

The student is usually introduced to the concepts of thermo-dynamics—the Carnot cycle, the principle of reversibility, the idea of entropy—in a way which does some violence to credi-bility. After having been taught science in terms of the highly abstract systems of mechanics and optics, and a science of heat concerned with the exact measurements of such abstract quantities as temperature and specific heat, he is suddenly asked to accept a mysterious and entirely unrealistic engine of an almost completely impracticable nature as a fundamental concept of science. The introduction of the idea of entropy is almost equally abrupt. How or why such a system of thought was invented is never explained: the student is asked to take it or leave it. The natural reaction on the part of all but the most remarkably able on the one hand and the most gullible on the other is, surely, to feel that if this represents scientific thought then it is something the student is entirely unfamiliar with; and this is a conviction that he may never lose. Of course, after the initial discomfiture one can surmount the hurdle, and learn to accept and use these difficult ideas as intellectual tools, effec-tively and efficiently. But the basic issue remains: it is surely undesirable that science should be imparted in any way that smacks of arbitrary authority.

For it is a simple fact that the men who created thermo-dynamics used the ordinary thought processes common to all men of science; and the materials they worked with and the problems they faced were all derived from the scientific and technological advances of the time. A knowledge of this histori-cal context can, I suggest, help to make the concepts of thermodynamics more readily intelligible. Just as the Newtonian system was put forward to resolve the problems of planetary motion, which observational astronomy had progressively clarified, so thermodynamics was formulated in response to the intellectual challenges that were posed by the rapid improvements made in the heat engine and the accompanying advances in the science of heat. This historical context is the missing factor in the textbook accounts, and I hope, therefore, that this book may be of some use to those whose duty it is to expound this mysterious yet fascinating science to coming generations of students.

This, then, is an account of a scientific revolution that took place between about 1790 and 1865, one of whose root causes was the rapid progress made in power technology during this period. It is not a general history of theories of heat; indeed I doubt if such a history could usefully be written in the present state of knowledge. Accordingly I have done little more than mention such topics as Fourier's analytical theory of heat and the contemporary speculations about the nature and causes of animal heat. At the same time I have made no mention of the doctrines of *Naturphilosophie* and their effects on a man like J. R. Mayer, nor of the later discussions about the theories of 'energetics'.

Nevertheless I have had to devote the middle chapters of this book to themes which at first sight do not appear to be related to each other. This is an unavoidable outcome of the fact that scientific advance means the unification of knowledge. If one attempts to describe a particularly important advance then one must describe the previously unrelated components which, by virtue of that advance, were welded together into one comprehensive system of thought.

Therefore, in spite of the omissions, both necessary and unintentional, in the following narrative, I hope it will be possible for the reader to find some hints towards further studies that should prove illuminating. The revival of German physics and the work of von Helmholtz, the development of chemical thermodynamics, the development of the concept of absolute zero, the significance of the Manchester School of science, and the disastrous collapse of British science in the second half of the nineteenth century are all topics which are well worth further study. For the rest I cannot do better than commend the collateral works by Dr Robert Fox (*The caloric theory of gases from Lavoisier to Regnault*: in press), Dr Richard Hills (*Power in the industrial revolution*: Manchester 1970), and Dr David Theobald (*The concept of energy*: London 1966).

I am very grateful to the following friends and colleagues who have read through the typescript, correcting errors and making valuable suggestions for the clarification of the argument: Dr Robert Fox (University of Lancaster), Mr Rodney Law (Science Museum), Dr Arnold Pacey (U.M.I.S.T.), and Dr Jerome Ravetz (University of Leeds). They are, of course, in no way responsible for errors and blemishes that remain. I am also grateful to my friends Dr Richard Hills (U.M.I.S.T.) and Dr Geoffrey Talbot (University of Belfast) for the use I have made of their theses, and to Mr Harry Milligan for the excellent photographic plates.

Manchester, 1970 DONALD CARDWELL

Chapter 1

The Origins of the Science of Heat

'FIRE' IN SEVENTEENTH-CENTURY PHILOSOPHY AND ARTS

The importance of fire and heat in the economy of nature and for the life and well-being of mankind has been universally recognised from the earliest times. But sun or fire worship do not constitute even primitive forms of science, while the ancient Empedoclean division of the 'elements' into Air and Water, Fire and Earth amounts to little more than a geographical ordering of the familiar world around us. The latter was the prevalent doctrine of heat up to the seventeenth century and together with the idea of four conjugate qualities—dry and wet, hot and cold—it indicates the extent of ancient and medieval knowledge of the subject. It is true that Plato's *Timaeus* and Lucretius' *De rerum natura*, among other writings, contained seeds which could have germinated in the right intellectual soil, but generally speaking it is impossible to agree with those historians of science who proclaim that there were, in ancient and medieval times, significant foreshadowings of modern theories of heat and thermodynamics. It might be supposed at first sight that Francis Bacon's judgement that heat is of the form of motion—expounded in 1617 with a scrupulous regard for the rules of evidence*—marked the beginning of modern scientific theories. But Bacon's assertion is basically Aristotelian: heat belongs to the group of phenomena that we commonly classify as 'motion'—just as we classify lions as members of the cat family. Now a developed Aristotelian theory of heat might be possible in principle, but such a theory can only be conjectural, for the actual course of events lay in quite a different direction. The contrast between Bacon's ideas and those of later science is made clear when we recall that Bacon had no notion

* Bacon was not a great lawyer for nothing.

I

of *quantity* of heat or of temperature, and necessarily therefore had no idea that a given amount of heat had different thermometric effects on different substances. His conclusions were not *logically* related to subsequent developments.

The most important step towards the establishment of a science of heat was the invention of the thermometer at the beginning of the seventeenth century.[1] Unfortunately there was for a long time doubt about what it actually measured. As a clinical instrument it certainly reduced the subjective element in the diagnosis of fevers, but beyond this the exact significance of its readings was uncertain and interpretation of them depended on the observer's philosophy of nature.

Galileo argued, in *Il saggiatore* and again in the *Discorsi*, that the sensation of 'heat' is caused by the very rapid motion of certain specific atoms. 'Heat' is thus an illusion of the senses, a product of the mental alchemy which transmutes the eternal scurrying of inert atoms of the 'real' world into our sensual world of colour and warmth, sweetness and bitterness, comfort and discomfort. There is therefore no such thing as 'heat' if there is no one present to experience it: in exactly the same way there is no such thing as a real pain at the tip of a dentist's drill, or a real tickle at the end of a feather.* Yet the sensation we call 'heat' is correlated with definite physical changes: bodies do expand as they seem to get 'hotter' and if they are 'hot' enough they change state, melting or vapourising as the case may be. Clearly, while we cannot talk about a quantity of 'hotness' (how can one quantify a subjective sensation?), we must try to account for expansion and change of state. This the Galilean atomic model can do, at least qualitatively, and if we have a measuring instrument, or thermometer, we can talk reasonably about a 'quantity of heat'. For the expression now indicates the total quantity of motion of all the specific atoms; the assumed cause of such (objective) phenomena as expansion and change of state.

The views of Rene Descartes (1596–1650) are more explicit than those of Galileo. Descartes tried to explain all physical phenomena in terms of extension and motion only. As the Cheshire Cat vanishes leaving only the abstract grin, so in Descartes' system specific substances and qualities vanish,

---

* These examples are taken from the writings of Sir Arthur Eddington and Galileo respectively.

leaving only a real world of extension and motion. A void or vacuum is impossible for Descartes as it would imply extension without body and for him extension *is* body. To account for the world as we know it Descartes requires three types of particles, which are distinguished from one another by their geometrical properties only—that is, by 'extension'. These are the small 'fire particles'—some of them are infinitely small—the intermediate '*boules*' and the 'gross material particles' which in aggregate constitute all material bodies. The first two sorts of particles form the all-pervading aether which circulates in a gigantic but invisible whirlpool or vortex about that large hot body, the sun. It is the motion of this vortex that carries the planets round in their circular orbits, and it is the pressure due to the aether flow round the earth—which has its own relatively small vortex—that accounts for the phenomenon we call 'weight'.

Light is a sensation caused by pressures transmitted through the aether to the retina of the eye, while the sensation of hotness is due to the vibrations of (probably) the material particles stimulating the nervous system. On the other hand the phenomena of expansion and change of state are caused by fire particles and *boules* penetrating between the material particles and forcing them further and further apart. It may even be that vibrations of fire particles and *boules* agitate material particles and are thus the first cause of the sensation of hotness. Dr G. R. Talbot has pointed out[2] that later Cartesians who did not share their master's anxiety about the void, or completely empty space, saw no reason for both fire particles and *boules* and so lumped them together as a 'subtle fluid': a rather hazy concept which developed in the course of the eighteenth century into the doctrine of material 'caloric'.* Dr Talbot remarks that while Descartes showed considerable insight in his observation that iron can contain more 'subtle matter' (fire particles plus *boules*) than can wood, he was unable to go on to discover the concept of specific heat, or heat capacity.[3] Specific heat is measured in terms of the weight of or quantity of matter in a specific body; but Descartes believed that weight is simply an

---

* J. F. Scott has remarked that Descartes refuted the view that heat was '. . . a material fluid (caloric) as was generally supposed'. This is quite erroneous. See J. F. Scott, *The scientific work of René Descartes (1596–1650)*, (London 1952), p. 66.

effect caused by the moving aether and is not directly related to the quantity of matter in a body.

A pioneer of the mechanical philosophy in England was Robert Boyle (1627–1691).[4] A disciple of Descartes, at least in the period when he did much of his best work on heat, Boyle wrote that '. . . heat seems principally to consist in that mechanical property of matter we call motion'. The motion in question is that of the constituent atoms, and it is random and very rapid; otherwise we should have to conclude that the wind or any large moving object must, *ipso facto*, be hot! Confirmation for this theory of the nature of heat came from the fact that it can be generated by friction or by percussion. Boyle shrewdly observes that a nail being hammered into wood does not become hot until it is driven right in and can move no further. Heat is a positive quantity; but is it entirely satisfactory to say with Descartes that cold is no more than a deprivation of heat? For if we do, how can we explain the immense force exerted when water freezes? If freezing suggests that 'frigorific' particles are entering the water, binding it into a solid mass, then there should be a detectable gain in weight when it has all been turned into a block of ice. But no increase in weight was found. On the other hand it was established that when bodies burned or were calcined there was a marked increase in weight, clearly indicating the absorption of fire particles; an observation which had its effect on the subsequent acceptance of the phlogiston theory. Boyle also reported that two blocks of ice melted more rapidly if rubbed together than if left to melt undisturbed. He noted, too, that a large quantity of ice took a long time to melt, but he drew no conclusions: the concept of quantity of heat—a prior necessity for the discovery of latent heat—had yet to be formulated.

With all this experimental work, much of it perceptive but little or none of it followed up by further research, Boyle's major contributions to the development of the scientific study of heat remain his association of it with chemistry, the publicity he gave to the famous law relating the pressure to the volume of an 'elastic fluid', or gas,* and his study of the thermometer. Having realised the desirability of an international thermometric

---

* This law, which in the English-speaking world is called Boyle's law and in Europe Mariotte's law, was first discovered by Richard Towneley. Boyle himself called it 'Towneley's hypothesis'.

scale Boyle proposed that the scale should be set by the expansion of a given fluid in a standardised thermometer bulb with a capacity of 10,000 volumetric units. The bulb should be filled at the boiling-point of water and degrees would represent unit increments or decrements of volume above or below that point: one degree would thus indicate a temperature slightly above the boiling-point of water and a volume of 10,001. In this way temperature measurements made anywhere in the world would be immediately intelligible to any observer.

Generally speaking, then, seventeenth-century theories of heat combined the idea of a subtle fluid with that of the motion of its constituent corpuscles or atoms. In this respect there was no essential difference between the views of Galileo, Descartes, Boyle and their numerous disciples. Nevertheless, from their rather diffuse notions two distinct and indeed rival theories could be deduced, which were later to become increasingly important. According to the first theory there was no 'subtle fluid', and the phenomena of heat were to be accounted for by some function of the motions of the atoms that constitute all material bodies. But the atoms remained scientifically inscrutable until Daltonian chemistry was accepted, and the unknown function of their motions was not understood until the concept of *energy* was established in the nineteenth century. The second theory ignored atoms and their motions and concentrated on the more tangible notion of a 'subtle fluid' whose presence accounted for the phenomena of heat. As we have remarked, this hypothesis developed in the course of the eighteenth century into the caloric theory. Mathematicians and physicists tended to favour the first theory while chemists came increasingly to accept the second. But we must remember that few men, if any, held wholly consistent theories throughout their scientific careers—or even for any considerable portion of them. Men's views and ideas change; they modify their hypotheses in the light of their own experiences, reflections and discussions. And men do not always write as clearly as the historian would wish; sometimes because they lack literary facility, sometimes —perhaps usually—because of confusion in their own thoughts and sometimes because they do not want to commit themselves too strongly to a theory about which they may have secret doubts.

If the seventeenth century was characterised by a variety of philosophies it was also conspicuous for the extent and precocity of its experimentation. The academicians of Florence, Paris and London cast their nets wide. They made use of lenses and concave mirrors to try to detect the heat of moonlight; they made lenses of ice and tried to ignite inflammable materials with them. The air pump and receiver were very popular, and were used to show that burning candles and glowing coals were extinguished in a vacuum, that heat was transmitted through a vacuum while sound was not, and that tepid or moderately warm water boiled as the air pressure was progressively reduced. They greatly improved the construction of thermometers—the Florentines were especially good at this—producing instruments of high craftsmanship and elegance; but there was continued disagreement about the best thermometric fluid to use and what 'fixed' points should be accepted. Some favoured the point at which water boiled, others that at which butter melted, while still others suggested the remarkably uniform temperature of deep cellars: after all, daily and even seasonal variations of temperature are imperceptible only a few feet below the ground. Finally there were experiments that were well in advance of any scientific theory. Thus Mariotte discovered that a sheet of glass cuts off the heat from a fire but has little effect on the heat from the sun; and several observers noted that the speed with which a body warms up in the sunshine depends upon its colour, being faster for dark colours than for light. The reasons for the phenomena of 'radiant heat', as it was later called, were at that time unknown. On the other hand there were the inevitable wrong ideas based on misleading experiments: one that had a long life was that heat tends to travel more readily downwards than upwards in a solid body.

Besides this extensive if random experimentation there were four major fields of inquiry and effort in which ideas about heat were important, or were soon to become so. These were technics, meteorology and the study of the atmosphere, chemistry, and medicine. The technical importance of the science of heat was not, in spite of Bacon's hopes, very great at first. Pit coal was only gradually, and to the disgust of many, beginning to replace wood as fuel in England, and its role in nascent industries was minimal; not until 1709 did Abraham

Darby succeed in using coke to smelt iron at Coalbrookdale.[5] These, however, were achievements in chemistry and metallurgy rather than in heat technology. In the meantime there were numerous speculations about the power of fire and of combustibles like gunpowder, but these did not become of practical significance until the end of the seventeenth century.

More immediately important were the developments in meteorology and studies of the atmosphere. The revelation that we live at the bottom of a finite ocean of air (the atmo*sphere*), which exerts a considerable pressure on everything underneath it, was one of the most important achievements of the seventeenth century. It caught the imaginations of the virtuosi and academicians; and with reason, for in the distinctive properties of the atmosphere and the void lay the keys to many fundamental advances in science. The elucidations of the role of the atmosphere, or rather of its main constituents, in sustaining combustion and respiration were to be major factors in the advance of chemistry and of physiology. The treatment of the atmosphere as an Archimedean fluid ultimately led to the formulation of Dalton's atomic theory in the early years of the nineteenth century.[6] Finally we note that the establishment of a science of meteorology presupposes knowledge of the great ocean of air and of the dynamics of its movements.

There were, of course, complicating factors: meteorology would be impossible without refined and standardised instruments such as thermometers, barometers and rain gauges[7] as well as detailed and collated information about the great wind systems, the rainfall belts and the pressure and temperature variations over the face of the globe. To put it briefly, extended field studies were essential for the development of the science. In this respect the growth of maritime trade was important, and nations which were sea powers were well placed to contribute to the science.

The phenomenon of the tides had intrigued seventeenth-century philosophers, Bacon and Galileo among them, until Newton had finally settled the matter. The problem of the trade winds, blowing steadily from the NE when one was north of the equator and from the SE when one was south of the equator, was not perhaps dramatically obvious or indeed so fundamental, but was still worthy of scientific study. Martin

Lister suggested that they were due to the dense vegetation of the sargasso sea-weed: the respiration varied diurnally, he believed, as the sun's heat governed the plant's activity. This would explain why the trade winds commonly fell away at dusk.[8] On the other hand George Garden argued, in more orthodox fashion perhaps, that the great wind systems of the globe must be explicable in terms of Cartesian vortices.[9]

Edmund Halley (1656–1742), the astronomer and geo-physicist, was more systematic and his work was altogether more satisfactory. He drew the first map to represent the wind systems of the world, showing the distribution of the trade winds and the complications introduced by the monsoons. He had no hesitation in ascribing the trade-winds system to the heating effect of the sun, which is at its maximum along the equator. The principle is that of convection (to use the word invented by Prout many years later): the heat of the sun rarefies the air, which expands and so, being lighter, rises:

> But as the Cool and Dense Air, by reason of its greater Gravity, presses on the Hot and Rarefied, 'tis demonstrative that this latter must ascend in a continued Stream as fast as it Rarefies, and that being Ascended it must disperse itself to preserve the *Aequilibrium*, that is by a *continuous current*, the Upper Air must move from those parts where the Greatest Heat is: So by a kind of Circulation N.E. Trade Winds below will be attended by a S.W. above, and the S.E. with a N.W. Wind above.[10]

The empirical fact that heated air rises was known long before Halley's time. But his application of this knowledge to explain the trade-winds system was as bold as it was original; so, too, was his postulation of counter-currents of air in the upper atmosphere. He was, however, unable to provide a satisfactory explanation as to why these winds have an easterly component and do not blow simply from the north and south. A convincing answer was given later by George Hadley,[11] who pointed out that air being drawn towards the equator from high latitudes, north or south, would necessarily have a slower rotational speed towards the east than points actually on the equator, which would be travelling eastwards at about 1,000 miles per hour; the net effect then would be to shift the trade winds from north or south to NE and SE. Thus the easterly component of the trade winds provides confirmation of the actual rotation of the earth.*

Halley was also interested in the problem of evaporation, a process which seemed at once paradoxical and fundamental. How was it, he asked, that an intrinsically heavy substance like water could rise through a light substance like air to form clouds? On Archimedean principles water vapour should rest in a dense layer just above the surface of the parent water. Halley's suggestion was that the water atoms actually expand on evaporation, increasing their diameters nine-fold until they have the same specific gravity as the air, whereupon they rise. These distended atoms are not homogeneous; they are spherules filled with 'finer air', specifically lighter than common air. Oddly enough this arbitrary and unsatisfactory explanation was accepted for some time. It was effectively criticised by J. T. Desaguliers in 1729[12] and, afterwards, by John Rowning.[13]

The fundamental importance of evaporation is that it provides a link in the chain of the 'hydrologic cycle'. Halley experimented on the rate of evaporation of water from a surface of given area at different temperatures and then calculated that the total evaporation from the Mediterranean sea should be much more than enough to account for all the water brought into it by all the great rivers. He had already computed the rate of flow of the river Thames, and he suggested that it would be reasonable to assume that each of the great Mediterranean rivers carried, on average, ten times as much water as the Thames. Even so, the amount supplied would be only one-third of that lost by evaporation. Thus Halley, like Perrault and Mariotte before him and Dalton later on,[14] had made a notable contribution to our understanding of the hydrologic cycle. The circulation of the waters—rainfall and snow, springs, streams and rivers—becomes a closed process when we take account of the barely perceptible but immensely important phenomenon of evaporation. There are no occult processes, no system of subterranean capillaries or syphonic caverns, as earlier speculators had supposed. The principle that perpetual motion is impossible was thus preserved; and

* This was the final answer to the long-forgotten objections to the Copernican theory raised by men like Tycho Brahe: why, if the earth is spinning, is there not a wind blowing from the east? The answer, briefly, is that there is such a wind.

ultimately, as we shall see, the whole process was understood in terms of the dynamical action of heat.

Another important meteorological problem that aroused Halley's curiosity was the differential heating effect of the sun's rays.[15] It is greatest at the equator and gets progressively less as you travel north or south. And even though the polar regions enjoy much longer summer days than the equatorial ones, this does not compensate for the extremely long nights and the low altitude of the sun in the sky during day-time. Halley asserted that the heat of the sun varies as the sine of the angle of incidence, and to some extent therefore he anticipated Leslie's discovery of the sine law of radiation, announced over a hundred years later. Halley did not consider the *emission* of heat by a body; he was concerned only with the insolation, or the solar heat received by a unit area as the angle of incidence varies. But the principles are closely related.

Halley was not a philosophical scientist and did not participate in the protracted metaphysical disputes of his contemporaries; he belonged rather to the tradition of the cosmographers and discoverers. He brought the science of meteorology to as advanced a state as was possible without such improved instrumental aids as balloons and radio and without further fundamental developments in chemistry and physics: notably the elucidation of the idea of quantity of heat, the discovery of latent heat, knowledge of the adiabatic heating and cooling of gases, and the invention of the concept of entropy. Accordingly, although Halley's theory of the trade winds is still put forward in popular accounts and school geography books as almost self-evidently true it is in fact an over-simplification, a first approximation.[16] Halley's lasting contribution was that he saw the atmosphere as a vast dynamical system, the 'go' of which resulted from the differential heating of the earth's surface. It was to be a long time before the problems of meteorology were studied by men whose insights rivalled those of Halley.

THE FIRE-ENGINE

The discovery of the atmosphere thus profoundly affected the development of science. As we shall see, it was no less important in its impact on technology.

Two aspirations which engineers, dreamers and philosophers had from time immemorial were the ambition to fly, or at least to make a flying machine, and the ability to harness the evident power of fire. From the time of the legends of Daedalus and Icarus, through the shadowy figure of Eilmer of Malmesbury[17] and the speculations of Roger Bacon to the more substantial plans of the Italian Renaissance inventors, men have sought ways to imitate the birds: by means of artificial wings flapped by the muscular efforts of the would-be aviator. But the dreams proved empty: the power/weight ratio was far too small. The Chinese achieved a degree of success with their ingenious invention of the kite, but it was strictly limited.

On the other hand there is no doubt that about two thousand years ago Hero of Alexandria had, or at least proposed, a model steam-engine, in the form of a reaction turbine,* that would actually work. The dream of the heat-engine, like that of the flying machine, was taken up by Renaissance engineers, and by the middle of the seventeenth century quite a number of reasonable suggestions had been put forward, among them being those of de Caus, Giovanni Branca, and the Marquis of Worcester. These suggestions were all based on the use of steam as the 'working substance', as it seemed the most effective way to harness the power of heat or fire. But there were insuperable snags, since the immense power of heat proved dangerous to generate and difficult to apply. As it happened, medieval and Renaissance Europe did possess one very effective heat engine: as Osborne Reynolds remarked, 'The combustion engine, in the form of the cannon, is the oldest form of heat engine.' But this engine required an immensely thick cylinder or barrel to contain the force, used a very expensive working substance

* This is an anachronistic use of the word 'turbine', which was coined in 1824 to describe a certain type of hydraulic machine designed in accordance with very advanced mechanical principles. Strictly speaking, the word should only be used to describe machines (water, steam or gas) designed in accordance with these principles, which were quite unknown to Hero.

(gunpowder) and worked on the inverted principle that the power generated was uncontrolled and destructive—at least as far as the point of application (the enemy) was concerned.

Thus by the seventeenth century two ancient ambitions, the flying machine and the heat-engine, were bogged down in practical difficulties. Men could see the possibilities clearly enough but they had no means of bringing them within the realm of the practicable. The discovery of the atmosphere by small but brilliant groups of Italian and French mechanical philosophers in the middle of the seventeenth century dramatically changed this situation. The two ancient technological dreams were given an entirely new twist—one which the old pioneers could never have suspected. If we must regard the air as an 'atmosphere', a finite ocean of gas or elastic fluid, having a finite and measurable depth, and if we can treat it as an Archimedean fluid, then there is no reason why we should not conceive of air*craft* or air*ships* to navigate across, or rather through it. Accordingly, as early as 1670 Father Lana Terzi, a Jesuit, proposed such a vessel: large but thin copper globes were to be exhausted of air, and so being specifically lighter than the surrounding medium would have enough positive buoyancy to lift a gondola and crew. The vessel was to be propelled and steered by means of large fan-like oars. Unfortunately it was impossible to make flotation tanks to Father Lana's exacting specifications, and his vacuum-aircraft remained a dream. But the basic principle was sound enough. Thanks to greatly increased knowledge of the properties of gases it was realised in the eighteenth century that while a vacuum was ideal it was not strictly necessary: a light-weight fabric balloon filled with a gas lighter than the surrounding air would provide enough buoyancy to lift a respectable payload, including an aeronaut. Thus the lighter-than-air craft was born.

The corollary of the displacement principle in hydrostatics is the law of the increase of pressure with depth. The credit for making the pressure exerted by the atmosphere entirely explicit belongs to Otto von Guericke who, in 1672, published the famous book in which he described his air pump and the experiments that he had made with it from the mid-1650s onwards.[18] A dramatic picture in this book, which was subsequently widely reproduced, showed two teams, each of eight horses, straining ineffectually to pull apart two small hemispheres just over a foot in diameter. These had been evacuated of air by means of a pump and were held together in the form of a sphere solely by the weight of the atmosphere. In this way von Guericke graphically illustrated the power that could be generated by the weight of the atmosphere. It should be added that either the sixteen horses were remarkably feeble or von Guericke was exaggerating. The latter is the more probable conclusion.*

If, however, von Guericke was exaggerating the 'power' of the atmosphere it was a pardonable, indeed laudable, offence; for almost immediately Huygens and Hautefeuille, independently of one another, set about devising methods of harnessing this power for useful purposes. A few years later Papin was experimenting with the same technique.[19] The actual 'breakthrough' came with the water-raising machine devised by Captain Thomas Savery in 1698. In Savery's engine (see Figure 1) a large metal cylinder is filled with steam from a boiler and then, the supply having been turned off, is drenched with cold water so that the steam condenses, a vacuum forms in the cylinder and water is sucked up a pipe from a well or reservoir, which is not more than 32 feet below the engine. When the cylinder is full of water the steam is turned on again and its pressure drives the water out of the cylinder and up the rising pipe to a storage tank or drainage channel. When the cylinder is once again full of steam the supply is turned off and the cycle repeated.

Exactly how many of these engines were made is unknown; whether any of them worked satisfactorily is uncertain. The difficulties and dangers of working with high-pressure and therefore high-temperature steam limited the maximum permissible lift to about 150 feet. The steam that was condensed and therefore wasted in heating up the cold metal cylinder and in pressing on the surface of the cold water meant that the engine was chronically inefficient. It is even doubtful whether it could have worked in the way in which Savery specified, for

* Von Guericke said that the hemispheres were about four-fifths of a Magdeburg ell in diameter. The ell was not a fixed and universal measure, so we must deduce its modern equivalent length. Guericke stated that a water barometer stood at a height of 19 to 20 ells, and as this must equal about 32 feet we conclude that the Magdeburg ell was about 17 inches. The spheres would be about 1 foot in diameter and the total atmospheric weight on them would therefore be about 1,500 lb.

'airlogging' may well have occurred. The seventeenth-century academicians had discovered that when water is boiled, the air dissolved in it is given off together with the steam. Unless there is some way of expelling the air it must accumulate in the cylinder until the engine eventually stops, 'airlogged', for no amount of cold water will condense air. It is interesting that Savery in his account of his invention at no point refers to the need to ensure that air is expelled from the cylinder if the engine

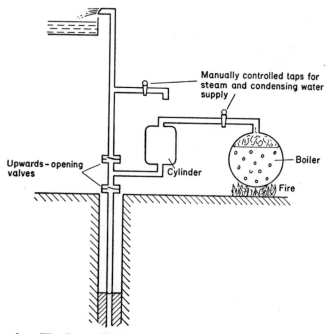

Figure 1.   *The Savery Engine*

is not to become inoperable through 'airlogging'. The obvious way to do this would be for the engine-man to flush the system every so often by allowing steam to pass right through the cylinder and up the rising pipe, thus clearing the engine of air. But Savery explicitly advises against this: '. . . it is much better to let none of the steam go off, for that is but losing so much strength and is easily prevented, by pulling the regulator some little time before the vessel forcing is quite emptied.'[20]

   The balance of probabilities is, so far as we know, that the engine may well have worked as a small demonstration model and that such full-sized specimens as were made may have seemed quite promising but in fact never worked satisfactorily for any length of time. Alternatively, engineers may have learned by experience to ignore Savery's advice, for they had found that the engine worked much better if they *did* allow the steam to 'go off' every so often.

The problem of the Savery engine evidently merits further study, backed up by experimental work. However, the defects of the engine as a 'miner's friend' soon became a matter of academic interest only. The need of the progressive mining industry—the growth industry of the period—for a means to cope with flood waters was met when, in 1712, Thomas Newcomen (1663–1729)[21] set up the first successful atmospheric pumping-engine. Once again a cylinder is filled with steam (see Figure 2) but one end is now open, and a piston moving up and down inside is connected by means of a substantial chain to one end of a great beam oscillating about a central pivot. The beam is counterweighted at the other end so that it tends to pull the piston up, drawing steam after it into the cylinder. When the cylinder is full the steam is turned off and cold water sprayed into it so that a vacuum is formed; the weight of the atmosphere drives the piston to the bottom of the cylinder, thus providing the powered half of the cycle. The condensing spray is then turned off, and the steam readmitted, the next cycle begins with the piston rising up the cylinder again.

In the Newcomen engine of 1712 explicit provision was made for the expulsion of air from the cylinder. The 'snifting valve' opened outwards in such a way that the rush of steam into the cylinder at the beginning of each cycle carried the accumulated air out with it through the valve. In this way, once a cycle, the engine made a wheezing noise—'like a man snifting with a cold'—as it cleared itself of air.[22]

The Newcomen engine was extremely important. It was perhaps the first major invention to be made by an Englishman,[23] and it ushered in a period of British technological supremacy that was to last about 150 years.* It was the first really successful prime mover to be put to work apart from the immemorial ones of wind, water and muscle. The momentous

---

* This is almost exactly correct, for it was at the International Exhibition of 1862 that the first signs became apparent that Britain's technological lead was threatened.

discovery of the atmosphere had thus made possible the first practicable and safe heat-engine, and although the actual physics of the process whereby fire or heat is made to produce useful work was somewhat obscure to Newcomen's contemporaries, their intuition that it *was* a heat-engine was sound enough, for it was commonly referred to as the 'fire-engine'.

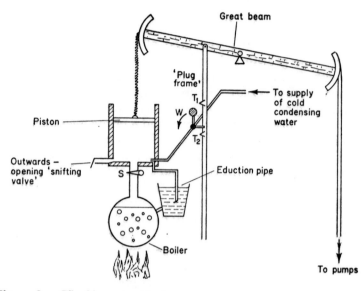

Figure 2. *The Newcomen Engine*

The movement up and down of the 'plug frame', a long board hanging from the great beam, causes the tappet, or plug, T_2 set in it to trip the weight-operated valve, W. When the plug frame moves in the opposite direction another tappet, T_1, resets the valve. A similar mechanism, not shown, controls the steam supply by means of the valve, S.

The 'eduction pipe' enables the condensed steam and the warmed condensing water to be returned via a well to the boiler, thus conserving heat.

The machine itself was magnificently adapted to its end. Notwithstanding the genius that went into its design, especially into the automatic valve mechanism that was operated by the motion of the great beam, it made no demands beyond the technical resources—the very limited technical resources—of the early eighteenth century. It was safe, reliable, and basically simple; it was powerful, and economically competitive with other means of pumping water. It belongs to that small but select group of inventions that have decisively changed the course of history.

PHYSICS AND CHEMISTRY: SOME BASIC CONCEPTS

The scientific study of heat and the dawn of recognition of its cosmic importance began roughly at the time when Isaac Newton, encouraged and helped by Halley, published his *Principia mathematica* (1687). One would naturally expect to find that the immensely successful Newtonian achievement influenced the development of the study of heat in much the same way that, according to Professor Cohen,[24] it influenced studies of electricity. But the basic Newtonian ideas such as mass, momentum, central forces etc., were not immediately relevant to the science of heat, and their importance was therefore mainly indirect.

Apart from the prestige that Newton conferred on all branches of science, he contributed several interesting hypotheses to the study of heat, some experimental work, and, indirectly, a number of fruitful analogies prompted by his system of mechanics. Fortunately he made no authoritative pronouncements about heat, and his interest was, so far as it went, cautious and eclectic. Thus he asserted in later editions of the *Opticks* that the cause of heat is the rapid motion of the *parts* of bodies, while elsewhere in the same work he proposed a kinetic theory of evaporation.[25] He suggested that the reason why black bodies heat up more rapidly than those of other colours is that they absorb the incident light, which is thereby 'stifled and lost' inside them,[26] but he does not expand this interesting idea.* He accounted for the thermal expansion of air by postulating that the constituent atoms vibrate with amplitudes that increase with temperature.[27] On the other hand, when he put forward a theoretical explanation for Boyle's law the model he used was essentially a static one: the atoms comprising an elastic fluid or gas are supposed to repel one another with a force that decreases with increasing separation.[28] Assuming that each atom repels only its immediate neighbours, Newton shows that air, or any elastic fluid so constituted, would

---

* Compare the suggestions of Lavoisier and Laplace. See page 62 below.

B

obey Boyle's law; but whether elastic fluids are really like this is problematical, as Newton carefully pointed out.

Newton's experimental work is represented by his attempt to define an extended temperature scale. In 1701 he published anonymously an account of his method of measuring high temperatures by extrapolation.[29] The limitation of conventional thermometers was that the liquid boiled long before really high temperatures could be reached. To get round this Newton invoked his 'law of cooling', which seems to have been inspired by his own demonstration that a moving body resisted in proportion to its velocity decelerates in proportion to that velocity.[30] Analogously, then, the rate at which a hot body cools down should be in proportion to the body's excess temperature over its surroundings.* If we heat a block of iron until it is red hot and then place small pieces of different but easily fusible metals on it, the pieces will melt, and each will solidify again only when the block cools down through the appropriate temperatures. All we need to do is to note the time at which cooling started, the times at which the metals solidified and, when the block is cool enough for its temperature to be measured directly, the time taken for it to cool from one measured temperature to another. Application of the cooling law will then give us those temperatures which were too high to measure directly, including of course the initial temperature of the block. However, we must add that Newton did not know that change of state takes place at *fixed* temperatures; he believed that it occurred over small ranges of temperature.

More important than the work of Newton was that of Guillaume Amontons (1663–1705). In the course of his comparatively brief career Amontons made a number of important discoveries and put forward some very valuable ideas. He seems to have been the first to have realised the importance of actually measuring the thermal expansion of elastic fluids, and the first to have studied the increase of *pressure* with temperature.[31] In the absence of an established thermometric scale, and without an accurate thermometer, he could do little more than

---

* i.e. $-\dfrac{dT}{dt} \propto T$; where T is the excess temperature and t is the time. This gives us $t \propto \log T + \text{const}$. If we know the time taken to cool from one measured temperature to another we can, given the cooling time, find any required temperature.

estimate that air expands, or its pressure increases, by about one-third between the temperatures of 'cold' and of boiling water. But it was a good estimate, and some historians even go so far as to claim that Amontons anticipated Gay-Lussac's (sometimes called Charles') law of gaseous expansion. This law states that all gases at the same pressure increase their volume equally if heated through the same temperature range. But Amontons could not possibly have arrived at this law: the chemical individuality of gases—oxygen, nitrogen, hydrogen, chlorine etc.—had not been discovered at this time.[32] All that was recognised was 'air' and its various modifications. He did, however, discover that unequal weights of air equally heated increased equally in pressure, and he went on to show that the relative increase in pressure was independent of the initial pressure and depended only on the rise in temperature; so that if 'cold' air at a pressure of 30 in. of mercury is heated to the temperature of boiling water the pressure increases by about one-third, to 40 in., while air at 60 in. increases to 80 in., and so forth. He was impressed by this discovery that the same small increase in temperature can produce progressively greater increases in pressure as the initial pressure is greater.* All this was novel and important; it represented a substantial step towards the establishment of the gas laws.

Amontons also noticed, as Halley had done,[33] that when water boils the temperature remains constant. Unlike Halley, he seems to have recognised the importance of this: boiling water provides an eminently suitable datum, or 'fixed point', for a thermometric scale. And as a thermometric fluid Amontons proposed air, for it is very expansive and can be used over a much wider range of temperature than any liquid. For these reasons Amontons developed a constant-volume air thermometer which measured temperature by increase of pressure. With its aid he found that Newton's law of cooling was seriously inaccurate.

Finally, he described a remarkable 'fire-engine' of his own invention. This consisted basically of a vertical wheel with hollow spokes and rim, fitted with internal valves. It was partly filled with water, and a fire was lit under one side so that

---

* i.e. in accordance with the now familiar equation, $\dfrac{p - p_0}{p_0} = \alpha(t - t_0)$. But this was, of course, not known to Amontons.

the air in that side of the wheel expanded, and acting through the valves drove the water to the other side. The wheel was thus imbalanced and made to rotate. The importance of this admittedly impractical engine was twofold. It can be fairly claimed that it was the ancestor of all those engines, such as the diesel engine, the gas turbine etc., which work by the thermal expansion of air and which were developed in the nineteenth and twentieth centuries.[34] It is true that the windmill is a kind of air engine, and a horizontal fan set in a chimney and driven convectively by the fire at the bottom[35] is, in a sense, a hot-air engine. But such machines were not ancestors of the modern air-engine in the way in which Amontons' engine was; for he had a direct insight into the possibilities of using the thermal *expansion* of air as a moving agent. In the second place, Amontons actually estimated the 'duty', or the work done for the consumption of a given weight of fuel, that his engine should have yielded, and went on to claim that it compared favourably with the cost of power from the traditional sources: wind, water and muscle. All this was quite theoretical, but he was entirely convinced of the practicability of the 'fire-engine'; the cannon, he pointed out, was just such an engine, and if the power of fire can be used to give violent action can it not also be used to move machinery?[36] This approach to the problem of the heat-engine was novel, for it showed that Amontons understood that general principles were involved, and also that an 'idealised' engine could be used to compute thermo-mechanical processes. In other words, he was taking the first steps towards the synthesis achieved by Sadi Carnot 125 years later, and he should therefore be regarded as one of the founders of the science of thermodynamics.

Amontons' achievements should be set against the back-ground of the more mundane problems of thermometry. Apart from the practical difficulties of making satisfactory glass instruments, there was a grave lack of basic data. The choice of boiling water and melting ice as determinants of 'fixed points' may seem obvious, but there were those who claimed that these points were, in fact, variable. Impurities in the water, variations in atmospheric pressure and bad experimental techniques would account for these; sooner or later the defective observa-tions would be eliminated, and it would be generally agreed that if distilled water were used at a given atmospheric pressure then the points would indeed be fixed. Less easy to cope with was the quite distinct argument that the fixed points actually varied with geographical location. William Derham and Petrus Musschenbroek thought that the freezing-point became lower the further north one went; and their views were by no means eccentric, or even implausible. It took time to clear up travellers' reports that the freezing-point in Siberia, on Hudson's Bay or in Peking was not the same as in Europe. It required the progressive accumulation of extensive and increasingly reliable data by increasingly competent observers to establish, during the first three or four decades of the eighteenth century, that the boiling- and freezing-points of water were, under standard conditions, invariable everywhere.* Once again meteorological, or geophysical, considerations played a part in the development of the science of heat: it could not have been developed in the laboratories and studies of Europe without reference to the larger world outside.

The immediate fate of Amontons' air thermometer was unpromising. It proved difficult to use and was bulkier and more fragile than the conventional liquid thermometer. Furthermore there were fundamental doubts, which lasted many years, as to its reliability. As George Martine, writing in 1739,[37] put it:

> . . . the dilatation of air is not so regularly proportional to its heat, nor is its dilatation by a given heat near so uniform as he [Amontons] all along supposed. This depends much upon its moisture: for dry air does not expand near so much by a given heat as air stored with water particles; which by being turned into steam increase vastly the seeming volume of the air.

Amontons' work was not however neglected. D. G. Fahrenheit was much impressed by his demonstration of the invariance of the boiling-point of water, and used it as a fixed point on his own famous scale.[38]

If two thermometric fixed points have been selected there still remains the question of how accurate the thermometer is in between the two. Do thermometric fluids expand uniformly? When the mercury column of a Fahrenheit thermometer stands

---

* As late as 1744 R. A. F. de Réaumur was delighted to find that the temperature of melting snow in the Andes was exactly the same as in Europe.

exactly half-way between the freezing-point mark at 32° and the boiling-point mark at 212°, is the indicated temperature 'really' 122°? A comparison with thermometers using other liquids proves nothing, for all the liquids might obey a non-linear law of thermal expansion; in fact the question as it stands is unanswerable. But a step towards its clarification was taken when, in 1723, Brook Taylor published an account of an experiment which was notable not so much for the results as for the implicit assumptions he made.[39] He mixed equal quantities of water at the boiling- and freezing-points and observed that the temperature of the mixture was the arithmetic mean of the higher and lower temperatures. Further experiments showed that when the proportions of hot and cold water were varied the temperature of the mixture was always determined by the ratio of the quantities and their temperatures. Taylor tried to make his containers as thin as possible, indicating thereby that he must have had a qualitative insight into what later became known as the 'water equivalent' of the container, or calorimeter. The significance of this was not so much that the expansion of the thermometric liquid was shown to be linear but that Taylor was working, at least implicitly, with the concept of *quantity* of heat, i.e. that the rise in temperature of a given mass of water must depend on the amount of heat imparted to it. But the point was not made explicit; it seems to have been accepted as a common-sense assumption. In any case, it is unlikely that Taylor was very original; in fact, he had collaborated earlier with Dr Desaguliers in carrying out very similar experiments.

Desaguliers tells us that he and Brook Taylor conducted experiments to show that the 'actual heat is to the sensible heat as motion is to velocity'.[40] There is some danger of confusion here, as the word 'heat' is being used, as it often was in those days, to denote both temperature *and* quantity of heat. We recall that Newton had, in deducing his law of cooling, considered temperature as an analogue of velocity, and we should therefore interpret this Newtonian analogy to mean that quantity of heat is to temperature as 'motion' (or momentum: mv) is to velocity (v). Just as two moving bodies in collision share their momentum according to a certain simple arithmetical rule, so the mixing of hot and cold bodies results in the sharing of heat between them according to an analogous rule.

Furthermore, the principle of conservation of motion (momentum) of Descartes and Newton implies a no less fundamental principle: that of the conservation of heat. Accordingly Desaguliers and Brook Taylor conclude as a general rule that the amount of heat that will warm a quart of water through so many degrees will warm two quarts through only half as many degrees.

Desaguliers, the popular lecturer and author of the first English textbook of physics, gives us a useful insight into generally acceptable ideas of the nature of heat before the middle of the eighteenth century. He accepted Newton's very tentative hypotheses to account for Boyle's law and, with the enthusiasm of the disciple, surpassed his master and flatly asserted that Newton had *proved* that elastic fluids are constituted of mutually repellent atoms. The outcome was that those who, in ever-increasing numbers, learned their Newton not from *Principia mathematica* but from other men's interpretations,* came to believe that Newton had given an authoritative demonstration that gases are atomic in constitution.

Unlike Newton, however, Desaguliers was very interested in meteorology. The composition and dynamics of the atmosphere, made up of different 'airs' and vapours, puzzled him. He accepted Newton's heterodox view that the ultimate particles of air may be heavier than those of water,[41] but as he had rejected Halley's ideas he had to provide his own explanation of evaporation and the rise of water vapour through the atmosphere. The atoms forming the surface of a pool of water are normally held together by short-range attractive forces but are exposed to the atmosphere; he suggested that they can therefore be electrified by atmospheric electricity. As this occurs they increasingly repel their immediate neighbours on all sides and underneath until they are ejected from the main body of water. On the other hand the conversion of water into steam by boiling results in an enormous increase in volume by something like 14,000 times (this was actually a gross exaggeration; as Watt later showed, the correct figure should be about 1,700). The

* cf. Robert Smith, who according to Professor Guerlac '. . . in his massive *System of optics*, published in 1738, made no distinctions between Newton's conjectures and his more rigorous discoveries'. See H. Guerlac in *John Dalton and the Progress of Science*, edited by D. S. L. Cardwell (Manchester, 1968), p. 62.

substantial reduction in specific gravity naturally allows steam to rise in air. To a Newtonian this increase in volume on vapourisation must be due to an equivalent increase in the repulsion between atoms, and this in turn must be due to the heat imparted to the atoms. The inference that heat increases the mutual repulsion of atoms had important consequences for the development of the material theory of heat.

Two agencies can therefore be invoked to explain the apparent repulsion between the atoms of elastic fluids: electricity and heat. Thus the cold condensing water injected into the cylinder of a Newcomen engine destroys the elasticity of the steam by depriving it of its heat; but it does not destroy the elasticity of the air since that depends mainly on electrical repulsion. This latter hypothesis did not prove acceptable for it raised more problems than it solved, and thus before long the elasticity of both vapours, like steam, and elastic fluids, like air, was ascribed to the repulsive force of heat acting between the constituent atoms.

In view of Newton's achievements perhaps it is not unduly insular to contrast English ideas about the nature of elastic fluids with Continental and Cartesian notions of fire particles and subtle fluids, to say nothing of occult entities such as atoms in the form of minute coiled springs, which were postulated to 'explain' the elasticity of air. The first inklings of a sounder, kinetic theory are to be found in the works of the Bernoullis and their school. Jacob Hermann, a pupil of Jean Bernoulli, wrote at the end of his *Phoronomia* (1716)[42] that '. . . other things being equal, heat is proportional both to the density of a hot body and to the square of the agitation of its particles': a statement which at first sight seems remarkably prescient but in fact leaves the most important questions unanswered. Jean Bernoulli later envisaged constituent atoms moving in a universal subtle fluid, in which heat caused vorticial motions so that the atoms were carried round in circular orbits, the resulting centrifugal forces giving rise to the pressure of the elastic fluid.[43] Boyle's law is accounted for in this manner, and Bernoulli asserted that the velocity of the particles is the measure of the degree of heat. But since he believed that the pressure of an elastic fluid is as the square root of the intensity of the heat, his view of the relationship between molecular velocities and temperature is not comparable with the modern one.[44]

In his *Hydrodynamica* (1738),[45] Jean Bernoulli's son, Daniel, put forward some very promising ideas. He treated the cannon as a heat-engine and referred, tantalisingly enough, to reports he had heard from England of a working fire-engine; but he does not seem, alas, to have had any direct experience of the Newcomen engine. His views on meteorology were disappointing compared with those of Desaguliers, or even Halley, but his famous theories of the nature of heat and of the constitution of elastic fluids were far in advance of the times. He may have been influenced by Hermann, for both were in St Petersburg in 1730–1 when Daniel was working on his book. In the new theory the fire particles and the subtle fluids are dispensed with, the constituent atoms being all that are required to account for the known physical properties of elastic fluids. He introduces the fundamental notion of the average distance that an atom travels between collisions—the mean free path, D—and he shows that the number of times an atom strikes the walls of the containing vessel must be proportional to $1/D$, while the number of atoms making such impacts must be proportional to $1/D^2$. The product of these two, $1/D^3$, gives him the pressure of the gas, and as D^3 is proportional to V, the volume of the gas, it follows that Boyle's law is obeyed. He was also able to show that if the effect of heat is to increase the velocity, 'u', of the atoms the pressure must vary as u^2/V, which is consistent with Amontons' discovery that the pressure of an elastic fluid varies with the temperature.

It is notorious that this very promising pioneer work by Daniel Bernoulli was neglected by succeeding generations of physicists and chemists. According to Drs Talbot and Pacey, there were three main reasons why this happened:[46]

(1) A great deal more physical and chemical knowledge of the properties of elastic fluids had to be obtained before the advantages of the kinetic theory could be appreciated. As things were, over the greater part of the eighteenth century the theory had little or no predictive value, and quantitative as it was, it accounted only for facts that were known and could be explained equally well by other theories.

(2) One of these other theories in particular—the (static) repulsion model—was backed by the great authority of

Newton. Furthermore, it offered some hope that the properties of elastic fluids could be accounted for in terms of central forces (e.g. gravitation). Newton, in a famous and eloquent passage, had written that:[47]

> . . . by propositions mathematically demonstrated . . . I derive from the celestial phenomena the forces of gravity with which bodies tend to the sun and the several planets . . . I wish we could derive the rest of the phenomena of Nature by the same kind of reasoning from mechanical principles, for I am inclined by many reasons to suspect that they may all depend upon central forces by which the particles of bodies, by some causes hitherto unknown, are either mutually impelled towards one another and cohere in regular figures, or are repelled and recede from one another. These forces being unknown, philosophers have hitherto attempted the search of Nature in vain; but I hope the principles here laid down will afford some light either to this or some truer method of philosophy.

(3) Atoms, molecules, corpuscles and particles remained in the last resort speculative. In an age of scepticism and positivism such entities would be suspect for many thinkers and the best scientific explanations would be those that dispensed with them. Not until after the work of John Dalton did the scientific status of atoms become generally respectable.

If, therefore, 'physics', as the subject was later to be called, had made quite respectable progress in the study of heat, there were still some formidable obstacles in the way, and the next significant contributions were to come not from that science but from a collateral one. This was not a setback or a defeat; after all, scientific knowledge rarely progresses in a strictly linear fashion. There is constant interplay between the different branches, and without such cross-fertilisation it seems probable that science would soon ossify. Accordingly we must now turn to the researches and speculations of the chemists of the period.

Hermann Boerhaave (1668–1738), Professor of Medicine, Chemistry and Botany at the University of Leyden, was a worthy upholder of the great tradition of medical teaching, which went back to the Middle Ages. Such was his reputation that students from all over the civilised world attended his lectures. When they returned to their own countries these men propagated Boerhaave's teachings, and it was in this way that the medical schools in Scotland and in Vienna were revivified, subsequently acquiring a distinction for medical science that they still retain.

However, Boerhaave is of interest for us because he was an able chemist and the first one, with the exception of Boyle, to contribute significantly to the science of heat. Like Desaguliers he had had the doubtful pleasure of seeing his works published in a pirated edition in England in 1727 and, again like Desaguliers, he was provoked into publishing an authorised version (1735).[48] A third version appeared in 1741.

Boerhaave's lucidly expounded ideas set the pattern for his contemporaries and successors. He was basically a Cartesian and his concern was therefore with 'Fire', which he held to be distributed throughout all bodies and throughout 'space'. He maintained that Fire is the great universal agent of change, and is abundantly present even in the coldest things under the coldest conditions; for it had been shown that one could strike sparks from a flint in the depths of an arctic winter. An ultimate limit of cold can be imagined, he says, when a body would be wholly deprived of Fire, but such a state cannot be reached either by nature or by technics: it is therefore not worth bothering about. In fact, Boerhaave's attitude throughout was one of common sense, so even when he was wrong, by later standards, he was never unreasonably or dogmatically so.

He sets out as an essential axiom the proposition that Fire is always conserved; he argues that it cannot be created *de novo*, as claimed by certain English philosophers, and that in all changes the total quantity remains unaltered. Now this statement was perhaps crucial, for it is very difficult to see how a science of heat could have developed without a basic conservation principle: such a principle was plainly essential for the formulation of the concepts of quantity of heat, of specific heat and of latent heat. Conversely, the lack of such a principle was a fundamental weakness of the embryonic dynamical theory of heat (as proposed by Bacon, Boyle, Newton, and others), and may well account for its comparative failure in the eighteenth century. Only with the establishment of the doctrine of the *conservation of energy* in the mid-nineteenth century could the dynamical theory come into its own.

Fire manifests itself in a number of ways but the only infallible test for it is, according to Boerhaave, the expansion of the body to which it is imparted. Fire seems to be quite weight-less and shows itself actively only when it encounters some resistance; otherwise it is transmitted without effect. Heavy substances like iron seem to have a distinctive ability to retain Fire. There can be intense heat at the focus of a large burning-glass, but if the sunlight is suddenly cut off it vanishes at once; on the other hand a piece of iron or of some other metal at the focus will retain the heat for a long time.[49] Judged by the criterion of expansion, rarer substances like air or water respond much more sensitively to Fire than do denser ones like mercury. Again, Boerhaave noted that a dense liquid like mercury can cool down hot bodies much more rapidly than a less dense one like water. Perhaps, then, different substances have different attractions for Fire depending on the nature of the substance and in particular on its density? Apart from the problem of conductivity, it might be thought that Boerhaave was within reach of the concept of specific heat. But he was considering the general causal agent Fire, and not heat; between Fire on the one hand and its thermal effects on the other the causal nexus was not necessarily direct or simple. As he put it: '. . . altho' we are able to discuss the Force of Fire by its sensible effects; yet we cannot from its Force make certain judgements of its quantity'.[50]

If, says Boerhaave, two equal quantities of water at different temperatures are mixed the resulting temperature will be the arithmetic mean of the two. But if we take equal volumes of water and mercury at different temperatures and mix them the final temperature seems to bear no relationship to the fact that mercury is fourteen times heavier than water; in fact it seems to be fairly near the arithmetic mean. To get a final temperature that *is* the arithmetic mean we have to mix three parts by volume of mercury with two parts of water. The ratio of 3 to 2 is so far removed from 1 to 14 that the theory that bodies attract Fire in proportion to their densities is decisively refuted; while 3 is sufficiently close to 2, at least for Boerhaave, to confirm the view that Fire distributes itself uniformly in space irrespective of particular substances.

Boerhaave was later criticised for drawing this conclusion. His defence would have been that he was concerned with *Fire* and not, as his critics thought, with *heat*,[51] for, as he had care-fully pointed out, we cannot make *certain* judgements about the quantity of Fire from measurements of its force. This would be quite fair, but if thermometer readings were reliable enough to refute the theory that Fire is attracted to substances in proportion to their densities, the difference between 3 and 2 is surely great enough to cast doubt on the rival doctrine that Fire is uniformly distributed by volume. Boerhaave himself called the thermometer an '. . . infallible instrument'. In short, although his critics did not do him justice on this point they had some reason on their side.

No less important, as we shall see, were Boerhaave's observations on the effects of geometrical form and scale on the heating and cooling of bodies:

> 'It is demonstrated by the Geometers that if Bodies remain the same in all other circumstances, the bigger they are the less surface they will always have in proportion to their solidity . . .'[52]

The result is that larger bodies retain their Fire longer and take longer to heat up because their surface areas are less per unit volume compared with smaller bodies. Dividing an iron bar into shorter lengths increases the total surface area, and the pieces accordingly heat up or cool down more rapidly. Water placed in a spherical glass flask will stay hot much longer than if it is poured into a shallow, flat dish. Although the argument is not rigorous in the mathematical sense, the intuition that geometrical factors determine the heating and cooling of bodies represents an extension to the science of heat of the principles pioneered by Galileo in the first section of *Two new sciences*. These observations on the effects of geometrical factors on the heating and cooling of bodies very probably influenced James Watt in his early investigations of the steam-engine; they may be said to have reached fruition with the publication of J. B. Fourier's *Analytical theory of heat* nearly a hundred years later.

Thus Boerhaave systematically explored the properties of the basic agency Fire with its associated thermal phenomena. His role was that of a minor Aristotle in that he established a morphology for the subject, and it is hardly an exaggeration to say that the most fruitful theories and experiments of the next fifty to a hundred years were proposed or carried out within the framework he had established. Indeed, one especially shrewd comment might have been thought pertinent more than a century later:

> 'Among all the Bodies of the universe, that have hitherto been discovered and examined, there never was yet found any one, that had spontaneously, and of its own nature, a greater degree of Heat than any other.'[53]

Boerhaave's immediate followers were much less interesting. Musschenbroek, a disciple of both Boerhaave and Desaguliers, published a popular and widely read *Essai de physique* in 1727 which was subsequently translated into English. He reproduced Boerhaave's theories and offered further evidence in support of the doctrine that bodies are increasingly sensitive to Fire as their densities are less. This caught the attention of George Martine (1702–41), an able young Scottish doctor who (between 1738 and 1740) published a number of essays on heat and thermometry.[54]

Martine pointed out that practical evidence did not support this plausible doctrine. He filled two identical glass flasks, one with mercury and the other with water, and placed them equidistant from a large fire. Each flask had a thermometer in it and temperature readings taken at regular intervals showed that the dense mercury warmed up much more rapidly than the less dense water. The converse experiment in which the heated liquids were allowed to cool down in a cold room confirmed that the thermometer in the mercury fell much more rapidly than the one in the water. There could be no doubt: substances are not increasingly sensitive to Fire as their densities are less.* We must not read too much into these experiments. They were purely 'operational', and concerned solely with the rates at which substances heated up or cooled down; they were not intended to throw any light on specific heats or upon the absorption of radiant heat. These things had not been discovered at that time.

The idea of quantity of heat is implicit in Martine's work and the metaphysical entity Fire is dismissed without comment. The correlative notion that the (finite) amount of heat in bodies must depend on the lowest possible temperature leads him to envisage the possibility of absolute zero. We are not yet,

---

* As Boerhaave would have put it. Martine referred only to the heating and cooling of bodies.

he wrote, in a position to determine this, but Amontons' experiments on the thermal expansion and contraction of air show us that it should lose all its elasticity at a temperature of 431 degrees below freezing-point ($-239°C$). Was this the absolute zero: the temperature at which air would be wholly deprived of its heat? Possibly, but Martine had grave doubts about the regularity with which air expands and contracts (see page 21).

This excellent work, carried out by a man whose career was tragically short, marked the beginning of the great Scottish school of science. Subsequent members of this school—Black, Watt, Leslie, Forbes, Kelvin, Rankine and Maxwell, together with several others—were to make major contributions to the development of the doctrines of energy and thermodynamics.

Summarising briefly the progress achieved by the middle of the eighteenth century, we recall that the fundamental axiom of the conservation of heat had been propounded: rather tentatively and on analogical grounds by men like Desaguliers and Brook Taylor; more explicitly but on metaphysical and rather misleading grounds ('Fire') by Boerhaave. At the same time the notion of quantity of heat was just emerging as a concept distinct from that of intensity or degree of heat. Curiously, no one claimed to be the inventor of this concept. It seems that a number of scientists started using it independently, more or less simultaneously, and without any fuss. It was a concept that led inevitably to the postulation of an absolute zero of temperature and, a little later on, to the discovery of specific and latent heats. Finally, we must mention the work of the two Baltic scientists G. W. Krafft and G. W. Richmann, who tried independently to discover a formula for the temperature of any mixture of two different quantities of water at different temperatures. The problem is, of course, complicated by the absorption of heat by the containing vessel and the thermometer as well as by losses during the experiment. Richmann (1747–8) was aware of this, but the concept of heat capacity, or specific heat, had not yet been formulated.[55]

Chapter 2

The Eighteenth-Century Contribution

THE SPREAD OF THE FIRE-ENGINE

However arbitrary and fragmented was the pattern of the political map of Europe in the middle of the eighteenth century, that of the scientific map was simple and uncomplicated. Paris was unquestionably the scientific and intellectual capital of Europe, and the important provinces were Scotland, Switzerland and the Scandinavian states together with an exotic colony in Russia (St Petersburg). For the rest, although able men like Cavendish, Priestley and Smeaton in England were to be found in most countries, they were not supported by active scientific institutions and societies. Germany was at the nadir of her political, economic and military fortunes, and even Holland, after Boerhaave, lost her eminence. Many of the ablest German scientists had to seek their fortunes abroad in cities like Paris or St Petersburg.

Mining was the most important growth industry and was the pacemaker for technology. The Newcomen engine was the most notable technological innovation of the time and its use had spread rapidly in the second quarter of the century: all over England, in Scotland and abroad. On his return to Sweden in 1726 Martin Triewald F.R.S., set up a Newcomen engine at the Dannemora mines, and subsequently provided the world with one of the first detailed accounts of the engine together with a careful illustration of it.[1] A year or two before the Dannemora engine, one had been erected at Schemnitz (Banska Stiavnica, in what is now Slovakia) which was at the centre of one of the leading mining areas of Europe and the seat of what was perhaps the first modern technical college, established by Maria Theresa in 1770.[2] It was used in France, Spain, Russia and the Americas. Two independent attempts were made—by

Gensanne and by the Portugese engineer de Moura—to revive Savery's engine by making it self-acting, but it could not compete with the triumphant Newcomen engine as a 'miner's friend'.

Later, in 1780, Gabriel Jars gave a revealing account of how five Newcomen engines had been erected at Windschacht, near Schemnitz.[3] They had been put up by an Englishman* who had demanded in payment for his services only the savings over the previous pumping methods that the machines achieved in ten years. Jars pointed out that horses provided virtually the only competition for these wood-burning engines and, he added, '. . . this Englishman made a considerable sum'. This was a time of growing power shortage, and as industry made increasing demands for this precious commodity, so men necessarily became very cost-conscious. There was no engineering industry to manufacture Newcomen engines and the man with the 'know-how' had to obtain payment for his services as best he could. Hence the novel system used at Schemnitz; a system which James Watt was later to use in Cornwall and which many people wrongly believe he actually invented.

In Britain John Smeaton was undoubtedly the leading engineer, and in the third quarter of the century he was able to approximately double the efficiency of a group of Newcomen engines on the north-east coast of England. He kept every component of the engine constant except for one, which he varied systematically. He noted which variation gave the best performance and then repeated the procedure for every other component in turn until he had produced an engine of optimum performance. The average 'duty' of the engines was raised from about 4·7 to about 9 million ft. lb. for the consumption of a bushel of coal. In this way Smeaton added a new component to the historical process of technological advance: that of systematic evolutionary improvement.

Smeaton's improvements, however, were too late in the day to save the original Newcomen engine; for already in Scotland the new school of scientists was doing work that was to transform the science of heat as well as the technology of power and, ultimately, to lead to the establishment of thermodynamics.

* Possibly Isaac Potter.

The first member of the Scottish school, apart from the ex-patriate George Martine, was William Cullen (1710–90). A teacher of medicine and chemistry in Glasgow and Edinburgh, he actually published very little. His single paper on heat, published in 1756,[4] dealt with the phenomenon of cooling by evaporation, particularly under reduced pressures, and he called attention to the fact first noted in the seventeenth century that water boils at lower temperatures as the pressure is reduced. The paper seems to have aroused some interest, and it influenced the course of James Watt's researches.

Although Cullen's pupil Joseph Black (1728–99) was one of the most influential scientific thinkers of the eighteenth century it is, as we shall see, difficult to assess his contributions to knowledge. In 1754, after submitting a brilliant thesis which won him the M.D. degree of the University of Edinburgh and which was soon recognised as a classic of chemistry,[5] Black settled down to a career as a university teacher of medicine and of chemistry, first at Glasgow and then from 1764 onwards, at Edinburgh. His career was similar to that of Boerhaave in that he was a professor of medicine who was also a distinguished chemist and a famous teacher. Indeed, in the last respect he may even have surpassed Boerhaave. Many men who were to become famous in different walks of life attended his lectures and subsequently testified to his abilities as a teacher. Lord Brougham—surely an expert witness?—remarked towards the end of his long life that Black was the greatest public speaker he had ever heard.[6] It is even possible that Black's success as a teacher limited his achievements as a scientist, for he published little during his career. We can think of several reasons for this,[7] but whatever the ultimate explanation may be it is a fact that Black never published his lectures. Our access to his ideas must therefore be indirect: through an anonymous account of his doctrines published in 1770, surviving lecture notes taken by students, the statements of his contemporaries, and Robison's edition of the lectures, published after Black's death.[8] These are sufficient to give us an insight into Black's ideas and teachings from about 1760 onwards.

Black was indebted to Boerhaave for the framework of his

ideas. But he had had the benefit of the advances made by men like Martine, and his philosophical standpoint was quite different from Boerhaave's. The latter had postulated Fire as the general principle of change but had been unable to bridge the gulf between familiar thermal phenomena and their postulated metaphysical cause. In contrast, Black's attitude smacked of the pragmatic positivism that sometimes characterised Newton: we do not know the cause of gravity but we can devise quantitative laws that account for the behaviour of bodies when gravity acts upon them. Accordingly, Black confessed himself agnostic on the nature of *heat*, and with a sound grasp of the realities of the situation ignored the elusive 'Fire' to concentrate on the operationally intelligible measures of temperature and 'quantity of heat'. He formulated the latter quite naturally, and was almost certainly unaware of the great advance that it represented. It may have seemed to him that in view of the work of Brook Taylor, Desaguliers and Martine it was not much more than a form of words to describe a measure that they had already used. The conservation principle is maintained but is tacitly transferred from Fire to heat.

Here we should remember the intellectual climate of the time and the place. Black's contemporaries and friends in Edinburgh and Glasgow included Adam Smith and David Hume. Generally speaking, the philosophical temper of these men was rational, sceptical, tolerant and positivistic. It does not matter for our purposes whether Hume, who was older than Black, actually influenced the younger man or whether his writings brought into sharp focus certain dominant ideas and beliefs which they all shared. We can certainly appreciate how Boerhaave's metaphysical notions of Fire as a causal agent would be unacceptable to anyone who sympathised with the philosopher who so destructively criticised the idea of causation, and who argued that there is nothing in any object, considered in itself, which can afford us a reason for drawing a conclusion beyond it. Lastly, we note that Black was a friend and supporter of James Hutton the geologist, whose 'uniformitarian' doctrine represented an application of positivist principles to the new science of geology.*

* The uniformitarian principle lays down that in attempting to account for the nature of the earth's surface we must assume that no agency acted in the past that we cannot actually observe at work today: frost, rain, wind,

If we have formed an idea of 'quantity of heat' it must soon appear that the same amount of heat exerts different thermal effects on different substances, warming some a great deal and others much less. This necessarily leads us to the concept of heat capacity. The Boerhaave–Fahrenheit experiments, as Black interpreted them, prove very clearly that heat does not distribute itself among bodies in proportion to their densities or to their volumes. It follows, then, that heat distributes itself according to some specific determinant for each particular substance.

Here we come very close to the concept of specific heat, quite untrammelled by metaphysical entities like Fire. But there is a difficulty with Black's conception.[9] He states that Martine's experiments show that the heat capacity of water is much greater than that of mercury, thus going far beyond Martine's own interpretation of his results. Black happened to be correct, but the facts did not justify the inference. The shiny surface of the mercury must have reflected a great deal of radiant heat; the transparent water must have transmitted a substantial amount; there was no assurance that the heat absorbed by the two liquids was the same. Black cannot be blamed for knowing nothing about radiant heat, which had not been identified at that time, but his scientific instincts must have been asleep, for it was well known that a glass flask of water can act as a burning-glass.

If we take two equal cubes, one of iron and the other of wood, heat them in an oven and then hold them in the hand, the iron cube will feel hotter for much longer than the wooden cube; whence Black infers that iron has a greater capacity for heat than wood. But, in fact, all this proves is that iron is a much better conductor than wood; had Black split the cubes open he would have found the wooden cube much hotter inside than outside. It was on experiences of this sort that Rumford later based some of his discussions of conductivity. Once again, Black is not to be blamed for ignorance of the concept of conductivity, which had not been distinguished at that time.

It seems, therefore, that Black's idea of heat capacity was, as it were, hazy round the edges. It was not quite the same as

avalanche, deposition of silt, earthquakes and volcanoes, coastal and river erosion etc. This, therefore, is a positivistic doctrine.

'specific heat', for thermal phenomena due to differing conductivities or absorptivities were attributed to it. It was perhaps a measure of the facility with which a substance absorbs or gives up heat. Black, we may note, explicitly asserted that heat enters all bodies—other than spongy ones with air spaces in them—at exactly the same speed, so that the rate at which a body heats up is determined solely by its heat capacity.

The concept of heat capacity, though imperfect, was fruitful, and destined in time to evolve into the concept of specific heat. Only a rigorous purist would deny Black the credit for this pioneer work. What does disappoint us is his failure to develop and apply the new idea. He does not offer us a list of the heat capacities of different substances compared with that of water, nor does he give us an account of the methods to be used in determining heat capacity. His attitude is, in fact, more qualitative than quantitative. According to some of the lecture notes, he mentioned by way of illustration the heat capacities of four or five common substances.[10] The Robison edition mentions one determination of heat capacity: a pound of gold at 190 degrees* is put into a pound of water at 50 degrees and the final temperature of the mixture is found to be only 55 degrees. As the gold has been cooled by 135 degrees and the water heated by only 5 degrees it is evident that, weight for weight, water has a much greater capacity for heat than has gold, the ratio being 19 to 1.

It may well be that Black had some intimation of the difficulties that we have mentioned, and these may have deterred him from taking his idea further. However, his other notable discovery, that of latent heat, was hedged with fewer ambiguities.

Whatever the combination of experiment and reflection by which men arrive at the idea of 'quantity of heat', its acceptance is logically prior to the discoveries of specific and latent heats. Once we start to think *quantitatively* about the simple process of boiling water we are forced into a dilemma, for which the only solution is the invention of the concept of latent heat. When water is boiling the temperature, as Amontons showed, remains constant, yet heat must be entering it all the time. Black calls

* Only in the second half of the eighteenth century did the now universally accepted symbol for the degree of temperature (°) become common; before then the expression for a degree was written, very often as 'gradus', or for short, 'gr.'.

this heat, which produces no sensible effect on the thermometer, the latent heat, and it is specifically associated with change of state. An exactly similar argument indicates that the melting of ice is associated with the absorption of latent heat.

Black gives us two examples from common experience which show how important this phenomenon of latent heat really is. A kettle of water at a temperature marginally or infinitesimally below the boiling-point would, on the common-sense view, require the addition of only a trivial amount of heat to raise its temperature *above* the boiling-point. But if this happened the water would be all converted into steam, technically into super-heated steam, with explosively disastrous consequences! This does not occur because a vast amount of latent heat is required if water is to be converted into steam. Again, on the common-sense view only a trivial amount of heat would serve to convert a vast snowfield just below freezing-point into (disastrous) flood water. This does not happen; snow is notoriously slow to melt, and on the northern slopes of the highest Scottish mountains it lingers on through summer. Once more the doctrine of latent heat provides the explanation. In this way an advance in the science of heat gave new insights into a range of meteorological and other 'transformation' processes. One is reminded of Halley's explanation of the trade winds.

The methods Black used to measure latent heats were ex-tremely simple. He compared the time taken to heat up a certain amount of water from a known temperature to boiling-point with the time taken for the water to boil away. Assuming that heat was entering the water at a constant rate, he computed by simple arithmetic the temperature that the water would have reached had it not been turned into steam. The ex-cess of this temperature above boiling-point is his measure of the latent heat of vaporisation of water. In other words it is the measure of the heat that has entered the water during the time it was boiling; the value he obtained for it was about 960 degrees Fahrenheit. His method for determining the latent heat of fusion of ice is essentially the same: he compares the time taken for a piece of ice at freezing-point to melt with the time taken for it, in the form of water, to heat up from freezing-point to a given temperature. The figure he gets for the latent heat of fusion is 140 degrees Fahrenheit.[11]

We must not interpret Black's idea of latent heat in terms of so many B.Th.U.s per degree Fahrenheit, or so many calories per degree centigrade, even though his results are numerically very close to the modern figures. There is a difference in interpretation. For Black the nature of heat is inscrutable, and he has no means of directly determining a given quantity of it; he has no 'heat meter'. But he can use temperature as a *measure* of quantity of heat, provided that he specifies the conditions of measurement so that quantity and temperature are not confused. The procedure is quite simple. A length is not the same thing as a velocity but we can, if we wish, use length as a *measure* of velocity. A body moving with a velocity v can rise to a height h, which is determined by the simple formula $v^2 = 2gh$. We can therefore use h as a *measure* of velocity, and if we are careful to specify that it is the (rising) height to which the velocity corresponds there need be no confusion. In the same way we can use temperature as a measure of quantity of heat, and there will be no difficulty if we remember that, the body being specified, a rise in its temperature is the measure of the quantity of heat entering it.[12]

Black's measures, then, are operational. His discoveries are independent of any particular theory of the nature of heat, whether it be a subtle fluid, an atomic vibration or undulations in an hypothetical aether. Discussions about the ultimate nature of heat are not, observed Black, very useful in the present state of knowledge.* And although a quantitative procedure suggests that we can conceive of bodies wholly deprived of heat, and therefore at an absolute zero of temperature, there is little point in speculating about such a remote state. Indeed, Black remarked, experience suggests that there is no body so cold that a suitable freezing mixture will not make it colder.

Difficult as it is to fault Black's positivist approach, it did carry an implication that was to cause trouble later on. If we assert that the heat released by a given weight of water in freezing would heat the same weight of water by 140 degrees, can we expect to find instances in nature in which the freezing of water in one place causes a rise in temperature elsewhere? Can we ever expect low temperature heat—the latent heat of

---

* This was his attitude in public; privately he later came to favour the material theory of heat, commending William Cleghorn's idea (1779) of heat as a subtle fluid composed of atomic particles, which repel one another but are strongly attracted to the atoms of ordinary matter.

fusion—to pile itself up to produce high temperatures? With the second law of thermodynamics behind us we can see that this is simply not possible; but in Black's day, when so many of even the simpler processes of nature were obscure, the answer was not obvious. There is, after all, nothing *clearly* absurd about the statement that one often finds it colder during a thaw because a vast amount of (latent) heat has to be absorbed to melt all the snow and ice.

Black's great achievements were to set up the fundamental concept of quantity of heat and, by considering its implications, infer the correlative concepts of heat capacity and latent heat. Having done all this by about 1760, and having also seen the importance of the two latter concepts in the processes of nature, Black did nothing further; he did not even publish his discoveries. If we agree that persuading others of the significance of his discoveries is a duty of the scientist, and when we recall the indifference and even hostility that, for example, J. P. Joule and J. R. Mayer suffered when their works were published and before their ideas were accepted, we must concede that Black's claims are a little tarnished; he did not play the game according to the rules: his friends played it for him. When, therefore, J. C. Wilcke published his discovery of the latent heat of fusion of ice in 1772, and later, in 1781, his discovery of what he called 'specific heat', he deserved full credit as an original and perceptive scientist, for he knew nothing of Black's work.[13] More, Wilcke gave precise details of his experimental methods, and drew up a list of specific heats of different substances, which had been determined by the method of mixtures. His work was carried on by the Finnish scientist Johan Gadolin.[14]

While Black was engaged in his desultory researches James Watt (1736–1819) was also at Glasgow University, employed as an instrument maker. In the comparatively open society of the eighteenth-century Scottish University, the relationship between the young craftsman and the distinguished professor was friendly, and as Black's work on heat and Watt's technological researches were collateral it was quite natural, perhaps even inevitable, for people to assume that Black's scientific discoveries directly stimulated Watt's technology. The exact relationship between the reticent, able but academically indolent professor and the inventor and applied scientist of genius has been a subject of discussion ever since and, at the same time, the source of a persistent but demonstrably false legend. The legend is that Black told Watt about his discovery of latent heat, whereupon Watt perceived the importance of economising the consumption of steam in the operation of the Newcomen engine, and to this end invented the separate condenser. The source of this legend was John Robison, who published it in a famous and in most other respects ably written account of the steam-engine,[15] and also in his edition of Black's lectures. Neither Black nor Watt was given to publishing their works and Robison was left as the best authority, for he had been closest to both the leading characters. And although, as the Rev. Vernon Harcourt pointed out a long time ago, Robison was a most unreliable historian,[16] and although Watt afterwards denied both accounts—in detail, authoritatively and on several occasions[17]—the legend has been more or less generally accepted up to the present day. The following quotations will confirm this.

Dr Black . . . published [*sic*] his important and beautiful discovery of latent heat. The knowledge of this doctrine led Watt to reflect on the prodigious waste of heat in the steam engine, where steam was used merely for the purpose of creating a vacuum . . .—Edward Baines, *History of the cotton manufacture in Great Britain* (London 1835), pp. 222–3.

In the 1850s J. P. Muirhead published extensive selections from the writings and correspondence of his kinsman Watt.[18] As these included Watt's written denials of the Robison legend it is hardly surprising that subsequent writers, such as Galloway and Goodeve,[19] paid less attention to it and gave good accounts of the history of Watt's inventions. However, in the dull closing years of the nineteenth century, the epoch which in Whitehead's words '. . . celebrated the triumph of the professional man', interest in the history of science and the history of technology waned; it did not revive until the end of the Great War. By then Muirhead's work was in the Victorian limbo and historians going back to their sources found—Robison! Sir William Ramsay, in his biography of Black (1918)[20] was perhaps one of the first to revive the legend. Other writers followed suit.

. . . Watt's improvement of the steam engine . . . was directly derived from Black's discovery of latent heat . . .—D. McKie and N. H. de V. Heathcote, *The discovery of specific and latent heats*, p. 122.

. . . Watt's impulse came from Black, the man who, besides his work on heat, began the chemical revolution . . .—H. T. Pledge, *Science since 1500* (H.M.S.O., London 1947), p. 110.

Watt (1765) applied Black's discovery (of latent heat) to his contrivance of the separate condenser.—Charles Singer, *A short history of scientific ideas* (Oxford University Press, 1961), p. 300.

The Doctor has had a good press. The best way of disposing of the Black–Watt legend, which has evidently misled able historians for too long, is to describe the actual sequence of Watt's discoveries and then to demonstrate that the legend makes no sense *in terms of basic physics*. The demonstration should then be as conclusive as it is possible to be in the history of science.

Watt's interest in the possibilities of generating power from steam appears to have been aroused before 1760. In those days he was an instrument-maker and a scientist, not an engineer. He had had no experience of Newcomen engines nor it seems of any other large-scale machines, and his knowledge was derived from readings of Desaguliers and Belidor.[21] Accordingly he took the obvious step, for a speculative inventor, of investigating the possibilities of using steam *pressure* to develop power; no experienced engineer, aware of the Newcomen engine's qualities and of the difficulties of using high-pressure steam would, one feels, have contemplated such a thing. But Watt was young, and with his friend Robison, who was then an undergraduate, he carried out experiments with a Papin's digester—an early form of pressure cooker—in 1761 or 1762. He used a syringe with plunger as a makeshift cylinder and piston, and he found that the pressure of steam from the digester was enough to cause the plunger to raise about fifteen pounds, a considerable weight. Promising though this might have seemed, Watt abandoned the whole idea as he, too, realised that it was impossible to make a boiler that would stand the necessary high pressures for any length of time. The main problem of the Savery engine had not been solved.

Watt's real opportunity came from quite a different quarter. In the winter of 1763–4 he was engaged in repairing the model

Newcomen engine that was used in Professor Anderson's natural philosophy class. When he had done this he noticed that the little engine could perform only a few strokes on load before it ran out of steam and stopped. This puzzled him, as the engine of which the little model was supposedly a scale replica worked perfectly satisfactorily; why then the disparity in performances between the large and the small engines? Watt knew that cold metal condenses steam and he was well aware that the little cylinder became very hot when full of steam. He therefore reasoned that as the volume of the model cylinder was much smaller in proportion to the interior surface area than it was in the case of the big cylinder, the steam in the little cylinder must have been exposed to much more cold metal surface, relatively speaking, than the steam in the big cylinder. It would therefore cool down and condense much more readily. Thus, owing to the 'scale effect' the small cylinder engine could not work as efficiently as the big one. This would have been a remarkable piece of analysis had it had no precedent. However, it seems very probable that Watt was in fact making a straightforward application of Boerhaave's prior ideas about the scale effect on the heating and cooling of bodies of identical form but different sizes (see page 29). Watt can hardly have failed to have known of Boerhaave's teachings on heat; either he read them for himself or heard them expounded by some friend or colleague in Glasgow.

Mr Rodney Law has recently made the important discovery that the model engine is not, in fact, a true scale replica of the big engine: the walls of the cylinder are disproportionately thick.* The relative excess of metal—and therefore of heat capacity—was almost certainly the true cause of the inferior performance of the little engine. It follows that as Watt's explanation cannot have been deduced from the experimental data it must have been based on prior knowledge, and the probability that it was derived from the influential Boerhaave is accordingly strengthened.

If the heating and cooling of the metal cylinder was a factor limiting the performance of a small Newcomen engine, did it seriously affect the efficiency of a big engine? And if so, could the waste it caused be eliminated or drastically reduced?

* See R. J. Law, *James Watt and the separate condenser*, Science Museum Monographs (London 1969).

Watt's next step was to measure the actual consumption of steam in every cycle of the engine and to compare it with the amount that would be required to fill a volume equal to that of the cylinder. He measured the volume of steam produced by evaporating a given volume of water and found that a cubic inch of water yields about one cubic foot of steam. Next, by measuring the amount of water evaporated from the boiler in a given time he could find the volume of steam actually supplied to the engine, and he could compare this with the amount theoretically required, which was merely the volume of the cylinder multiplied by the number of cycles performed. The first volume turned out to be several times greater than the second: the excess of steam evidently representing the quantity required merely to heat up the cylinder after it had been cooled down by the cold condensing water. The waste was, therefore, very serious.

The obvious solution to the problem was, in terms of contemporary physics, to make the cylinder of some material which, while strong enough to stand the wear and tear of continued use, had a much lower heat capacity than iron or brass. After he had carried out a number of tests to find the heat capacities of different substances* Watt concluded that wood was the best material, and he therefore made a wooden cylinder which he soaked in linseed oil and then baked dry. He found that little steam was required to fill a wooden cylinder —after all, wood is a poor conductor of heat and a little steam was enough to warm up the surface layer to about 212 degrees. An economy had therefore been achieved, but the wooden cylinders were not mechanically satisfactory, for they soon warped and cracked. And then another difficulty arose. For maximum economy the cylinder must not be cooled down; only enough cold water to condense the steam should be injected since any excess will cool the cylinder down and so waste heat. The problem, therefore, is to determine the optimum amount of condensing water.

* Dr Talbot points out[22] that this suggests that Watt cannot have heard of heat capacity from Black; had he done so he would not have needed to carry out these tests. It may also be significant that Watt, in his letter to Brewster (May 1814) wrote: 'It was known to some philosophers that the capacity or equilibrium of heat as we then called it, was much smaller in mercury and tin than in water'.[23] But why no mention of Black as the discoverer of heat capacity? Why the plural 'philosophers'?

At this point an evident paradox struck Watt's analytical mind. On the accepted law of mixtures an ounce of steam at 212 degrees mixed with an ounce of water at 52 degrees should yield two ounces of tepid water at 132 degrees. But Watt knew very well that in every cycle much more condensing water was required than was evaporated from the boiler in the form of steam. The law of mixtures in fact broke down. Watt had stumbled on the *phenomenon* of latent heat. He went on to measure the excess heat in steam by bubbling it through a known weight of water at room temperature until it boiled; the gain in weight of the water represented the steam condensed. This steam gave up enough heat, merely by being condensed into boiling water, to warm up a much greater weight of cold water to the boiling-point. The weight of the cold water multiplied by the rise in its temperature gave Watt a figure for the latent heat of vaporisation of water. It was close to that which Dr Black had obtained.

Watt was very puzzled by this breakdown in the law of mixtures and he took the problem to Black, who (as Watt later remarked) '. . . then explained to me his doctrine of latent heat, which he had taught for some time before this period, (summer, 1764) . . .'

The admitted fact that Black explained the doctrine of latent heat to Watt is the ultimate source of the legend that Black's discovery stimulated Watt's invention. But the legend rests on a simple misunderstanding of the word 'explained'. A scientific explanation is a demonstration that a unique and puzzling event is really no more than a particular manifestation of a very common and relatively simple natural occurrence. So we *explain* an eclipse of the moon by showing that it amounts to nothing more than the shadow of the earth crossing the face of the moon; the unique and primitively alarming phenomenon is thus seen to be only another instance of the familiar principle that opaque objects cast shadows. In the same way Watt was puzzled by the unique phenomenon that he had come across experimentally; Black was able to explain it by pointing out that it was consistent with a whole range of familiar occurrences such as the boiling of kettles and the melting of snow. Black's approach had been philosophical and general and he was therefore able to identify the particular as one instance among many others, while Watt had given no thought to the general.

Watt's measurement of the latent heat of evaporation enabled him to compute exactly how much water should be injected to condense all the steam without cooling the cylinder down. But he found that when the engine was operated at this theoretical maximum of efficiency there was a serious loss of power. The engine was now economical but no longer powerful: why was this?

The thermometer confirmed that the temperature of the condensing water coming out of the cylinder was very high. Watt knew, from Cullen's experiments, that tepid water would boil *in vacuo* and that in doing so it would generate steam or water vapour, thus vitiating the vacuum. The problem, then, was to determine the pressure of steam that would be generated by the hot condensing water inside the cylinder. The simplest way to do this was to use the digester to find the law of increase of pressure for temperatures above the normal boiling-point at 212°F, and then to extrapolate back to find the pressure associated with the temperature of the condensing water. Watt actually presented the results which he obtained in this way in the form of a graph: a very unusual procedure for those days:[24] 'From these elements I laid down a curve in which the abscissa represented the temperatures and the ordinates the pressure, and thereby found the law by which they were governed sufficiently near for my then purpose.'

These results are shown in Figure 3 with a graph based on modern data for comparison. The sharp rise in pressure with increase in temperature may, incidentally, have confirmed his judgement that high-pressure steam was impracticable as a source of power, but more immediately the curve told him that the steam pressure due to the hot condensing water in the cylinder would be strong enough to detract seriously from the power of the engine by opposing the weight of the atmosphere on the piston. Watt was thus on the horns of a dilemma. He could not, at the same time, keep the walls of the cylinder boiling hot, have a pool of near-boiling water at the bottom and a near-perfect vacuum inside. In such—hypothetical— circumstances the hot water *must* boil, generating steam and so destroying the vacuum. The water would go on boiling until the appropriate pressure was reached, and the more 'economical' the engine the greater this opposing pressure must be. In fact we can sum up the dilemma quite simply: for maximum

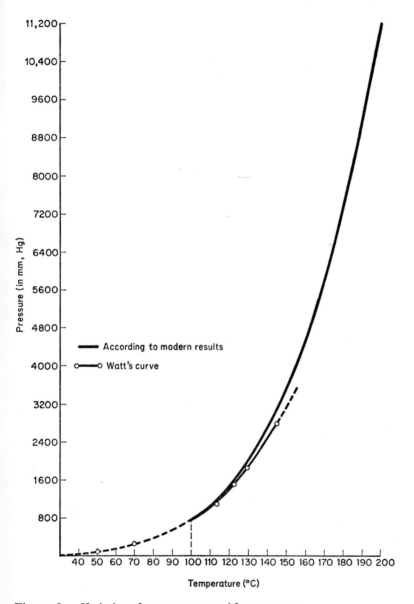

Figure 3. *Variation of steam pressure with temperature*

economy the cylinder must be kept hot all the time, but for maximum power it must be cooled down once every cycle. These desiderata are quite incompatible, for a cylinder cannot be both hot and cold at one and the same time (Figure 4).

Or rather they can be compatible if we have *two* cylinders in place of one; the first to be kept hot all the time, the second cold. Having thus resolved the dilemma in a moment of insight, Watt was immediately able to sketch out the practical details of his improved engine: the true steam-engine. The first or working cylinder, in which the piston moves, is connected at the bottom by a pipe fitted with a valve to the second or condensing cylinder (Figure 5). When the steam under the

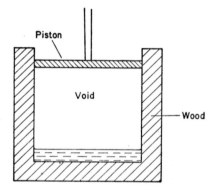

Figure 4.   *The physically impossible 'ideal' engine*

piston is to be condensed the valve is opened so that the steam can flow into the (void) condensing cylinder, or condenser, there to meet a spray of cold water. More steam will follow, and by this Gadarene process all the steam in the working cylinder will be condensed and there will be a near perfect vacuum in both cylinders, although the first remains boiling hot all the time and the second cold. In this way power and economy are reconciled.

Atmospheric-engines of this sort were made, and indeed the last one was working as recently as the early nineteen-thirties. But Watt, a perfectionist, had from the beginning seen the drawback: cold atmospheric air driving the piston down must cool the hot walls of the cylinder and so cause unnecessary waste of heat. To avoid this he proposed to use steam at atmospheric pressure to drive the piston. The modification it involved was simple and is indicated by the dotted lines on Figure 5. After the steam has driven the piston to the bottom of the cylinder, valve V_1 is closed and V_2 is opened, the counterweighted beam causes the piston to rise and the steam is driven via V_2 to the bottom of the cylinder. When the piston has reached the top of the cylinder V_2 is closed, V_1 (which provides access to the condenser) is opened, and the piston is driven to the bottom again. All the valve actions are of course automatic, as in the case of the Newcomen engine.

This then was the remedy which Watt proposed (in May

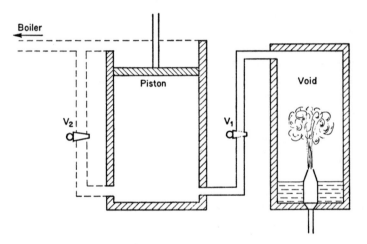

Figure 5. *The Watt Engine*

1765) for the basic defect which he himself had revealed in the operation of the Newcomen engine. The very difficult business of transforming this laboratory solution into an actual working engine, which required a number of auxiliary inventions such as an air pump in place of the snifting valve, does not concern us here.[25] The fortunate intervention of John Roebuck* and later the partnership with Matthew Boulton belong rather to economic and business history than to the history of technology or of science.

There are however two general points concerning Watt's researches that we must make. In the first place Watt's invention

* Roebuck had studied under Boerhaave.

c

of the condensing-engine, based as it was on scientific research and technological insight, could not in principle have been achieved by the Smeatonian method of evolutionary improvement (page 33). This method, used by Smeaton himself, indicated that cooling the cylinder right down and then heating it up again was such a slow process that the power of the engine was seriously reduced. Maximum power could be obtained if the cylinder was kept warm—that is, if some steam was left in it after 'condensation'—and heat losses could be reduced to a minimum for the most economical working by lining the outside of the cylinder with wood. In other words, quite apart from Smeaton's abilities as an engineer, his method logically implied the perfection of the Newcomen atmospheric-engine and nothing more. Watt's revolutionary invention belonged to quite a different order from that of Smeaton's evolutionary improvement.

In the second place it must be emphasised that Watt's investigations were always guided by cost considerations; he was very careful to establish that the alternate heating and cooling of the cylinder *did* cause a big consumption of heat—this was by no means obvious—and then that this particular diseconomy was *the* major one in the operation of the engine. Inefficient combustion might easily have been a more serious and a *more easily remediable* cause of waste. Indeed James Brindley had already tried to reduce the coal consumption of a Newcomen engine by putting the furnace and flue inside the boiler instead of underneath it. Brindley had even experimented with wooden cylinders. But he lacked Watt's scientific knowledge and insight; he was a millwright and was naturally prejudiced in favour of wood, a relatively cheap and easily worked material which he used whenever possible.*

There remains the problem of the relationship between Watt and Black. Surely there is nothing improbable in the supposition that two men whose interests overlap might work in the same university at the same time? We know, for example, that much later on Kelvin and Rankine were both working more or less independently at the University of Glasgow on the fundamental

* Samuel Smiles, *Lives of the engineers*, third edition (London 1874), Vol. 1, pp. 149–52. See also John Farey, *A treatise on the steam engine* (London 1827) and C. T. G. Boucher, *James Brindley, engineer 1716–1772* (Goose and Son, Norwich 1968), pp. 39 ff.

theory of thermodynamics. Later still, at Cambridge, J. J. Thomson was working on the cathode ray 'corpuscles', while Joseph Larmor was studying 'electrons' a few hundred yards away. They were both studying the same thing but were substantially ignorant of each other's activities. Watt was a man of genius, Black was a man of the highest abilities; and there is no reason why they should not have worked out their ideas independently of one another, even though Watt may have benefited from the occasional discussions he must have had with Black.

The case against the Black–Watt legend can, however, be put on other grounds besides circumstantial probability and historical documentation. The legend in fact violates the elementary principles of physics. Anyone with access to a table of physical constants and a knowledge of the substance and weight of a Newcomen engine cylinder can easily confirm that by far the greater part of the heat supplied must be used to warm up the cool cylinder at the beginning of each cycle. The heat loss in a Newcomen engine is demonstrably in and through the heavy metal cylinder: the latent heat of vaporisation is wholly irrelevant, for the loss would be exactly the same if the latent heat were zero. Things might be even worse, for in this hypothetical case about six times as much steam would be needed to heat up the cylinder, and this might have posed impossible problems for the engineering of those days. We therefore reach the paradoxical conclusion that, contrary to the legend, the relatively high latent heat of vaporisation of water was a positive *advantage* in the working of the Newcomen engine.

It was characteristic of Watt's analytical mind that he should speculate about the waste of power represented by the Gadarene rush of steam from the cylinder into the condenser.* When the piston is at the top of the cylinder the exhaust valve, V_1, is opened and all the steam rushes out of the cylinder with gale force into the condenser, thus wasting power. That this *is* a waste of power is quite clear, for if we were to put a small windmill or turbine in the exhaust pipe the rush of steam could be made to do useful work. How, then, can we economise the power at present running to waste every time the exhaust

* Did this perhaps show the influence of Smeaton? Smeaton had recognised the waste of power represented by water leaving a waterwheel with excess velocity (see below, p. 69).

valve V_1 is opened? Watt's solution was to turn off the supply
of steam from the boiler before the piston had travelled more
than a fraction of the way down the cylinder.[26] The steam
would go on driving the piston down, though with steadily
diminishing pressure as its volume increased, until when the
piston reached the bottom of the cylinder, the steam pressure
would be only marginally above the residual pressure in the
condenser. The much expanded steam is now shunted, via V_2,
under the piston, but when V_1 is opened so that the next cycle
can begin this steam is so attenuated and its pressure is so low
that there can be no wasteful gale of steam into the condenser.
In this way Watt indicated the ultimate limit of efficiency—the
extraction of the last drop of 'duty' from the last wisp of steam.

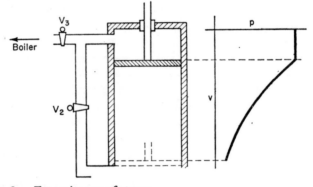

Figure 6.   *Expansive use of steam*

By his invention of the expansive principle (1769) this meticulous Scotsman foreshadowed the progressive improvement of
heat-engines and the postulation by Sadi Carnot of a general
theory of the motive power of heat. With astonishing insight
Watt had laid one of the cornerstones of thermodynamics. The
next steps in this direction were to be the realisation that there
is a relationship between the amount of heat supplied and the
work done; and that for a given amount of heat there is a
maximum of work that cannot be exceeded.

All this did not, however, exhaust Watt's contributions to the
formulation of the concept of the heat-engine and the establishment of thermodynamics. There was also his work on 'total
heat' which led to the formulation of the important, if inexact,
'Watt's law'; and lastly there was the invention of the indicator

diagram with which his name is associated. We shall discuss the invention of the indicator diagram in Chapter Three.

The fact that water under reduced atmospheric pressure boils at progressively lower temperatures suggested to Watt that great fuel economies might be achieved by boiling *in vacuo*. Economies of this sort might be significant in such processes as distilling, and might even be of marginal advantage to the steam engine.* Accordingly Watt lost little time in trying to measure the amount of heat required to warm up and then vaporise the same amount of water under different pressures. He was disappointed. In 1766 Black was telling his students that Mr Watt had discovered that, contrary to what might be supposed, steam generated at low pressures and temperatures imparted just as much heat to the refrigeratory, or still, as steam generated at normal atmospheric pressure; in other words, that the 'total heat'—the sum of the sensible and the latent heat—was about the same in both cases.[27]

Generally, Watt found that as the temperature and pressure of steam increased so the latent heat decreased. Was the increase in sensible heat exactly compensated for by the decrease in latent heat, so that the sum of the two was always constant? Watt did not commit himself on this point; for him it was enough that boiling under reduced pressure did not lead to a tangible economy of heat and therefore of fuel. As we shall see when we consider the work of John Southern, there are grounds for believing that Watt himself doubted whether the total heat was always constant. For the present we note that the late Professor Partington was misled when he asserted that J. A. de Luc, working in Watt's Birmingham laboratory in the Spring of 1783, made experiments on the latent heat of steam '. . . on which the so-called "Watt's law" is based'.[28] As Black's remarks indicate, experiments of this sort were made by Watt himself long before 1783; and even more conclusive is de Luc's own testimony. No one, wrote de Luc, knows more about the properties of steam than does Mr Watt, who in his (de Luc's) presence had carried out a series of experiments to demonstrate that as the temperature of steam varied, above or below

---

* This, of course, was very problematical. A reduction in the steam pressure would mean a loss of power in the engine, and it was by no means certain that this loss would be more than compensated for by an enhanced heat economy.

212°F, so the latent heat decreased or increased.[29] Furthermore, Watt showed that a very large amount of latent heat is required for the evaporation of water at normal temperatures. In short, it seems that Watt established the fundamental idea of 'total heat', and showed that as the sensible heat varied so the latent heat changed in the opposite sense. The assertion that the total heat, or the sum of the latent and sensible heats, was constant ('Watt's law') was probably made by others despite Robison's ascription of it to Watt; in any case it was later shown to be inaccurate.

Let us now summarise the two most important features of Watt's experiments and inventions.

In the first place, no 'common-sense' appreciation of the heat losses involved in the operation of the Newcomen engine would have justified Watt's inventions. What was needed was the measurement of the actual amounts involved. This Watt was able to provide, for he belonged to one of the most active scientific groups in the world; a group which was, moreover, pioneering the scientific study of heat. The engine he envisaged was, in a sense, an ideal engine: one that was to develop the maximum power with the maximum efficiency. Since it was necessarily much more expensive than any other engine its construction could only be acceptable in those places—like the Cornish mining area—where power and economy were both equally important.

In the second place, Watt's engine was unambiguously a *heat*-engine. The Newcomen engine had been recognised as a 'fire-engine', but the central role of heat was obscured by the mechanical action of the earth's atmosphere. In Watt's engine, on the other hand, the presence of the condenser indicates (as Dr Pacey points out) that *heat* is the basic principle and that power is generated by the flow of heat from the furnace to the cold condenser. It would have been very difficult to draw such a fruitful inference from the operation of the Newcomen engine— or even from a non-condensing high-pressure steam engine.

The extent of Watt's intuitive insight into thermodynamics —the science and the name were undreamed of in those days—is clearly revealed by the words he used to describe the *principles*, as he termed them, of his engine in his patent specification of 5 January 1769: 'First. That vessel in which the powers of steam are to be employed to work the engine, which is called the cylinder in common fire-engines, and which I call the steam vessel, must, during the whole time the engine is at work, be kept as hot as the steam which enters it . . .'

These ideas, coupled with the expansive principle, constituted the basis of thermodynamics; but nearly sixty years were to elapse before cognate scientific knowledge and technological practice had advanced sufficiently for Sadi Carnot to present them in one great synthesis.

In a rather tentative fashion, knowledge of the work of Watt and of Black was spread abroad by authors like de Luc, Magellan[30] and the anonymous writer of the 1770 volume. As was perhaps inevitable with second-hand accounts, ambiguities were brought into the narrative and, thanks to the efforts of Robison, some spleen too. The actual development of the doctrines of latent and specific heats was undertaken by two men whose scientific standing was inferior to that of Black and Watt: William Irvine and Adair Crawford.

Irvine (1743–87) was a pupil of Black, and another reluctant author.[31] His particular theory was basically very simple in that he took the concept of quantity of heat to its logical conclusion. Each and every body, according to Irvine, contains a certain absolute quantity of heat, which is fixed by its heat capacity and its absolute temperature. If for any reason the heat capacity of a body should change then it must either emit or absorb heat; thus the heats of combustion, of chemical reactions in general and the latent heats of fusion and vaporisation are merely the consequences of abrupt changes in the heat capacities of the substances concerned. In fact all productions or absorptions of heat indicate changes in heat capacity. Now this theorem enabled Irvine and his followers to calculate the absolute zero of temperature. To illustrate this let us suppose that the specific heat of ice at 0°C—for the sake of simplicity we use modern measures—is 0·9 and that that of water at the same temperature is 1·0. If, on Irvine's theory, the latent heat of fusion is 80 calories this must be the difference between the absolute heats of water and of ice at 0°C. Assuming that the temperature of melting ice is $x°$ above absolute zero, we can express this as:

$$(1·0)x - (0·9)x = 80$$

which makes the absolute zero $-800°$.

Figure 7 illustrates the principle on which this calculation is based: the analogy with water in tanks of the same depth but of differing cross-section (= heat capacity) is very plain.

The plausibility of Irvine's theory rests entirely on the assumption that specific heats are always smaller after the production of heat and greater after its absorption. It also requires that absolute zeros of temperature calculated by the

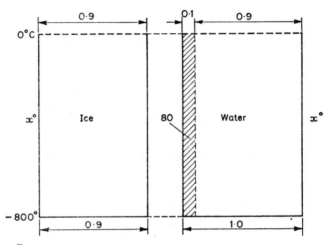

Figure 7

above method should be reasonably consistent with one another. As we shall see, these desiderata laid the theory open to immediate experimental test.

Irvine's theory did not, however, arouse the controversy that the work of his immediate disciple Adair Crawford (1748–95) did. Crawford, a London surgeon, visited Scotland and probably attended Irvine's lectures in 1776. As a medical man he was interested in the problem of animal heat, and he saw in the teachings of the Scottish school the possibility of elucidating this puzzling phenomenon. In 1779 he published an account of his experiments and observations in the form of a book on animal heat.[32] Crawford measured the specific heats of a large number of inorganic and organic substances and, something quite new, the specific heats of common air and 'fixed' air (carbon dioxide). This he did by filling a bladder with the two airs and comparing the heating and cooling effects they had on a given amount of water at a known temperature. He concluded that the specific heat of fixed air was very much smaller than that of ordinary air, only 1/67 in fact. Hence as in respiration one breathes in ordinary air and exhales fixed air with a much reduced heat capacity there must, on Irvine's principles, be a substantial emission of heat as a result of the change. Perhaps this is the real cause of animal heat? In his second and much expanded edition of the work (1788) Crawford described experiments using brass containers in place of bladders for the measurement of the specific heats of gases. The results, of course, could hardly have been accurate, and he had no appreciation of the fact that in his first experiments, in principle at any rate, he had been measuring the specific heats of gases at constant pressure (C_p) and in the second at constant volume (C_v). The importance of this distinction was not realised until much later. For the present it is enough to note that Crawford, following Irvine, had developed and applied the ideas of specific heat which Black had done nothing about for twenty years or more.[33] However, because he gave more credit for the basic ideas to Irvine than he did to Black, Crawford stirred up an angry controversy and brought down much criticism on his own head.

THE ESTABLISHMENT OF THE SCIENCE OF HEAT

Despite the caution of Black and his followers in the second half of the eighteenth century, there was a hardening of opinion in favour of the material theory of heat. The difficulties of the kinetic theory which was favoured by mathematicians and physicists were still unresolved, while the chemists who came increasingly to dominate the study were, perhaps by tradition (in the footsteps of Boerhaave), perhaps by metaphysical preference, predisposed to favour the substantial theory. An exception was P. J. Macquer,[34] but he had few disciples.

In 1776 Bryan Higgins put forward a solution to the problems of the elasticity and expansion of gases.[35] If, as nearly everyone agreed, gases are composed of material atoms that attract one another then, according to Higgins, their elasticity is accounted for by atmospheres of fire, or heat, which, he suggests, surround all atoms. These atmospheres are themselves atomic but fire atoms are unusual in that, while attracted to ordinary material atoms, they repel each other strongly. This very simple theory accounted satisfactorily for the thermal expansion of solids and liquids as well as for the elasticity and expansion of gases. At about the same time Torbern Bergmann was propounding the same idea and, a few years later, very similar theories were put forward by Cleghorn and by Lavoisier.

The notion of material atoms surrounded by self-repellent atmospheres of fire or heat could be commended on several grounds. For one thing, it could easily be reconciled with the received accounts of Newton's views. Indeed in this respect the theory could fairly be described as vulgar Newtonianism: Newton had not explained the mechanism by which the atoms of an elastic fluid repel one another (Proposition 23, Book 2; *Principia mathematica*); was it not reasonable to suppose that this elasticity was due to the self-repellent material of heat?

In the second place the theory accounted for the phenomenon of the heating (or cooling) of a gas when it is suddenly compressed (or expanded): adiabatic heating or cooling as it was later called. This had first been noticed in connection with experiments on the air pump.[36] Cullen mentioned it but had no insight into its significance. Bryan Higgins also referred to it, and J. H. Lambert pointed out that when air enters an evacuated vessel the temperature rises.* His explanation was that even 'empty' space contains the matter of heat, so that the entry of air carrying more heat must cause a rise in temperature;[37] and he went on to suggest that suddenly reducing the volume of a void should have a heating effect. More perceptive was Erasmus Darwin, who argued that since the evaporation of ether produces great cold, and boiling off water requires considerable latent heat, it would seem to be generally true that expansion into (or of) an elastic fluid requires a supply of heat if it is not to be cooled: 'I was led to suspect that elastic fluids when they are mechanically expanded would attract or absorb heat from the bodies in their vicinity and that when they are mechanically condensed the fluid matter of heat would be pressed out of them and diffused among the adjacent bodies.'[38]

The picture of heat being squeezed out of a compressed gas as water can be squeezed out of a sponge is a convincing one.

---

* This particular experiment later played an important part in the work of Gay-Lussac and of Joule.

It is, in essence, modest in the demands it makes on our credulity, on our willingness to accept conceptual models put forward to explain phenomena. Thus, early on, the material theory of heat was woven into the explanation of one of the most important phenomena of heat and thermodynamics: the adiabatic heating and cooling of gases. Darwin's suggestion that the conversion of a liquid into a gas requires much heat because of the associated expansion in volume was extremely important: it evidently influenced John Southern, and through him the development of thermodynamics (see pp. 161–2). In all this Darwin was not original; he had in fact been anticipated by Lavoisier and Laplace, as we shall see. What does distinguish this paper is the way in which he applies these ideas to meteorology and to technology.

Darwin points out that the summits of the equatorial Andes are always snow-covered and the foothills are tropical, while there is a temperate region in between. The stability of these sharply contrasting climates in close proximity is ensured by the fact that ascending hot winds are cooled by expansion while descending cold air from the summits is heated by compression. He goes on to observe that the formation of clouds and the precipitation of rain are caused by the cooling of moist air consequent upon its expansion, and he tries to relate the variation of barometric pressure with warm and cold weather. Once again, as in the cases of Halley and of Black, the study of heat had contributed to a deeper understanding of meteorological and geophysical processes.

In the sphere of technology Darwin solved a famous problem when he pointed out that the formation of ice and snow which accompanied the puff of exhaust air from the 'Hero's engine' at Schemnitz was simply due to the cooling of expanded moist air.* He also commented that in the operation of the Watt-type

---

* This water-raising engine driven by compressed air was erected in 1756. The air was compressed by the weight of a column of water 260 feet high and was used to force flood waters from the lowest level in the mine, 104 feet further down, up to the intermediate level, whence both the raised water and the 'driving' water could drain out of the mine through an adit, or sough. The process was cyclic, and necessarily therefore the driving air had to be greatly expanded at the end of the cycle, when it was expelled from the pressure vessel. The resulting formation of snow and ice interested many scientists at the time. The short account of the engine in *Philosophical Transactions* lii (1762), p. 547, wrongly but understandably ascribes the

steam-engine the rush or expansion of the steam into the condenser should cause a cooling of the vapour, with consequent rapid condensation. This raises interesting speculations about Darwin's relationship with his friend and fellow-member of the Lunar Society, James Watt.[39] But what is immediately important is that the expansion of steam—the 'working substance' —is now associated not only with a fall in pressure but also with a fall in temperature, and that the latter factor is held to be important for the economical operation of heat-engines.

Shortly after Darwin began to work out these ideas an obscure Swedish apothecary published a remarkable essay entitled *On air and fire* (1777). In a few short paragraphs the author, Carl Wilhelm Scheele (1742–86), a chemist of great ability, made a fundamental contribution to the scientific study of heat. By an argument that was at once simple, in the elementary facts on which it was based, and lucid in the reasoning employed, he established the concept of what he called—and we still call—'radiant heat'.[40]

It is quite natural to regard light and heat as essentially the same. The rays of the sun or from a hot fire provide illumination and at the same time warmth. This has been the commonsense view from time immemorial. It is true that some seventeenth-century investigations might, as we saw, have led to a deeper knowledge, but it was not until Scheele's work that light and radiant heat were clearly distinguished. There is, observed Scheele, some confusion about fire (shades of Boerhaave!), light and heat. If a red-hot charcoal is placed at the focus of a concave metal mirror, is it the light that ignites inflammable material at the focus of another, conjugate mirror? Or is it the heat? Or are both acting together? If we sit before a hot stove on a cold winter's day with the front door wide open we can feel the heat five or six feet away from the stove while, at the same time, we can see our breath in the cold air. Evidently the air is not warmed by the heat passing through it. Further tests show that a strong draught in the room does not reflect the heat travelling horizontally from the stove, and a candle flame is quite undisturbed by it. This sort of heat does not cause a beam of light to produce that characteristic quivering effect

---

freezing phenomenon to the presence of sulphur and nitre in the water which could thus supposedly form a freezing mixture.

that heat *rising* from a hot body does; and it can be completely cut off by a sheet of glass, which has no detectable effect on light passing through it.

From these and other simple observations Scheele concludes that the heat that rises above a stove is quite different in kind from the heat that travels horizontally into the room. The latter is similar to light: it travels in straight lines, it obeys the optical laws of reflection and it does not 'combine' with the air through which it passes. But it is not the same as light, for a sheet of glass which freely transmits light will absorb it: furthermore, we can see the light of a fire much further away than we can feel its heat. Scheele then distinguishes the familiar convection process as a mode of heat-transfer quite different from what he calls 'radiant heat'.

Thus in a few brief but cogent observations which could (in theory) have been made at almost any time, for he used no more elaborate apparatus or concepts than were available to Archimedes, Scheele distinguished the two dominant modes of heat transfer. The new radiant heat could very easily be harmonised with the material theory of heat as we have outlined it above. A very hot body would naturally tend to throw off its excess heat, which, being a very subtle, self-repellent substance, would travel very rapidly outwards, rectilinearly from the body. Such was the idea of 'radiant caloric' as it was later called. In contrast, the kinetic theory which involved the additional hypotheses of undulations or pulsations in an invisible, intangible aether, must have seemed very forced and improbable. When we recall that the material theory gave a very plausible account of the nature of the expansion of solids, liquids and gases, as well as of 'adiabatic' heating and cooling of gases, we must surely concede that the material theory satisfied Occam's principle that entities must not be multiplied unnecessarily and that it was, therefore, at this time and for many years to come, the more scientific of the two rival theories.

Robertson Buchanan, writing in 1810, was less than fair to Black when he wrote: 'Dr Black's theory led to vague, indistinct and inaccurate notions on the subject; nor was the word *Capacity* well chosen. The term *specific heat* is that which is more approved by later writers, particularly Dalton and Leslie.'[41]

But he had a point. The achievements of the Scottish school at this time had hardly been commensurate with the abilities of its individual members. The Scandinavians, in spite of the talent of Wilcke and the genius of Scheele, were too isolated to do much more, and so it was left to two supremely able Frenchmen, Lavoisier and Laplace, to put the science of heat on a satisfactory and systematic basis. This they did in their historic *Mémoire sur la Chaleur*, published in 1783.[42] It was appropriate, and indicative of changes to come, that the great chemist should have as his collaborator a man who was later to become a distinguished theoretical physicist. The result of their collaboration was a work in which precise definition, theoretical clarity and experimental insight are all equally prominent: as well, we must add, as certain axioms whose subsequent modification marked fundamental stages in the development of thermodynamics.

They begin by noting that there are two main theories about the nature of heat: one that it is just the *vis viva* of the atoms that comprise all material bodies, the other that it is a subtle material fluid. The first theory satisfactorily explains the well-known fact that the sun's rays, which consist of streams of fast particles (cf. Newton: *Opticks*), heat bodies up but exert no appreciable force on them; for the *vis viva* is proportional to mv^2 while the force is proportional to mv (m, the mass of the particles is very small and v, their velocity, correspondingly large). The *vis-viva* theory also explains why seemingly unlimited quantities of heat can be generated by friction.* On the other hand the material theory offers more plausible explanations of certain other thermal phenomena. They do not presume to choose between the two, perhaps because Laplace, the theoretical physicist, had not yet been converted to the material theory;[43] but in any case it did not matter, for on both theories the total quantity of heat is always constant during the simple mixing of bodies. Now this was important: it was a direct assertion of the axiom of the conservation of heat: '*The conserva-*

* Lavoisier and Laplace were, at this time, clearly ignorant of Scheele's work. If one accepts the Newtonian (particle) theory of light and also Scheele's demonstration that radiant heat is similar to but not the same thing as light, then the probability is that radiant heat is caused by streams of characteristic particles. But this is to concede the material theory of heat. Lavoisier and Laplace's reference to 'M. Vilke' suggests that they were quite unfamiliar with the work of the Scandinavian school.

tion of free heat, in the simple mixture of bodies, is thus independent of all hypotheses on the nature of heat . . .'

They then put forward a basic principle which they claim is also common to both theories: 'If in a combination or in any change of state whatsoever there is a decrease of free heat, this heat reappears entirely when the substances return to their original state . . .' Generally: *'All variations in heat, real or apparent, which a system of bodies undergoes in changing state are reproduced in inverse order when the system returns to its first state.'*

This can be, and usually is, taken to mean that when a body heats up, melts or vaporises it absorbs exactly as much heat as it gives up when it cools down, solidifies or condenses. In this form the statement is not much more than an assertion of the principle of the uniformity of nature. But we should not take too restricted a view, for the statement is also quite general in its scope. It can be, and usually *was*, taken to mean that if a certain amount of heat is required to warm up and vaporise enough water for one working-cycle of a steam-engine then *exactly the same amount of heat* was delivered to the condenser at the end of the cycle. And this, as events showed, turned out to be a very debatable matter.

Lavoisier and Laplace then proceeded to set out the now familiar equations of the method of mixtures, the elementary book-keeping of the science of heat. The change in temperature that a body undergoes multiplied by its mass and the specific heat equals the quantity of heat that it absorbs or gives up: as they put it, $mq(a-b)$, where m is the mass, q the specific heat and a and b the two temperatures. But the familiar method of mixtures presents certain difficulties: chemical, if the two substances react chemically; physical, if they have incompatible mechanical characteristics like mercury and oil; theoretical, in the assumption that the specific heat of any substance is the same at temperature a as it is at temperature b. There is also the problem of the heat capacity of the container, or calorimeter as they call it. To get round these difficulties they advocate the use of the ice-calorimeter that they have invented. This avoids these difficulties and, although its use requires great experimental skill, it can be made to yield accurate determinations of specific heats as well as heats of chemical reaction and animal heats.

With the ice-calorimeter they submitted Irvine's hypothesis

to searching tests. Their measurements of specific heats indicated that, according to Irvine's method, the absolute zero must vary between $-600°$ and $-14,000°$. While they admit their findings are not absolutely conclusive, it is clear that they regard them as effectively destroying Irvine's hypothesis; and most scientists agreed with them. Furthermore, Irvine's hypothesis was open to question in two respects. In the first place Irvine assumed that all the heat entering a body must, if there is no change of state, have a thermometric effect. But Lavoisier and Laplace suggested that while some of the heat in a body was free and so affected the thermometer, some was chemically combined and so had no sensible effect. Irvine's hypothesis made no provision for this second possibility, so its basic assumption could be queried. But even if we ignore the possibility of 'combined caloric' it still does not follow that the specific heat of a substance remains the same at all temperatures. It may well vary according to some undetermined law, in which case the whole argument is again in jeopardy. Indeed:

> Since expansion, fusion and vaporisation are equally effects of heat we can confidently presume that in the production of the first of these effects as in the other two a quantity of heat is absorbed and which consequently ceases to be sensible to the thermometer; but in the passage of a body to different states of expansion by imperceptible steps one can only know the amount of heat absorbed by the increments in the specific heat, it is thus very probable that the specific heats of bodies increase with their temperature, but each one following a different law . . .

Irvine's hypothesis was thus impugned by experiment and called in question by reasoning based on commonly accepted principles. But more than this; Lavoisier and Laplace called attention to the absorption of heat due to the expansion of a body. This phenomenon was hardly important in the case of solids and not obviously important in the case of liquids, but it was eventually found to be extremely significant in the case of gases.

The authors point out that the different states of a body can be defined in terms of heat, for it is heat that determines the difference between solid and liquid, liquid and vapour. They refer respectfully to Adair Crawford, while they criticise his method of determining the specific heats of gases. Unfortun-

Plate I. Halley's map of the wind systems of the world. (*Philosophical Transactions of the Royal Society,* 1786–7)

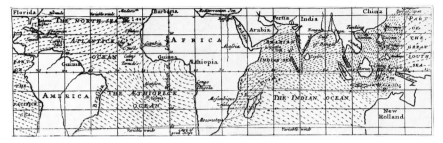

Plate II. The traditional source of power: a simple undershot water-wheel. This mill, at Rossett near Chester, was built in 1661. (*Photograph by the Author*)

Plate III. 'Fairbottom Bobs'. Erected in about 1760, this Newcomen engine was used to pump out a mine at Fairbottom, between Oldham and Ashton-under-Lyne. It lay derelict for many years until, in 1926, it was dismantled and removed to the Ford Museum at Dearborn, U.S.A. The boiler can be seen behind the cylinder on the right-hand side. (*By courtesy of the Manchester Central Reference Library*)

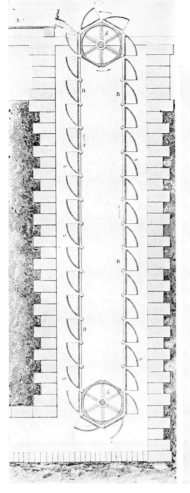

General Design of Mr Westgarth's Statical Engine

Contracted from Mr Westgarth's Original Design by J. Smeaton 1767

Plate IV. An early eighteenth-century machine intended to develop the power of falls of water too high for a single waterfall. (*J. T. Desaguliers*)

Plate V. Drawing of the column-of-water engine erected by William Westgarth at the Allenheads lead mine in Northumberland. (*Journal of the Society of Arts*)

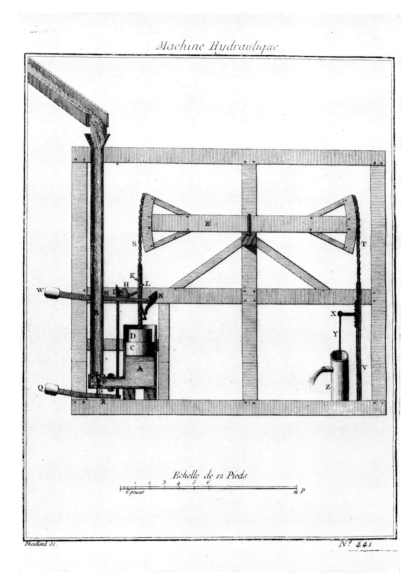

Plate VI. Design for an early column-of-water engine by Gensanne.
(*Machines et Inventions Approuvées par l'Académie,* 1741)

ately they are unable to propose a better system themselves, or offer better results. Finally they refer to 'M. Vilke', but do not mention Black. This last could perhaps be held against them, but their defence would be simple: Black had published nothing on the subject, so why should they mention him?

Lavoisier's establishment of the new system of chemistry and the publication of his *Traité élémentaire de chimie* (1789) set the seal on the acceptance of the material theory of heat.[44] Lavoisier, of course, set 'calorique' and 'lumière' at the head of his list of 'simple substances' or elements. Although he affected indifference as to whether caloric was ultimately a substance or *vis viva*, there can be no doubt where his real opinions lay, together with those of the majority of his contemporaries, so that the acceptance of the new chemistry meant, almost as part of the deal, the acceptance of the material theory of heat. It is perhaps ironic that ultimately the great debate that preceded the establishment of thermodynamics in the middle of the nineteenth century was not so much between supporters of the kinetic theory and supporters of the material theory as between those who accepted and those who rejected the axiom of the conservation of heat. Lavoisier and Laplace and their contemporaries assumed that this axiom was common to both theories; but the nineteenth-century thinkers had to replace it with the doctrine of the conservation of energy.

The achievement of a quantified science of heat before the end of the eighteenth century leaves us with one detail to mention. The discovery of the *concept* of conductivity was a more pedestrian affair than Scheele's brilliant discovery of radiant heat. It has always been appreciated that certain materials—textiles, straw, wood etc.—do not transmit heat very readily. Black himself conceded as much. But in the case of solid materials it could be, and was, argued that in all cases heat flows through them with exactly the same velocity, differential heating effects being due solely to differences in their heat capacities. The idea that different solid bodies may have different thermal conductivities came about as a consequence of advances in the new science of electricity. Copper is a good conductor of electricity, ivory and glass are very bad conductors; perhaps then these substances differ also in their capacities to conduct heat? Considerations like these led Ingen Housz to try to measure the thermal conductivities of different substances. He used uniform wires made of different metals, coated them with wax and then noted the speed with which the 'melting point' travelled along the wires when one end was heated.[45] This method involves an error which is the converse to that made by Black: it ignores the different specific heats of the various metals used and ascribes all the differential thermal effects to conductivity alone;[46] it also leaves out of account the heat transmitted from the metal bar to the surrounding air, or the external conductivity. But in the context of science at the end of the eighteenth century these mistakes could be, and soon were, corrected.

The way ahead was, Dr J. R. Ravetz points out,[47] cleared by J. B. Biot, who in 1804 measured the steady-state temperature of an iron bar heated at one end. The results enabled him to formulate a simple theory of heat-flow that took account of internal and external effects. It was on this work by Biot that Joseph Fourier began to construct his analytical theory of heat. Later still, G. S. Ohm, inspired by Fourier, commenced his study of the conduction of electricity; the wheel had come full circle.

But the establishment of the concept of thermal conductivity by the beginning of the nineteenth century calls our attention to an important (if vaguely formed) hypothesis that was in the back of men's minds at that time. Good conductors of electricity are also good conductors of heat. Electricity and heat are mutually related in that the passage of electricity generates heat and that both heat and electricity can be produced in apparently limitless quantities by friction. Electricity is unquestionably a subtle fluid—one can store it in jars—and heat, too, is almost certainly of the same nature. May they not therefore be different but related manifestations of some common, basic substance?[48] Here at any rate there appeared to be the likelihood of important new scientific advances in the closing decades of the eighteenth century and in the opening decades of the nineteenth.

Chapter 3

The Rival Power Technologies: Steam and Water

POWER TECHNOLOGY AT THE END OF THE
EIGHTEENTH CENTURY

The first extensive use of the steam-engine had been in the mining industry. In districts like Schemnitz, Cornwall, the Midlands and the north-east coast of England it had quickly established itself, and in doing so had helped to concentrate and define the communities of skilled engineers that had long been associated with mining areas. Among these conspicuous engineers were men who, like Richard Trevithick, were capable of developing the steam-engine right through into the age of the railway (we must not forget that the first railways were used in mines) and the steamship.

Although mining continued to be in many respects the pace-maker for technology, the historian whose concern is with energy and thermodynamics must at least take note of the quite phenomenal development of the textile industries in eighteenth-century England. Traditionally, spinning threads from easily broken animal and vegetable fibres had been a woman's occupation,* and the craft required a practised eye, good judgement and very sensitive fingers. The endeavour to get clumsy and insensitive machines to carry out these delicate tasks, to mechanise them in fact, seems when we look back to have been extraordinarily bold, almost rash; and yet it was successful. In technical, if not in strict chronological succession a number of able inventors analysed the craft into its component processes and then produced the machines for carding, roving and spinning; the last—the famous 'water frame'—was patented by Richard Arkwright in 1769, the year of Watt's condensing-engine. Although it is a fascinating subject, a

* cf. 'Spinster'.

67

discussion of the motives and ideas—economic, social and technological—of the men who led the textile revolution is outside
the scope of this book.[1]

The mechanisation of the textile industries demanded not
only new machines to replace skilled fingers, but also new forms
of social organisation, improved means of transport (this was
a great period for canal-builders) and new mill structures which
could carry the new machines and at the same time be both
reasonably fireproof and tolerably warm in winter.[2] Lastly
there had to be enough power available to drive the machines,
day in, day out. Windmills being unsuitable, water-wheels and
horse-wheels were left as the two main sources of power.[3] The
technology of water-power had, by the latter part of the
eighteenth century, attained a respectably progressive, indeed
scientific state. Prominent among the external reasons for this
was the textile revolution, which in effect demanded that every
inch of fall on every reasonably big river or stream should be
harnessed for production; and so it came about that textile
mills were spread out, like beads on a thread, along the rivers
of the textile counties. During this phase the industrial revolution was essentially rural in its location; concentration into
great industrial cities and conurbations came later, as a consequence of the wider use of the steam engine.

That the resources of water power could, in great measure,
satisfy the demands of the new industries was due to the efforts
of the scientific engineers who transformed what had been a
craft at the beginning of the eighteenth century into an
advanced and viable technology by the end of the century. It
all began with a classical paper by Antoine Parent (1704) in
which he tried to calculate the conditions under which the
maximum power (or *effet* as he called it) could be derived from
a given stream of water. If a stream was simply directed
upwards to make a fountain, then the weight of water rising
per minute multiplied by the height to which it rises represented the total available power (*effort*) in the stream. The
problem amounted to this: if the stream, instead of rising,
strikes the blades of an undershot water-wheel, what proportion
does the greatest load that the water-wheel can hoist to the
same height in one minute bear to the weight of water flowing
in the stream per minute? By using the new mechanics and the
recently invented calculus, Parent deduced that at best the

wheel could generate only $\frac{4}{27}$ of the total available power, or 'effort' of the stream, and that to do this it would have to rotate with a peripheral velocity of $\frac{1}{3}$ that of the stream.

He had, however, made a gratuitous assumption and a simple, if excusable mistake which combined to vitiate his results. If these are corrected the maximum efficiency and the optimum speed both become $\frac{1}{2}$ in place of $\frac{4}{27}$ and $\frac{1}{3}$. As it happened the discrepancy was soon noticed, in practice if not in theory, for Déparcieux in France and Smeaton in England both showed that actual water-wheels could easily exceed the theoretical maximum of efficiency. By a combination of practical observations 'in the field' and careful experimental work with models, Smeaton demonstrated that the maximum power to be derived from an undershot wheel is about $\frac{1}{3}$ that of the stream, while for an *overshot* wheel it is about $\frac{2}{3}$ (1759). The surprising difference between these two results was, Smeaton later concluded, due in part to the waste of power resulting from the turbulent impact of the water against the blades of the undershot wheel: power is, in fact, consumed simply in 'altering the figure of the stream'. An overshot wheel, on the other hand, does not waste power in producing turbulence, for the water can be made to fill the buckets smoothly and without impact, and if the wheel rotates sufficiently slowly—which implies large buckets—its maximum efficiency can approach 1. The last condition is necessary, for if the wheel rotates with appreciable velocity, the water leaving it must evidently carry off, as it were, appreciable power. It is impossible to exceed the efficiency 1; for if more power could be generated than is required to restore the *status quo*, simple perpetual motion would be achieved. The maximum possible efficiency of an undershot wheel is, Smeaton inferred, $\frac{1}{2}$.

Déparcieux's argument (1752) was simpler and less authoritative but nevertheless ingenious and informative. If, he said, we imagine a mechanically perfect overshot wheel driving an identical one in reverse (to constitute, as we would say, a scoop wheel) only a very small excess weight of water in the driving wheel is sufficient to set the pair revolving. Thus the limit of efficiency of water-wheels must, if we include the overshot variety, be 1 and not $\frac{4}{27}$, as Parent had argued.

This work had important consequences—scientific, technological and economic. In the first place we note that although the application of *Newtonian* mechanics will yield perfectly correct figures for the maximum efficiency and the optimum velocity of an undershot wheel it gives us no insight into the reasons for this inferior performance; save, perhaps, the common-sense (and correct) one that the water leaving the wheel still has an appreciable velocity and is therefore capable of driving another water-wheel. After all, momentum (mv) is conserved in the impact and there is, from this point of view, no problem. On the other hand the rival mechanics, that of Leibnitz and Huygens, does indicate a source of waste: that due to the loss of *vis viva* or 'living force' (mv^2) that accompanies the inelastic or turbulent impact of the water on the blades. The rival system of mechanics therefore offers us a deeper and more fruitful insight into the transformation process than does Newtonian mechanics, although both are equally 'correct'. Accordingly the *vis-viva* doctrine in its application to machines was soon put on a more systematic basis by the efforts of men like Borda and Lazare Carnot.

The study of the conditions under which energy is transformed began in this way in the middle of the eighteenth century.

In England the weight of Newtonian orthodoxy prevented any theoretical development of Smeaton's work and of the related *vis-viva* doctrine. Newtonianism had, however, little relevance to the development of the textile industries, and it could hardly inhibit the practical application of Smeaton's precepts. The water-wheel provided direct and smooth rotative motion; its prime cost was small and in the hands of scientific engineers it was soon made to develop very substantial power, much more in fact than any steam-engine could provide until well into the nineteenth century. Following Smeaton's recommendations the large industrial water-wheels that provided power for the big textile mills were either overshot or 'breast wheels' (the water entered the machine at axle height) and the greatest care was taken to maximise the power obtained from the available river or stream as well as to minimise the losses in transmission and distribution inside the mill. In this way the technology of water-power held the steam-engine at bay until well into the nineteenth century.

On the other hand the law of diminishing returns ensured

that the best sites on the best rivers were soon snapped up, and newcomers to the textile industries had to be that little bit more efficient or more enterprising than their established rivals, for they had to make do with less abundant power. The word 'abundant' is, however, entirely relative, and with the rapid growth of the industry even the best-sited mills had to pay urgent attention to the power problem if they were to increase production to keep pace with the market.* Plate 1 indicates the high degree of refinement that water-power technology in its application to the textile industries had reached by the beginning of the nineteenth century; the contrast with the rustic water-wheel of folk-lore and popular tradition is complete.[4]

The indications are that the 'take-off' into self-sustained economic growth which England achieved at the end of the eighteenth century was accompanied—inevitably—by a serious power shortage that was particularly acute in the new growth industry: textiles. Not only were there strong incentives for the improvement of water-power technology, but supplementary aids and alternative sources of power were kept under constant review. Arkwright had used horses in his first mill at Nottingham,[5] and thereafter horse-wheels seem to have been fairly common. Dr R. L. Hills remarks that in 1788 '. . . there were about twenty five mills in the whole of Oldham parish, eleven of them in Oldham itself while by 1791 there were eighteen in Oldham alone. Because the best water sites had been taken most of these mills were driven by horses.'[6] But there were fairly obvious limitations on the use of horses, limitations which became more apparent as the cost of fodder rose towards the end of the century.[7] The alternative was, of course, to use the steam-engine either as a booster, pumping water back over the water-wheel to meet peak loads and cope with summer droughts, or as a direct source of power. Used in the first capacity the steam-engine was inefficient, since its own deficiencies were added to those of the water-wheel. But as it was functioning only as an auxiliary, efficiency was a minor

* Thus Quarry Bank Mill, established by Samuel Greg in 1784 at Styal (Cheshire) expanded so rapidly that by 1796 the water-power system had to be entirely redesigned by Peter Ewart. Shortly after this a Boulton-and-Watt engine was installed to supplement the power derived from the river Bollin.

consideration, and it was sound financial policy to buy the cheapest possible steam-engine that would serve to compensate for the shortcomings of the site. On the other hand there were some serious drawbacks to using the steam-engine as a direct source of power. It was costly, and it was very difficult to transform the clumsy oscillations of the great beam into the necessary smooth rotary motion.[8]

The solution of the problems of the rotative steam-engine, involving the development of the double-acting principle and the inventions of the parallel motion and the sun-and-planet gear, was one of Watt's greatest achievements. The novel features of the new machine were kinematic rather than thermodynamic and we are not, therefore, directly concerned with it. But one point is of interest. As we have already noted, Watt's method of showing the advantages of and charging royalties on his (pumping) engine had been to claim a fixed portion of the savings which it yielded compared with the less efficient Newcomen engine. This method was well suited to the mining industry where there was detailed knowledge of and general agreement about the costs of running a Newcomen engine. But direct comparisons of this sort were not possible in industries like textiles which had little experience of the New-comen engine. On the other hand the horse was not only the conventional measure of strength, it was commonly used in the textile industry and mill-owners could be expected to have a good idea of the duties it could perform. Accordingly Watt expressed the capacities of his rotative engines in terms of so many horse-power, taking as his measure of one horse's power the ability to lift a given weight a given height in a given time. Of course he was not original in standardising the power of a horse in this fashion—others had done it before[9]—but the authoritative figure of 33,000 ft. lb. per minute postulated by Watt was (and still is) generally accepted by virtue of the wide currency he gave it through his business, technological and scientific contacts.

Indeed, Watt so dominated the technology of steam power in the last two decades of the eighteenth century that other engineers, some of them very able men, were frustrated by his authority, and a certain *fin de siècle* dullness became apparent. Such was Watt's genius that virtually every aspect of steam-engine application and development was covered by his patents. Another engineer might invent a brilliant new improvement to the engine only to find that if he wanted to market it in a practicable form he would have to use some of Watt's patents: sun-and-planet gear, parallel motion, condenser, etc. One could hardly get away from Watt's commanding figure: his genius and the business abilities of his partner Matthew Boulton had seen to that.

The remarkable success of the beam engine, introduced by Newcomen and confirmed by Watt, has focused attention on that particular form of engine to the exclusion of practically all others. But it would be a serious distortion of history if the wisdom of hindsight blinded us to the great ingenuity that eighteenth-century engineers displayed in exploring every possible way of deriving mechanical power from heat. In fact, the triumph of the beam engine and of the reciprocating principle in general was the final outcome of a process of natural selection from which only the very fittest survived. We have already mentioned one of the earliest and most interesting of these alternative machines—Amontons' fire wheel—and this was followed by a varied succession of others. Indeed, we can draw up a morphology, classifying heat-engines into separate families, or groups.[10] In this way we can set out, in the chronological order in which they were invented, the separate groups of heat-engines that were investigated by eighteenth-century engineers. It is important to remember that each group includes more, often many more, than one individual invention. The general principles of these engines are illustrated in Figure 8.

(1) The steam-reaction engine. A horizontal pipe is mounted like a compass needle so that it can spin round; the opposite ends of the pipe are bent at right angles and terminate in nozzles so that a powerful jet of steam emerging from the two ends causes the pipe to rotate about the vertical axis. Hero of Alexandria's engine was of this sort.

(2) Steam mills such as the one proposed by Giovanni Branca in 1627. A powerful jet of steam drives a paddle-wheel in the same way that a stream of water drives an undershot water-wheel.

(3) Imbalancing wheels (see Plate XVIII). These were based on the same principle as Amontons' fire wheel. A wheel with hollow spokes and rim and fitted with appropriate valves is half

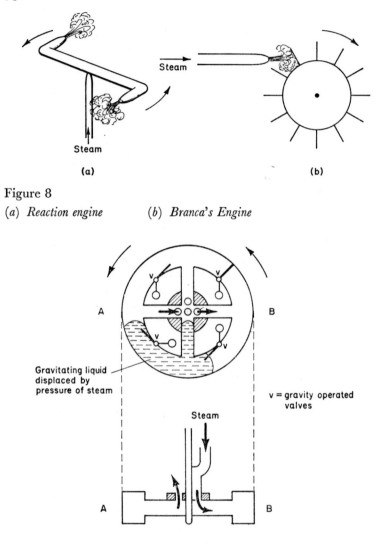

Figure 8

(a) *Reaction engine*   (b) *Branca's Engine*

(c) *Simplified diagram of Masterman's steam wheel (1821). A development of the engine developed by Amontons. Each hollow spoke has a circular hole in it near the axle. Two holes on opposite sides of a fixed, vertical disc, through the centre of which the axle passes, allow steam to enter each spoke as it passes and, at the same time, allow used steam to exhaust from the opposite spoke.*

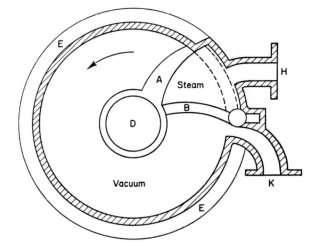

(*d*) *Watt's direct rotative engine, 1782.* (*Farey*)

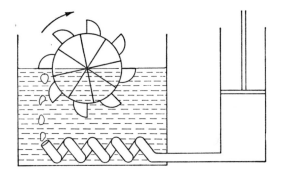

(*e*) *Buoyancy engine* (*Parker, 1792*)

full of a heavy liquid. The application of steam pressure or the thermal expansion of a gas in one side of the wheel displaces the heavy liquid to the other side so that the wheel, being imbalanced, must rotate.

(4) Savery engines. It has been shown that a form of Savery engine enjoyed a brief but limited revival towards the end of the eighteenth century.[11] These inexpensive if inefficient engines were apparently used to give the occasional boost to the water-wheels of cotton mills.[12] In these later versions of the Savery engine the pressure phase was dispensed with and water was

raised by suction only. The engine was also self-acting, the valve mechanism being operated by the rotation of the water-wheel, and provision was made for flushing the air from the condensing vessel. In view of these modifications it may be doubted whether the engine should still be called a 'Savery' engine rather than a novel invention in its own right.[13]

(5) Beam-engines.

(6) Direct rotative engines. These originated with Watt's patent of 1782 and were extremely popular in that every notable engineer seems to have experimented with this form. Two concentric drums are arranged so that the outer one is fixed while the inner one is free to rotate. A radial paddle or vane fixed to the inner drum and touching the outer one acts as a piston moving in an annular cylinder. A retractable baffle hinged to the outer cylinder constitutes the top of the cylinder so that steam injected behind the piston causes the inner drum to rotate. The baffle has to be retracted once every cycle to enable the piston to pass round, and it was the practical difficulty of arranging this that defeated every attempt to make this appealingly simple engine work.* The principle of direct rotation was, of course, ultimately attained with the invention of the steam turbine.

(7) Buoyancy-engines. These bulky and evidently inefficient engines were surprisingly popular with many inventors, possibly because they were cheap, simple and foolproof. A stream of expanded bubbles rising through hot water, or some other liquid, is caught in the downwards-facing buckets of a 'water'-wheel immersed up to its axle, thus causing the wheel to rotate. The bubbles can be either of steam, generated by boiling the water, or of air expanded by merely *warm* water; in the second case part of the power generated by the wheel is used to overcome the hydrostatic pressure of the water.[14]

(8) Finally there were desultory proposals to harness the seemingly 'irresistible' force of the thermal expansion of solids and liquids. These came to nothing, for the basic assumption is clearly unsound.[15]

The great majority of these engines were, it will be seen, intended to produce direct rotative motion. Steam, with its

---

* The direct rotative steam-engine bears a curious outward resemblance to the N.S.U. Wankel engine. Whether the latter will be more successful than its steam-driven predecessors were remains to be seen.

tremendous expansion (about 1,800-fold on vaporisation), was the favoured 'working substance', and the first suggestions (apart from (3) above) of an engine worked by the thermal expansion of air were not made until the very last years of the eighteenth century. On the other hand a number of engines in categories (1), (2), (3), (5) and (6) were described as versatile in that they could be driven by the pressure of water as well as by steam, and it was often additionally specified that driven in reverse the engine could be used as a water-pump. The commercial advantages of such a versatile machine were quite obvious. We should note here that Amontons had described a rotary pump.[16]

Nevertheless, after great efforts and the expenditure of much money, year in, year out, engineers were forced to the conclusion that the beam or reciprocating engine was the best that could be devised. What seemed at first sight to be an excellent idea always broke down in practice, and sometimes even before that. Let us consider for the moment the case of Baron von Kempelen's engine. This was patented in 1784, and Watt first learned the details of it from Joseph Priestley.[17] The engine worked on the same principle as a Barker's mill, which put it, theoretically at any rate, into category (1) above. Watt was at this time busy developing the rotative steam-engine and this widely-discussed possible competitor worried him. He therefore set out to compute its efficiency.

If we assume that the horizontal pipe rotates in a vacuum and if we assume that the static pressure of steam is 14 lb. per square inch, then according to Watt the steam will rush into the void with a velocity of 1,800 feet per second. Watt now assumes (wrongly) that the performance of such an engine will be governed by exactly the same principles as those which apply to the undershot water-wheel: '. . . the proper velocity for it is $\frac{1}{3}$ of y^e above, i.e. 600 ft. pr. sec and y^e effect will also be about $\frac{2}{3}$ of the power, or 10 lbs × 600 = 6,000 lbs 1 ft. high pr. second. The quantity of steam employed will be 1,800/144 = 12·5 cubic feet of steam'.[18] Compared with this Watt can show that one of his own engines would lift over 15,500 lb. 1 foot high for the consumption of 12·5 cubic feet of steam; thus von Kempelen's engine is markedly inferior to his own. He goes on: '. . . by another theory it comes out that the effect might . . . be greater but this second theory is so complicated that I cannot say I understand it thoroughly, nor do I think there is occasion because 600 feet p. sec seems to me an impossible velocity on account of the necessary friction in the machine . . .'

There are a number of interesting points about these observations.* For one thing, Watt not only uses the old Parent theory some twenty-five years after Smeaton had experiment-ally refuted it, he even uses the *language* of the theory: 'effect' for *effet* and 'power' (static pressure) for *effort*. The explanation must be that Smeaton had provided no alternative theory and Watt had perforce to use the only extant one he understood, even though it was defective. In the second place it is extremely significant that Watt endeavours to generalise water-power theory to include the performance of a heat-engine: this was almost certainly the first time this was done. Lastly we must wonder what was the 'second theory' to which Watt referred: could it have been Borda's?

If Watt had, in the event, little to fear from von Kempelen a much more dangerous and, from our point of view, more interesting rival appeared when Jonathan Hornblower patented his compound engine in 1781. In this machine the steam, having done its duty in a small, high-pressure cylinder is then ad-mitted to a much larger, low-pressure cylinder. As it drives the piston down in this larger cylinder it necessarily expands and thus, in effect, uses Watt's expansive principle. In fact Horn-blower, a clever Cornish engineer, had patented this device before Watt patented his expansive principle. But to be really effective the Hornblower 'compound' engine had either to operate at a reasonably high pressure, which in 1781 was hardly feasible, or it had to use Watt's condenser. Hence it fell between two stools. The original idea may have been formu-lated as early as 1776,[19] but the first Hornblower engine was not erected until 1782. In all some nine or ten of these machines were made, but they were not very successful, for the small advantage of using expansive operation from an initially low pressure hardly compensated for the cost of the extra cylinder

---

* The late Dr H. W. Dickinson commented that these observations were actually on the subject of the steam turbine, adding that it (the turbine) 'received practical and theoretical attention at such an early date as this . . .' But this is quite wrong. The theory of the turbine was built on the ruins of Parent's theory, and the application of the latter to steam-power cannot possibly be called turbine theory.

and the more complex mechanism. Hornblower, for all his gifts, was hamstrung by Watt's patents and by the partners' inflexible commercial policies. When he tried to have the duration of his patents extended in 1792 he was opposed by Boulton and Watt and lost the case. We are not concerned here with the quarrels and bitternesses of the past, nor are we concerned with apportioning exact measures of credit for the major inventions of the time. We are, however, very interested in certain aspects of Hornblower's work, for they have an importance that trandscends all the bickering.

Hornblower had a friend and adviser in Davies Giddy, the high sheriff of Cornwall who was later M.P. for Penzance and President of the Royal Society. This man was an Oxford mathematician and a wealthy landowner, who when he married in 1817 changed his name to Davies Gilbert, the name by which he is usually known and which we shall use henceforth. Although Gilbert's views on certain social matters such as education were reactionary,[20] he rendered good service to science and was a steadfast friend to a number of able Cornish engineers and scientists of the day; among them being Richard Trevithick, Humphry Davy—and Jonathan Hornblower.[21] Hence the evident alarm with which Watt wrote, in 1792, to his talented assistant John Southern about the progress of the litigation with Hornblower: '. . . they have brought Mr Giddy, the high sheriff of Cornwall, an Oxford boy, to prove by *fluxions*, the superiority of their engine, perhaps we shall be obliged to call upon you to come up by Thursday to face his fluxions by common sense . . .'[22]

This passage has been quoted by several modern writers, but it is actually more revealing than they realise. Its significance lies in the archaic and technically misleading term 'fluxions'. Indeed, from MSS in the Royal Institution in Truro it is apparent that Gilbert had obtained a mathematical expression for the expansive power of steam that was formally similar to that of Daniel Bernoulli.[23] This would, therefore, have been the first time it had been realised that in the case of an actual, expansively operated steam-engine the work done (in his calculations Gilbert denotes this as the 'efficiency') is given by the area under the pressure/volume curve: in other words the area of the indicator diagram.* In fact this involved integration

* In modern symbols the expression for this is, of course, $\int p dv = W$.

and not differentiation, so Watt should have referred to 'fluents' rather than 'fluxions', but this is a technical quibble.

What followed Gilbert's disclosures in court was intriguing, furtive and slightly comic. In the case of a simple Watt or Newcomen engine the desideratum was, as we have seen, to get as good a vacuum as possible in order to derive the maximum 'duty' or work from the machine. To this end a systematic engineer like Watt could fix a manometer to the cylinder in order to ensure that when the cylinder was full of steam the pressure was at atmospheric level; when condensation had taken place it was at, or near, zero. The duty performed by the engine would therefore be the product of the pressure times the area of the piston times the length of the stroke (force times distance). If, however, the pressure of the steam fell steadily during the motion of the piston the simple rule of 14 lb. times area times stroke would not give the answer. What, in fact, was needed was the area of the indicator diagram, the area under the pressure/volume curve (see Figure 6).

The practical solution of this problem is very simple and was found by John Southern in 1796. As one of Boulton and Watt's business acquaintances wrote: 'I am like a man parch'd with thirst in the Expectation of relief, or a woman dying to hear (or tell) a Secret—to know Southern's mode of determining Power . . . '[24] It was simple enough. A sheet of paper is pinned to a board, which moves to and fro with the reciprocating motion of the piston; at the same time a spring-loaded pressure-gauge causes a pencil, held in a lever arm, to move at right angles to the motion of the board. As a result the pencil automatically traces out the curve, relating the pressure of steam in the cylinder to the change of volume: that is to say it traces out the indicator diagram; and Gilbert's analysis showed that the area under the curve must be proportional to the work done or the power being developed by the engine.

Watt made no secret of the fact that he used manometers to determine the variation of steam pressure in the cylinder, but the indicator he kept secret for a long time. The first public mention of it seems to have been in a short letter in the *Quarterly Journal of Science* in 1822.* But it was a diffident introduction and John Farey, for example, first learned of it when he saw one being used on an engine in Russia as late as 1826.[25] Two

* I am grateful to Mr Rodney Law for reminding me of this letter.

reasons are sufficient to account for this extraordinary and effective secrecy; patents could hardly protect such a simple and easily hidden device from extensive piracy and, in any case, it was reasonable to suppose that any attempt to patent it would have been immediately challenged by Hornblower and Gilbert on the grounds that it was based on a procedure clearly sketched out in public: i.e. before the courts.

But if the actual device could easily be kept secret, the basic theory could not, for it was already public knowledge. In the 1797 edition of the *Encyclopaedia Britannica* Robison's famous article on the steam-engine, to which we have already referred, included an account of the theory of expansive operation.

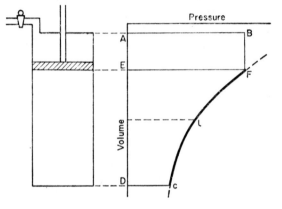

Figure 9

Indeed, Robison's analysis soon came to be much admired by contemporary writers on the steam-engine, even though he clearly knew of Gilbert's work and was therefore unoriginal.[26] Robison's account is rather more interesting to us for what is left out than what is said. 'The accumulated pressure during the motion of the piston from EF to DC will be expressed by the area EFcDE and the pressure during the whole motion by area ABFcDA. Now it is well known that the area EFcDE is equal to ABFE multiplied by the hyperbolic logarithm AD/AE = L.AD/AE and the whole area ABFcDA = ABFE (1 + L.AD/AE). It having been granted that Flc is a rectangular hyperbola' (because the expansion follows Boyle's law).

The term 'accumulated pressure' is more than a periphrasis for the work done, or Gilbert's 'efficiency'; it established the whole argument in the Newtonian tradition: duty is to be measured by *force*, and the insight afforded by the relationship between the work done and the area of the indicator diagram is, if not obscured, hardly clarified. Again, the coy avoidance of any reference to the use of calculus ('. . . it is well known that . . .') gave no loophole for Gilbert or Hornblower. Robison does in fact mention Hornblower, but only to argue that his compound engine was no more efficient than an engine using Watt's expansive principle, and also to praise the high quality of Hornblower's workmanship; but the latter, in Robison's scale of values, meant that Hornblower was being put firmly in his place as a worthy artisan.

So far we have considered the development of the expansive principle and the indicator diagram solely from a theoretical point of view. However, Boulton and Watt were not in the engine business in order to advance thermodynamics; they were in it to make money. They were selling power, and while their patents still ran the premiums they charged on their engines were at the rate of £5 per horse-power per annum (£6 in the London area). But the problems of matching the power of an engine to the machines in a mill proved difficult. Some mill-owners even put up their mills and started to install the machinery without giving any thought to the engine that would have to provide the power.[27] Most mill-owners, lacking experience of steam engines, changed their minds several times in their dealings with Boulton and Watt.[28] Others, more far-sighted perhaps and looking forward to the growth of the industry, wanted to put in larger engines than they immediately required. Should they be expected to pay £100 p.a. on a 20 h.p. engine if they could only run it at 5 h.p.?[29] It was problems like these that made some means of measuring the power developed by an expansive engine, or an ordinary engine run at less than full throttle, absolutely imperative; hence the invention of the indicator diagram.

The man who bore a good deal of the brunt of the difficult task of applying steam-power effectively in the new textile industries was Peter Ewart, Boulton and Watt's representative in the Manchester area.[30] As a man of scientific ability Ewart certainly equalled and probably surpassed John Southern. He found the scientific atmosphere more congenial in Manchester than in Birmingham, for Manchester had its famous Literary and Philosophical Society, whose *Memoirs* can claim to be the oldest scientific periodical in Britain, apart of course from the *Philosophical Transactions*, and there was a wide and deep concern in the area for scientific and technological education.[31] John Dalton was, or was shortly to become, a leading scientific figure in Manchester, and Ewart became his close friend. The great chemist and meteorologist and the able Scottish engineer had gifts that were in some ways complementary; Ewart certainly adopted the chemists' notion of caloric and was quick to accept Dalton's atomic theory. On the other hand he told Dalton about the principle of total heat and about Watt's law as well as winning his respect for the great paper 'On the Measure of Moving Force'.[32] Whether he told Dalton about the indicator diagram we do not know—he was privy to it from the beginning—but it *is* safe to say that the long friendship between Ewart and Dalton had some significance for the development of thermodynamics.

Let us now sum up. Watt had envisaged the principle of expansive operation as early as 1769, and Hornblower, quite independently and at some unknown later date, had hit on the same thing. Hornblower had been the first to patent the principle, in the form of the compound engine, and Gilbert had been the first to analyse it theoretically. The theory of the indicator diagram, which was formulated by Gilbert in 1792, led to the practical invention of the instrument by John Southern in 1796. And once again Robison fogged the historical issue and acquired for himself the unjustified reputation of being the first man to analyse expansive operation. It is well that the credit should be fairly apportioned, for the concept of expansive operation and the invention of the indicator diagram were of basic importance, as we shall show. But history returned her own verdict on the matter: the prestige of Cornish engineering was soon abundantly confirmed when the Boulton and Watt patents finally lapsed in 1800.[33]

At the turn of the century a new generation of Cornish engineers took over. The young Richard Trevithick, an intuitive genius if ever there was one, pioneered the use of high-pressure steam, much to the elderly Watt's disapproval. Once again Davies Gilbert had a contribution to make, for he acted as scientific adviser to Trevithick in much the same way that he had done for Hornblower. The advantages of high-pressure steam were obvious enough: the higher the pressure the smaller the cylinder needed to generate the same power. Moreover, the higher the pressure the less necessary the condenser became; to do without one and exhaust the steam directly into the air meant, for an engine working at an *excess* pressure of five atmospheres, losing only one-sixth or about 17 per cent of the total available pressure difference, while gaining enormously in simplicity, cheapness and mobility. Without the cumbersome condenser there was no real obstacle to putting a small and powerful high-pressure engine on wheels. Hence the railway locomotive was born from the supersession of Watt's low-pressure condensing engine by Trevithick's high-pressure engine. Further and incidental advantages were that the high-pressure engine, being small, might be expected to lose less heat by convection, conduction and radiation than the big low-pressure ones; and having fewer moving parts—for example no air-pump was required—it would lose less power through friction. It was therefore reasonable to expect that it would be markedly more efficient than the low-pressure engine.

If Trevithick's high-pressure engine represented a new and radical departure from the traditional Newcomen–Watt line of development, the great 'Cornish' pumping engines[34] of the following fifty and more years were well within that tradition. Large, condensing beam-engines, they were progressively developed until they reached an acme of perfection that could hardly be surpassed.

STEAM-POWER AND WATER-POWER

We have already pointed out that many heat-engines were, it was claimed, designed in such a way that they could be driven by the pressure of water as well as by that of a gas or vapour. In striving for such versatility inventors were confirming the immense scope that there was for the development of water-power, as well as the fact that in many parts of the world it was more readily available than was cheap fuel. This was particularly true of most European countries at that time. Water-wheels were, of course, commonly used as sources of power, and we have already discussed the theory and practice of these machines, but they by no means exhausted the list of devices for harnessing hydraulic power. There was, in fact, considerable and constant cross-fertilisation between the technologies of steam- and water-power; and this cross-fertilisation led not only to new and improved techniques but also to *new and profound insights into the natural processes of energy transformation.* Full discussion of the theoretical aspects of this subject must be deferred to a later chapter; for the present we are concerned mainly with the technical details.

It is easy to draw up a morphology of water-engines to parallel the one for heat-engines set out on pages 73 to 76.[35] In one very important respect, however, hydraulic engines differed from heat-engines in that the former all shared a common measure of efficiency. The criterion for undershot and overshot water-wheels, Barker's mills, hydraulic rams, etc., was the capacity of the machine to restore or recover the *status quo ante*; a machine whose efficiency was rated as 60 per cent, or 0·6, coupled to a perfect pump, could generate enough power to restore $\frac{6}{10}$ of the driving water to the source whence it had fallen, or, on Galilean principles, $\frac{6}{100}$ of the driving water to a height ten times that of the source. In other words, the duty of water-engines was commonly expressed as a fraction, which indicated the capacity of the engine to restore the motive agent (water) to the source. This mode of expression had been adopted by Parent and subsequently accepted by all other writers on the subject.

From our point of view the most important and interesting of the water-engines was the hybrid commonly called the column-of-water engine, or water-pressure engine.* In essence these were atmospheric or steam-engines driven by the weight or pressure of a column of water. Although they have a history which can be traced back to the seventeenth century, the characteristic form of the engine emerged only in the middle of the eighteenth century in Schemnitz (Banská Štiavnica, in Slovakia), in France, and a few years later in England. These early engines represented quite independent inventions, and all show quite clearly the influence of the Newcomen design.[36] As the atmospheric engine progressed, as it became the true steam-engine and as metal increasingly replaced wood in its

* *Machine à colonne d'eau, Wassersäulenmaschine.*

construction so the design of the column-of-water engine followed suit. And, at the same time, the mechanical refinements of the steam-engine were incorporated in the column-of-water engine, which was soon made in a double-acting form using parallel motion. Illustrated accounts of the engine were given by Delius,[37] Jars[38] and John Smeaton,[39] who thought it was '. . . the most important invention since the fire-engine'. Jars, like Delius, ascribes the invention of the Schemnitz column-of-water engine to Joseph Höll, the man who invented the famous air-engine (page 59), and goes on to give details of the performance and cost—as we have noticed, Jars was very cost-conscious—of the various column-of-water engines working at or near Schemnitz. He notes that all the mechanical details of these machines have been taken from the fire-engine* and observes that if there is a sufficient fall of water '. . . one should always prefer these machines to water-wheels which are more frequent in mines for one does not lose a drop of water'. This advice was accepted by other engineers and it soon became a general, or textbook rule that if the fall of water much exceeded forty feet (or about ten metres) a column-of-water engine was better than a water-wheel.

The column-of-water engine thus followed in the footsteps of the steam-engine, sharing the latter's improvements step by step, with the steam-engine acting always as the pacemaker. In fact Ewbank, the historian of hydraulic machinery, was merely echoing what many had previously observed when, in 1842, he wrote[40] that the column-of-water engines '. . . exhibit a very striking resemblance to high-pressure steam-engines. Indeed the elemental features of steam- and pressure-engines are the same, and the modes of employing the motive agents in both cases are identical.'

These engines were not, as it happened, extensively applied in Britain. Engineers like Trevithick, who erected about six of them in Cornwall,[41] were certainly impressed, and it has been suggested† that Trevithick's early experiences with column-of-water engines, operating at high pressure, led him to conceive the advantages of high-pressure steam-engines. But as it happened the combination of plentiful, cheap and good coal with an absence of substantial falls of water reduced the scope

* 'Toute cette mécanique a été prise sur celle de la machine à feu.'
† By Mr W. N. Slatcher (verbal communication).

for applying these machines in Britain. It was otherwise on the Continent, especially in Germany and Austria–Hungary, where these engines were more systematically developed.

The conditions for the development of the maximum power from a given fall of water, worked out by a succession of (mainly French) theorists culminating with Lazare Carnot, had been that the water must enter the machine without shock or violent impact and leave without appreciable velocity. Given good design the column-of-water engine could approximate to these requirements, at least in theory, and therefore deliver enough power to restore a substantial proportion of the driving water to the source, thus indicating high efficiency. It was, after all, from one point of view a pump driven backwards; and in the limiting case, when it works 'ideally' without frictional or other losses, it must generate enough power to enable a perfect pump to restore all the motive agent to the source.

This was not an entirely abstract argument, for as the engineer Guenyveau put it:[42] 'The column-of-water engine is . . . the inverse of a force pump and several German writers use the same formulae for both.' Guenyveau is also informative about the design of engines, to satisfy Lazare Carnot's two conditions for maximum efficiency. If when the piston reaches the end of the cylinder the full pressure of the driving water is behind it, not only will the machine receive a violent shock ('water hammer') but the *vis viva* of the column of water will be wasted. To obviate this he points out that the supply of water can be progressively throttled as the piston travels down the cylinder, being finally cut off as it reaches the end. Alternatively, an air-vessel can be attached, through a suitable valve, to the high-pressure pipe, thus giving the motive agent a certain degree of elasticity.[43] In this way the column-of-water engine in its advanced form incorporated a partial analogue of the expansive operation of the steam-engine.* But in this case there was no question of cross-fertilisation, for Guenyveau, although he discussed the steam-engine, did not mention Watt's expansive principle; this was not surprising, since it was quite unknown in France or elsewhere on the Continent at that time. The partial analogue of the expansive principle applied to the working of

* The analogue is partial since water is, of course, quite incompressible and inelastic. It is, however, an analogue since the insight into what we now know as the energy principle is the same in both cases.

the column-of-water engine was therefore arrived at by independent, if parallel, considerations of basic *vis-viva* principles that are common to both water- and heat-engines. The perceptive engineer of that day might well have wondered if any additional insights could be obtained through the study of advanced power technology.

One other charateristic feature of the column-of-water engine was that it was much more flexible than any of its hydraulic rivals. Unlike the water-wheel it could generate power efficiently, not merely from falls of thirty feet or so, but from hundreds of feet; indeed the only limit to the height of fall was the practical one set by the mechanical strength of the pipes, valves and cylinders. No great knowledge or experience was necessary to show that the power of a 120-foot fall could be more efficiently developed by using one column-of-water engine than by employing four 30-foot water-wheels in cascade, one above the other. Thus in a number of ways the column-of-water engine represented the most advanced and interesting form of water-power at the turn of the century, and one which could and did arouse the interest of the leading engineers of the time.

Chapter 4

The Beginnings of a
New Cosmology

If one looks back over the development of water- and heat-power during the eighteenth century, a contrast that strikes one immediately is that while from the time of Parent onwards a knowledge of the amount and velocity of the water flowing and of the useful work performed was enough to enable the efficiency of a water-engine to be computed—1, or complete recoverability, marking the ultimate limit—no such calculation was possible for the 'fire'-engine. Although fire was recognised as the motive agent and although engineers did measure the power-output of fire-engines, there could be no overall measure of their efficiencies; still less was it possible to regard such engines as *transforming* 'effort' from one mode into another. In so far as efficiency could be measured it was expressed practically as the 'duty', or the work done for the combustion of one bushel of good coals.

No progress was possible until Black, his contemporaries and immediate successors had established the quantified science of *heat*. Equally important were the researches of Watt and the inventions he based on them. For Watt, in effect, established the true nature of the heat-engine, and indicated for the first time what the efficient operation of such an engine implied. Dr Pacey points out that the invention of the condenser clarified the idea of the *flow of heat through the engine* as an essential feature of its operation. It made it apparent that a hot *and a cold* body were necessary if the engine was to work. The need for the cold body, and the concomitant flow of heat, was concealed from the understanding (and the imagination) in the working of the atmospheric engine, and it was not very evident in the case of the subsequent high-pressure Trevithick engines.

As far as heat-engines were concerned, the requirements at the beginning of the nineteenth century were, apart from piece-meal improvements, the elucidation of the laws of heat-transfer (radiation and conduction) and increased understanding of the physics of gases and vapours. Therefore in this and the following chapter we shall be concerned with the study of heat, no longer as a part of chemistry but as a branch of the new science of *physics*. This will require that we pay some attention to geophysics and meteorology as well as to speculations about the ultimate nature of heat. We shall then be able to discuss how it came about that the convergence, practical and theoretical, of the power technologies in the opening decades of the nine-teenth century helped to revolutionise the science of heat and to change man's idea of his universe.

GEOPHYSICS AND METEOROLOGY

While engineers and entrepreneurs in England were solving the many problems of applying steam-engines to new industries, and while engineers everywhere were seeking the optimum conditions for the generation of power, the geophysicists were pushing ahead with their exploration of the natural processes of the earth. The old seventeenth-century 'sacred' theories, such as those of Whiston and Woodwarde, had been superseded by the rationalist theories of Buffon and others, one consequence of which was the establishment of a general science of geology and the work of men like Werner and Hutton. J. H. Lambert tried to extend his photometric technique to radiant heat (or 'obscure heat' as he called it) in order to compute how much heat the earth received from the sun.[1] The problem was: did the earth receive more or less heat from the sun than it lost by cooling? And behind this question lay the more general one of whether the earth was heating up or cooling down, and hence, ulti-mately, the problem of the age of the earth. But an answer to this cosmic question was quite outside the scope of eighteenth- or even nineteenth-century physics: it was not to be forthcoming until after the discovery of radioactivity.

In one particular respect the horizons of human experience and scientific understanding were being widely extended at this time. There was a curious and very rapid development of interest in the great central mountain-chain of Europe, the Alps. Although there had been exceptions in all ages up to the eighteenth century the generality of men, even of men of education and culture, had disliked or feared the mountains. The reasons were simple and obvious: they are patently inimical to life, and extreme inaccessibility encourages all sorts of fears of hostile forces that may lurk among them. If this was no more than a superstitious peasant's attitude, there were also rational or sophisticated objections. How can one admire a barren wilderness? Natural beauty is that which evinces the taming or civilising hand of man. It is best seen in the formal beauty of the great parks and gardens of Europe or in the more subtle work of the English landscape-gardeners. Could any man of discrimination seriously compare a mountain torrent with, say, the fountains and cascades of Versailles—or Chatsworth? Now this attitude is a matter of taste, and taste only; it is as tenable today as it was at the time when Thomas de Quincy lowered the blinds of his coach to shut out the unpleasant prospect of the Lakeland mountains through which he was being driven.

It is usual to ascribe the growth of appreciation of the beauties of mountain scenery to the influence of Rousseau in the first instance and, more generally, to the great Romantic movement which began towards the end of the eighteenth century. It is not our purpose to question this thesis,[2] but it must be pointed out that there is some difference between the passive appreciation of mountains and actually trying to explore and to climb them. In short, it seems that to account for the rise of mountaineering we have to look elsewhere than to the romantic movement; in fact, to the contemporary stage in the development of science.

The association of physical science with mountains goes back, as far as we need be concerned, to the seventeenth century; to Perier's ascent of the Puy de Dôme in order to demonstrate that the height of the barometer fell as the altitude increased, and to Richard Towneley's ascent of Pendle Hill when he was discovering his 'hypothesis', still wrongly called 'Boyle's law'. In the middle of the eighteenth century Réaumur and Bouguer carried out geophysical investigations in the Peruvian Andes, but it was the developments towards the end of that century that brought the study of mountains to the forefront of science.

The rise of geology focused attention on the fantastic rock-formations of the high Alps, for as that pioneer of scientific alpinism, de Saussure, put it: '. . . it is above all the study of mountains which can accelerate the progress of the theory of the globe . . .'[3] Mineralogists, crystallographers, botanists and zoologists found much to interest them in the Alps, but it was perhaps the physicists who had most to learn. For here was, in effect, a great and natural physics laboratory in which physical processes were being carried out on a gigantic scale. The conversion of snow into ice, resulting in the formation of glaciers, the motion of these ice-rivers down to the warm valleys, and the appearance of streams from underneath the ice at the very ends of the glaciers, posed an abundance of problems. It was, we recall, phenomena of this sort which had led Joseph Black to formulate the concept of latent heat. Again, the mountains offered an obvious route to the study of the composition of the atmosphere at different heights, the formation of clouds, the fall of rain, snow and hail; as well as atmospheric electricity and the properties of radiant heat. It would scarcely be an exaggeration to say that as the solar system was the celestial model for Newtonian mechanics, so the high Alps were to provide a good deal of the laboratory material for eighteenth- and early nineteenth-century physical science.

Accordingly we find that men of science were intimately concerned with mountaineering and mountain exploration long before it became an international sport. That these early pioneers should actually enjoy mountaineering and love the mountains they explored in pursuit of knowledge is in no way surprising. It is a profound mistake to assume that scientists are detached, unemotional observers of nature, mere correlators of phenomena; on the contrary, the man of science seems to have an affection for and an involvement with the objects of his study. For without this commitment he will not be able to formulate his theories; as indeed Aristotle pointed out.

The first scientific mountaineers were, naturally enough, Swiss. De Luc, whom we have already come across, was a pioneer of scientific mountaineering, making the first ascent of the Buet near Chamonix (though perhaps not a very challenging one). Another and roughly contemporary pioneer was Father Placidus à Spescha, a geologist and mineralogist who organised the first ascent of the more impressive Tödi, or

Russein, in the Grisons. After this came de Saussure who similarly organised the ascent of Mont Blanc itself; thereafter came a flood of scientific mountaineers. Practically all those who played major parts in the subsequent development of studies of heat and thermodynamics seem to have carried out scientific expeditions in the Alps; these men included Pictet, Prévost, Leslie, Rumford, Biot, J. D. Forbes (who wrote extensively on glaciers), Joule and Tyndall.

De Saussure's name is perhaps not an important one in the history of theories of heat. He accepted Lambert's theory that the void contains Fire, which accounts for the rise in temperature noticed when air rushes into a vacuum, and which suggests that if the volume of a void space is suddenly reduced—one can hardly say compressed—there should be a rise in temperature. He also believed that the evaporation of water is essentially a solution process—a view which, as we saw, de Luc refuted but which had a surprisingly long life. On the credit side, he carried out experiments on radiant heat using conjugate metal mirrors.[4] Jointly with Pictet he showed that a hot, but not glowing hot, bullet at the focus of one mirror produced a rise in temperature at the focus of the other. To confirm that light in no way entered into the process Pictet repeated the experiment using a glass vessel full of boiling water instead of the hot bullet; again a rise in temperature at the focus of the other mirror was detected.

More important than de Saussure were his compatriots J. A. de Luc, Marc-August Pictet and Pierre Prévost. De Luc wrote extensively on the physics of the earth and on meteorology, and carried out some excellent experimental work when he established that the coefficient of expansion of liquids varied with the air dissolved in them. He then went on to show that the expansion of mercury was to all intents and purposes uniform.[5] De Luc's theory of heat shows the influence of Boerhaave: Fire is the universal cause of heat phenomena and is the most active physical agent on our globe. Heat, the consequence of Fire, has three principal effects: expansion, fusion and vaporisation.

Marc-August Pictet (1752–1825) made his mark with one excellent volume, *Essai sur le feu*,[6] which was published in 1791 and almost immediately translated into English. Like Lavoisier and de Saussure, Pictet accepts the material theory of heat. All hot bodies are, he says, in a 'forced' state in that the self-repellent matter of heat is retained only by the surrounding, ambient heat. If heat is not 'cohibited' in this way it will diffuse itself rapidly outwards. The two modes of heat-transfer, conduction and radiation, are distinguished, and Pictet points out that the heating-up of a body depends upon both its heat-capacity and its conductivity. The latter can be separated from the former by comparing the times which bodies of equal mass take to heat up with their specific heats. Pictet is very fair in apportioning the credit for the discoveries of specific and latent heats to Joseph Black, although he says that the first hint of the former was given by de Luc in 1772.[7] He has heard of Wilcke, whom he calls Wilkie.

Pictet's most important contribution was to the study of radiation. As we saw, he collaborated with de Saussure in experiments on radiant heat using conjugate mirrors. He even tried to measure the velocity of radiant heat, but all he could conclude was that it covered at least 69 feet in no measurable time at all. So much for radiant heat; what then of radiant *cold*? With a piece of ice at the focus of the first mirror he found that there is a *fall* in temperature at the focus of the second one. But this was no problem for him: every heated body is in a forced state, and the thermometer bulb is no exception. The heat from the thermometer will tend to leave it for the cold ice; thus it must indicate a fall in temperature. There is really no difference in principle from the case of the hot bullet, or the boiling water; there is no radiation of cold *per se*; only radiation of heat. In the first case the radiation is from the bullet, or hot water, to the thermometer, in the second from the thermometer to the ice.[8] In this way a difficult, seemingly paradoxical experimental result is no sooner obtained than it is satisfactorily explained in terms of the received theory of heat, without the necessity of inventing any additional concepts. This could be taken as further confirmation of the caloric theory; and conversely, as a refutation of the rival kinetic theory.

Pictet concludes his book by discussing the heat generated by friction.[9] This he sees as a difficult problem; but he goes on to suggest that as both electricity and heat can be generated by friction and as they are related in many other respects, perhaps they share some common factor.

Pictet's discovery of the transmission and reflection of cold had an important bearing on a theory that was being worked out by another Genevese scientist, Pierre Prévost. Prévost had studied Daniel Bernoulli's kinetic theory of gases and he knew of Le Sage's subsequent speculations on the same subject.[10] He therefore used these ideas, or concepts, to develop the theory of what he called 'mobile equilibrium'. Caloric is a subtle fluid whose atoms are tiny compared with their separation and which move very rapidly in straight lines in all directions. All hot bodies therefore tend to emit caloric atoms; the hotter the body the faster the atoms travel. If, then, we have two adjacent bodies, one hot and one relatively cold, both will be emitting caloric atoms and each will be receiving such atoms from the other. Inevitably, by this process of dynamical exchange the cold body will be heated up and the hot body cooled down. And this explains, still within the caloric framework but even more neatly, the phenomenon of the transmission and reflection of cold discovered by Pictet. In this way the famous and fundamental 'Prévost's theory of exchange' was born: the theory which states that all bodies are always absorbing and emitting heat, and which therefore constitutes a landmark in the history of our insights into natural and thermodynamic processes. The circumstances of its discovery indicate clearly enough the explanatory power of the caloric theory and also that, on the margins at any rate, the caloric and kinetic theories were not so antithetical as later writers have tended to assume.

RUMFORD; THE FALSE DAWN OF THE KINETIC THEORY

At the beginning of the nineteenth century the material theory of heat in the form which Lavoisier had formulated it was well established in France and, by virtue of the high prestige of French science, was increasingly accepted elsewhere. The merits of the theory were considerable, for it provided a very convincing explanation of the process of thermal expansion in solids, liquids and gases; it accounted for the latent heats of fusion and of vaporisation and it harmonised very well with the phenomenon of compressive heating or expansive cooling of a gas—indeed the picture of material heat being squeezed out of a gas was a particularly persuasive one. Finally, it provided a convenient framework for Prévost's theory of exchange; a theory which was soon recognised as marking a profound insight into the nature of heat transference.* Beyond all these advantages, the caloric theory held out some hope that a major correlation of human knowledge might be achieved if the phenomena of heat, electricity and light could all be shown to be the effects of some ultimate substance.

And yet it was at this time of increasing acceptance and confirmation, when the material theory of heat seemed more plausible than ever before, that a champion entered the lists on behalf of the kinetic theory. This was Sir Benjamin Thomson, Graf von Rumford, a man whose name is mentioned in all intermediate textbooks dealing with heat and thermodynamics[11] and one of whose experiments has its place in the folk-lore and mythology of contemporary science. This was the experiment by which, it is often said, he decisively refuted the caloric theory, and at the same time established the dynamical theory of heat in its place. The source for this is the famous paper which he wrote as a result of his observations on the boring of a cannon in an arsenal at Munich.[12] A great deal of frictional heat is generated during this operation. Rumford was impressed (as others had been before him), and by immersing the cannon barrel in a tank of water and using a deliberately blunted borer he showed that in less than two-and-a-half hours the water could be brought to the boil merely by frictional heat; in fact the supply of heat seemed inexhaustible. This, Rumford argued, was quite inconsistent with any material theory of heat. If caloric is flowing outwards in all directions from the source in the cannon barrel, how can the supply be kept up? One cannot, he urged, suppose that caloric flows *inwards* from the water and the main body of the cannon at the same time that it is flowing *outwards* from the boring head. Nor can we suppose that it results from any physical change during the boring operations, for tests showed that the specific heat of the metal filings and swarf is exactly the same as that of the metal from which it was abraded.† Rumford therefore concluded that heat cannot be

* This fundamental theory was in due course adapted to the dynamical theory of heat without difficulty.

† This, of course, was evidence against Irvine's theory, but it did not, as some modern writers seem to think, refute the caloric theory. Lavoisier and Laplace had already produced weighty evidence against Irvine's theory some fifteen years earlier. This did not prevent Lavoisier's supporting the caloric theory.

material, and he insists that the only thing communicated to the barrel has been MOTION [*sic*].

Some modern writers have professed, unwisely, to deplore the prejudice as they see it that led practically all Rumford's contemporaries to reject his arguments. There is, of course, some truth in the popular view that this 'experiment' refuted the caloric theory, but there is much more error, and this is unfortunate for several reasons. It is less than just to Rumford's critics, among whom were men of the highest abilities; it misinterprets the necessary conditions that must be fulfilled before a scientific theory must be thrown out and a new one put in its place and, lastly, it obscures and confuses Rumford's very real contributions to science. We shall try to clarify the situation by recounting the development of Rumford's main theory as it was put before the scientific public, and then by accounting for the reactions of that public to it.

Benjamin Thomson (1752–1814) was born in New England. After the end of the great war of independence, during which Rumford fought against his fellow-Americans, he came to England, where his talents were quickly recognised. He became an Under-Secretary of State and was later given a high military command; he was elected F.R.S. and duly knighted. After this he went to Bavaria, where he added to his laurels as administrator, technocrat and scientist. From about 1799 onwards he divided his time between England and France. His second marriage, to Lavoisier's widow, was not a happy one.

Rumford's first scientific papers were concerned with gunnery and explosives. These led him, quite naturally, to studies of heat, but his first efforts in the new field were abortive: he devised a simple method for measuring what we should now call the specific heats of solids only to find (in 1783) that he had been anticipated by Wilcke, whom he calls Wilkin. His first notable work was on the transmission of heat through various fluids: air under different pressures and with different humidities, other gases, water and mercury. His apparatus was extremely simple and his results no more than qualitative.[13] He found, as we should expect, that mercury was the best transmitter of heat, followed in order by moist air, water, air under lower pressures, and finally a Torricellian vacuum, which transmits heat although not very readily. It is perfectly clear that at this time (1786) Rumford had not yet distinguished the different modes of heat-transmission: conduction, convection and radiation.

In his next paper (1792) Rumford is able to take his investigations considerably further.[14] His motives are frankly practical,* and this particular study is concerned with the effects of familiar insulators, like wool, fur and feathers, on the transmission of heat. He finds that these substances 'conduct' heat less effectively than air because they reduce the free movement of the air of which they are mainly composed, in terms of bulk. His conclusion is startling; air, like all gases, is essentially a perfect non-conductor of heat; the sole mode of propagation of heat through air is by means of what we now call convection. The particles of air, on being heated by contact with the walls of the containing vessel, expand, and as their specific gravity is accordingly reduced they tend to rise, colder particles sinking down to take their place and be heated up in turn. There can, in an elastic fluid, be absolutely no communication of heat between particles, and this represents the essential difference between gases on the one hand and solids and liquids on the other. We can now see why materials like wool, fur, feathers, hair, etc., are such excellent insulators, and moreover we can now understand the reason for the distribution of fur on the bodies of animals that live in cold climates. A beneficent Providence has arranged that the fur is thick on the creature's back and fine and thin on the undersides, so that as heat is dissipated upwards in air, the animal is kept as warm as possible in the coldest weather.

His next paper (in 1797) almost inevitably extends his proposition to include liquids as well as gases.[15] All liquids, even mercury, are absolute non-conductors of heat, and the only way in which heat can be transmitted through a liquid is, as with a gas, by convective circulation. For Rumford the touchstone was his experimental demonstration that no heat seems to be conducted downwards through an absolutely still liquid; hence, he argues, there can be no communication of heat between the particles of the liquid. From the time of this paper onwards Rumford carried on a long debate with the other members of the scientific community who were interested in the study of heat. Are liquids conductors of heat or are they

* '. . . the confining and directing of heat are objects of such vast importance in the economy of human life . . .'

not? The experimental difficulties of proving the case one way or the other were clearly formidable; many different sources of error can confound the issue. True, dramatic experiments can be set up to demonstrate that water is a very poor conductor of heat, but that is not the same thing as proving that it is an *absolute* non-conductor. Indeed, to prove Rumford's case one would necessarily require an absolutely sensitive, or perfect, thermometer, and such an instrument was not forthcoming. It is hardly surprising, therefore, that Rumford made no converts.

Among those who joined in this discussion was John Dalton. He pointed out (1804) that if heat could not be communicated from particle to particle then a mixture of hot and cold water must, according to Rumford's proposition, sooner or later separate out into two layers with the hot water on top, the cold water underneath. To this Rumford replied, rather evasively,[16] that the viscosity of the water would prevent the separation of the hot and cold particles. But this was very unsatisfactory; how could liquids ever heat up if the constituent particles could neither move nor communicate heat to one another? Rumford's suggestion was basically unsound, and it is difficult to deny that John Leslie was quite justified in remarking that it '. . . really deserves no serious discussion'.[17]

Why, then, was Rumford so tenacious in holding to this palpably absurd theory? One possibility was that it was peculiarly apposite to the grand contemporary doctrine of the 'argument from design'. Rumford was evidently not influenced by the cool scepticism of Hume, Black, Hutton and the other members of the Scottish School; similarly the revolutionary ideas of Darwin, Wallace, Huxley and Tyndall were some fifty years in the future. His was the world of Archdeacon Paley, of natural theology and the Bridgewater Treatises. That fresh water is unique in that it reaches its maximum density four degrees (centigrade) above the freezing-point is really most convenient and considerate. It explains exactly why ice forms on the surface of lakes and ponds and why the temperature of deep lakes is always, as de Saussure found, just above 4°C. The formation of ice on the surface prevents winds agitating the water below and so, as water is a non-conductor, the loss of heat must stop, or at least be greatly reduced. Thus lakes do not freeze solid; fishes survive and the economy of nature continues.[18]

A benevolent and omniscient Deity has designed the Universe, created matter, constituted the physical laws that material things must obey and given life to all living things. It follows that the laws of physics must be so formulated as to protect and conserve life; hence the subtle wisdom which ensures that water reaches its maximum density at 4°C and is, at the same time, a non-conductor of heat. So, too, the oceans act as great reservoirs of heat, and the oceanic currents like the Gulf Stream distribute that heat so that otherwise frigid continents have mild and equable climates. In fact, observes Rumford, the simplicity of this contrivance for the preservation of heat '. . . deserves to be compared with that by which the seasons are produced . . . every candid observer . . . will agree with me in attributing them both to the SAME AUTHOR' [sic].[19]

For our purpose, however, it is important to remember the years in which Rumford enunciated and subsequently defended his doctrine that liquids are absolute non-conductors of heat. He first stated it in 1797; he was defending it vigorously in the early 1800s. Now it must be agreed that the mechanism for the propagation of heat in liquids and gases that Rumford describes is quite inconsistent with the dynamical theory of heat as put forward by Daniel Bernoulli and as elaborated by Joule and Krönig, Clausius and Maxwell in the nineteenth century. There is, on the latter theory, no reason at all why molecules should not move downwards in a liquid or gas. We must conclude therefore that in 1798 Rumford had not envisaged anything like the dynamical theory of heat; yet it was in this year that he published his famous inquiry into the source of heat generated by friction and made his apparently prescient remarks about MOTION. What are we to make of this?[20]

In the first place, the concluding paragraphs of that famous paper contain very significant reservations, which are usually left out in popular or textbook accounts of Rumford's works. We do not know, we may never know, says Rumford, what the ultimate nature of heat is; but we may study its effects on material bodies, just as Newton was able to study the effects of gravity and to formulate laws about it without knowing the ultimate nature of that agency. It seems, in fact, that at that time Rumford's view of the nature of heat was agnostic, or at least negative. He was not a philosophical scientist so it is unlikely that he was influenced by the positivism that characterised the Scottish School; it is more likely that he wanted to suggest that motion could, in some unspecified way, be taken as a *measure* of heat, or else in the last resort his nerve failed and he wished to provide a way of retreat for himself.

Secondly, it is worth remarking that Rumford nowhere in this paper suggests that heat is the measure of the *vis viva* of the constituent atoms. This hypothesis had, as we saw, been mentioned by Lavoisier and Laplace and, more recently, by Armand Séguin in his long critical essay of 1789. On the other hand Rumford made no attempt to measure the amount of work done to produce a given amount of heat, even though the Smeatonians in England and the practitioners of *mécanique industrielle* had provided appropriate measures of duty, or work, or *force-vive latente*. In fact in this paper Rumford uses quantification only for purposes of illustration: the power of two strong horses, the heat produced by six good wax candles burning steadily. In brief, there is no attempt to correlate the amount of work expended with the quantity of heat produced.

If, however, Rumford did not establish the dynamical theory, did he destroy the caloric doctrine? According to his contemporaries he did not. For one thing the properties of caloric were not, as William Henry pointed out,[21] so well known or so clearly defined that the doctrine could be falsified by Rumford's simple experiment. Again, if the fact that a seemingly inexhaustible supply of heat is available so long as friction continues proves that heat is of the form of MOTION, does not the very similar generation of electricity by friction prove that electricity, too, is motion? But at that time practically everyone[22] accepted, with good reason, that electricity must be a subtle fluid.* If, then, Rumford's argument proves the case for the proposition that heat is motion, it must prove the same case for electricity, and this makes the argument unacceptable; so, at any rate, reasoned Thomas Thomson,[23] Berzelius[24] and, later on, Becquerel.[25] Finally, we must add that to extrapolate from the experience of two-and-a-half hours to eternity—as is implied by the word 'inexhaustible'—was extremely bold, if not rash. Needless to say, before very long experimenters came forward with arguments and evidence to show that after a certain time the yield of heat from percussion or friction showed signs of falling off.[26] Others, like Haldat,

* Perhaps we still believe electricity to be a 'subtle fluid'?

found that the yield varied considerably when different materials were used to generate frictional heat.[27] Haldat further claimed that the yield was significantly reduced if the whole frictional apparatus was electrically insulated from earth.

In short, Rumford's challenge to the accepted theory of heat stirred up a hornet's nest of counter-arguments and experimental refutations. The consensus would seem to be that while the generation of frictional heat was rather difficult to explain on the caloric theory it was not necessarily impossible to do so; and in any case there were grave objections to any alternative theory. If heat can be created *de novo*, what happens to the fundamental axiom of the conservation of heat on which the whole science rested? No scientific theory is ever flawless, and the caloric doctrine was no exception. Rumford pointed to a weakness in the doctrine, but that was by no means the same thing as refuting it. Accordingly, most men of science (Dalton, Avogadro, Thomas Thomson, William Henry, Leslie and the great majority of Frenchmen) continued to accept the caloric doctrine in its various forms; and beyond that many of them, like Dalton, looked forward to an eventual union of the theories of heat, electricity and light in terms of one fundamental substance.

Rumford's famous paper was not so much a piece of science, in the proper sense of that word, as an instance of special pleading; a mustering of evidence for a particular point of view. As such, its methodological status is suspect: it is Baconian rather than Galilean. In this respect his paper on the weight of heat (1799) is much more impressive.[28] Following on the previous attempts to weigh heat, or (optimistically) to show that it was weightless, Rumford weighed a block of ice just before it began to melt and then after it had entirely melted. There had been no significant change of temperature, so a major source of experimental error was eliminated while, at the same time, a considerable amount of heat—latent heat—had entered the ice as it had melted. Rumford found that there was no significant change of weight. This was an impressive demonstration, and his simple calculation that the heat required to melt the ice would, if applied to the same weight of gold, raise the metal to a *bright red heat* illustrated the point picturesquely. Rumford did not explain how low-temperature heat—at 0°C—could be piled up, as it were, to produce temperatures of about a thousand degrees centigrade; but then he can hardly be blamed for not having foreseen the second law of thermodynamics.

From the time of these two papers onwards Rumford was passionately convinced that the caloric theory was false. He lost no opportunity, in paper after paper, of asserting that heat *cannot* be a material substance. His vision was consistent; he was the leading campaigner against caloric. All the known facts had to be made consistent with his rejection of caloric. He denied that anything could ever be emitted by hot bodies when they cool down, as plainly the caloric theory required. But then how *can* a hot body cool down if it does not emit, or lose, anything; or, for that matter, how can a cold body ever warm up without gaining something? Rumford's answer is revealed in his criticism of Pictet's experiment with conjugate mirrors: 'It is not possible', wrote Rumford, 'that caloric has an actual existence. The communication of heat and the communication of sound seem to be completely analogous. The cold body in one focus compels the warm body (the thermometer) in the other focus to change its note.'[29] The analogy between sound and radiant heat is clearly false; there is no effect in the transmission of heat analogous to that of resonance in the transmission of sound. But perhaps we should not be too critical of Rumford on this score. His view is that bodies cool down in consequence of *external* action, of the effects of rays that fall upon them.

Some time shortly after 1800 Rumford explicitly adopted the old Bernoulli doctrine that heat is merely the *vis viva* of the constituent atoms of all material bodies, and at about the same time he accepted the idea that radiant heat, just like light, is an undulation propagated in an all-pervading aether; indeed he thought that radiant heat and light may well be of the same nature. This was a really fundamental proposal. It was due however not to Rumford but, in all probability, to Thomas Young of the Royal Institution, who had put forward the idea, tentatively, in a paper read to the Royal Society in November 1801. Young had proposed the undulatory theory of light in 1800 and again, after some bitter criticism, in his Bakerian Lecture of 1803.[30] Rumford had been instrumental in appointing Young to his position at the Royal Institution, and thereafter had become one of the first supporters of Young's new theory of light; for Rumford the undulatory theory fitted in very well with his long-held convictions about the nature of heat.

Advanced as Rumford's position now was, there were still some formidable difficulties about his denial of the existence of caloric. His assertion that all bodies heat up or cool down in consequence of only *external* action led him to a most unfortunate statement about the existence of calorific and *frigorific* rays; the latter being for him as real as the former. It was not as reactionary as it appears, for according to Rumford the rays emitted by a body at an intermediate temperature will be calorific to—i.e. will heat up—a cold body, but will be frigorific to—i.e. will cool down—a hot body. Nevertheless the choice of words was unfortunate and laid him open to misunderstanding and misinterpretation. In fact the belief that a ray of heat may be calorific or frigorific is the unavoidable corollary of his *rejection of caloric together with his inability to intuit and apply the principle of energy*. This principle was, as we shall see, shortly to be extended by French engineers, such as Hachette and Christian, to include a range of thermal phenomena; while on the other hand Young, Wollaston and others often used the word 'energy' to denote the 'strength' of different rays of light. The possibility of an intuitive jump to the realisation that heat equals energy was certainly there but Rumford, alas, never showed much interest in the science and the technology of mechanics.

Thus Rumford's total hostility to the material theory of heat led him to some awkward, even reactionary positions. He asserts the reality of cold radiations;* he denies the possibility of an absolute zero of temperature since it was a notion which had come to be closely associated with the material theory of heat, and lastly he rejects Prévost's theory of exchange: bodies do not emit or receive *heat* and it is absurd to suppose that they can do both at the same time. We pause here to note that Prévost's theory of exchange is, in logical terms, inconsistent with the basic argument in Rumford's paper of 1798, namely that it is wrong to suppose that heat can flow *towards* the boring head in the cannon barrel at the same time that it flows *away* from it (see page 96). Thus to accept Prévost's theory neces-

---

* Thus he believed that negroes had dark skins designed to absorb such few frigorific radiations as may be available in the tropics; arctic animals had white fur to reflect such radiations.

sitates rejecting the force of Rumford's argument, and vice versa; a point that textbook writers might bear in mind.

We have discussed Rumford's works at some length mainly because of the traditional estimate of his importance as a founder of the dynamical theory of heat. We have found, on the contrary, that his experience revealed the grave difficulties that confronted the dynamical theory at that time, and which had to be resolved before it could be established. How, then, do we assess Rumford? His career indicates that he was a man of considerable force of character and with great powers of persuasion. He wrote with some elegance, he knew very well how to plead a case and he was always courteous in discussion, a gift which was by no means common at that time. But he was also rootless, both spiritually and intellectually. He had not enjoyed the discipline of the Scottish School, which was based on a higher education with a strong component of philosophy and logic;[31] he had not been subjected to French influence, with its emphasis on high and rigorous professional competence with a strong mathematical background; and, most surprisingly, he had little contact with the engineers who were at that time developing the steam-engine: easily the most important instance of the application of heat to practical purposes. Richard Trevithick relates: 'We found him [Count Rumford] a very pleasant man, and very conversant about fire-places, and the action of steam for heating rooms, boiling water, dressing meat & C, but did not appear to have studied much the action of steam on pistons & C.'[32] Trevithick's son Francis wrote, rather patronisingly, 'Count Rumford was not quite up to the idea of the new steam engine to be worked solely by the pressure of steam on the piston; still he gave his opinion of a proper fire-place for the boiler of the steam carriage.'[33]

It is therefore hardly surprising that as a scientist Rumford's vision is somewhat narrow and his method sometimes suspect. He is prone to argument rather than demonstration and is usually reluctant to discuss the case against any theory he has accepted. Again, his reasoning is sometimes seriously defective in that he uses circular arguments and is apt to draw sweeping conclusions from quite inadequate data.* But these are faults

* As an instance of circular argument we may cite his calculation (after the cannon-boring experiment) that in 21 days, 19½ hours the cannon would have been ground to powder, or filings, but enough heat would

to which most scientists are liable; more serious in Rumford's case was the fact that he showed little awareness of the science of mechanics and had only the haziest idea of quantification.

In Rumford's defence it can be argued that methodological considerations are not the sole determinant of scientific merit.[34] He was entirely consistent in his hostility to the caloric theory and in this he was eventually if posthumously justified, for his well-publicised opinions provided an effective and much-needed precedent for those who, like Joule, were later to establish the dynamical theory of heat. He deserves, of course, full credit for his work on the problem of the weight of heat and for his very fruitful suggestion that radiant heat is propagated by undulations in an aether and is therefore of the same nature as light. But perhaps his best contribution was the general one that he insisted upon the cosmic role of heat as a universal natural agency. Everything, everywhere, is subject to the determining influence of heat; it is one of the fundamental factors in the operation of the universe. This was later to become an article of faith for the founders of thermodynamics.

As a pendant to our discussion of Rumford we must mention the famous, or rather notorious, experiment which Davy said that he carried out in refutation of the material theory of heat. He said that he pressed two smooth blocks of ice together and, by means of clockwork, caused them to rub against each other with a reciprocating motion; as a result of friction the ice melted and this, Davy said, refuted Irvine's theory in particular and the caloric theory in general. However, Andrade has pointed out[35] that if the ice did melt it was most probably due to the conduction of heat through Davy's apparatus; alternatively one might reasonably infer that the melting of the ice proved, not that the caloric theory is wrong, but that the melting point of ice is lowered under pressure—as indeed it is (see page 243 below). For our part we find it impossible to imagine how the 'experiment' was carried out. If it was performed at a suitably low temperature then the melted ice would immediately re-freeze and the effect would not be apparent; if, on

---

have been generated to have melted the cannon sixteen times over. This, he thought, confirmed his anti-caloric argument. As an instance of a sweeping conclusion from inadequate data his assertion that liquids and gases are absolute non-conductors of heat would be difficult to beat.

the other hand, it was performed at a temperature near the melting-point there could be no assurance that any water formed was due to the frictional effect and not to simple heat conduction or radiation from the surroundings.

RADIANT HEAT

John Leslie (1766–1832) who was Professor of Mathematics at Edinburgh University from 1805, was, like Rumford, also interested in the problems of heat, his *magnum opus* being the famous *Experimental enquiry into the nature and propagation of heat* (London 1804). Unlike Rumford he was rather a quarrelsome writer and affected a portentous prose style. His Scottish university background ensured that in intellectual terms he was technically better equipped and philosophically more sophisticated than Rumford; indeed, he was one of the first notable British scientists to use the Leibnitzian calculus notations in place of the outmoded Newtonian fluxions; it is also worth noting that he regarded himself as a disciple of David Hume.[36] Certainly he was appropriately cautious: he had no time for the aether hypothesis and he dismissed any idea of linking heat with motion as '. . . a shapeless hypothesis' which '. . . explains nothing; it throws out a delusive gleam and then leaves us in tenfold darkness.'[37] Leslie's approach to the problem of the nature of heat required the application of Occam's razor: why, he asked, multiply entities unnecessarily? Radiant heat is no more than the calorific effect of light, for light and heat are one and the same thing, a highly elastic fluid, which, when 'projected', manifests the phenomena of light, when 'combined', of heat. The radiant heat from hot but dark bodies, which Scheele described, is accounted for by an undulatory pulsation which, rather like sound, is propagated through the air. The alternate compressions and expansions of the air carry, by some mechanism which Leslie does not venture to describe, the caloric particles outwards from the hot body. This theory explains why radiant heat from an obscure or dark source is not transmitted through glass while the light of the sun together with its calorific effect is.

Leslie invented a particularly sensitive thermometer, the 'difference thermometer' as he called it,[38] with which he studied the effects of different surfaces on the radiation and absorption of heat. He describes these experiments in some detail in the *Experimental enquiry*. But perhaps his most important discovery was, as he put it, that 'All the preceding investigations concur to establish this simple proposition—*That the action of a heated surface is proportional to the sine of its inclination.*'[39]

And so, over a century later, the logical corollary of Halley's sine law of isolation was discovered. It was not neglected, as Halley's discovery had been, for almost at once it was taken up and made a basic axiom of Fourier's theory of radiant heat (see below, pages 116–20). But Leslie's idea of the nature of radiant heat was not generally accepted, possibly because of the obscurity at the heart of it. He himself admitted a serious difficulty: the 'pulsatory discharge' of heat from a dark hot body does **not** appear to be affected by the rarefaction of the surrounding air, and it appears to be unchanged when the body is immersed in different gases.[40]

Leslie distinguishes between the three modes of propagation of heat; for conduction he uses the word 'abduction', for convection 'recession', for radiation 'pulsation' and, logically, he should add a fourth term ('lumination'?) for the mode by which heat is propagated through empty space. He shows experimentally that the cooling power of a stream of air varies with its velocity and suggests that a suitable thermometer could accordingly be used as an anemometer.[41] He frequently refers to the 'energy' of heat transmission but unfortunately does not bother to explain what he means by 'energy'. He accepts the notion of an absolute zero of temperature, which he puts at −750°C, and tends to favour Irvine's theory of heat capacity. He is critical of Dr Black because: '. . . the theory of *capacity*, at once so luminous and comprehensive, was reared without his participation'. And he gives the credit for it to Irvine and Wilcke, followed by Crawford, Gadolin and others. This, of course, is a judgement which must be respected; it is not possible to refute it.

Like Rumford, he is convinced of the cosmic importance of heat. But a curious, indeed unique, consequence of his theory is that since the earth does not emit light, or rather only the trivial amounts generated by a few volcanoes and the fires lit by men, and since calorific pulsations cannot be transmitted through empty space, it follows that this and all the other planets must be getting hotter all the time! He did not say whether, in his view, this startling inference had any theological implications.

Leslie and Rumford were, then, pioneers of a new science of heat. Neither of them was or had been a chemist and it was the chemists who had previously been the leaders in the scientific study of heat. Possibly the chemists had tended to think of heat as a passive, static substance which could be doled out, rather like a chemical, thereafter to remain in equilibrium. Now, with the increasing interest in the transmission of heat, a new aspect was coming to the fore. And the study of heat-in-motion transferred the subject from the realm of the chemist to that of the physicist, as we should now call him.

We have referred on several occasions to 'physicists' of the eighteenth century, but strictly speaking it was anachronistic to do so, for this science hardly existed before the middle of the nineteenth century. In 1800 surprisingly few sciences were 'established': mathematics, *mécanique rationelle* (Newtonian mechanics in England), planetary astronomy, *mécanique industrielle* (Smeatonian technology in England), together with the very new science of chemistry virtually completed the range of solidly based sciences; all other studies were, as Whewell put it, in a 'progressive, empirical' stage. When we recall that the general concept of energy, thermodynamics, field theory and the study of atomic structure had still to be thought of we can appreciate how very shadowy, how devoid of familiar contents the field of physics was in 1800. There were also lacking at this time the familiar methods, techniques and standards of the physicist. The use of graphical method was still very rare; the theory and method of dimensions was only just being founded;[42] the idea of physical constants had not been grasped,[43] and appreciation of experimental error and order of accuracy was rudimentary.*

The study of the transmission of heat constituted an important phase in the establishment of physics as an autonomous science; independent of chemistry on the one hand and much more general than classical mechanics on the other. The studies of radiant heat and the conduction of heat lent themselves

* For example, Rumford, knowing the value of π to five places of decimals, calculates the capacity of a glass globe 1·6 inches in diameter to be 2·14466 cubic inches.

readily to mathematical analysis, transcending in subtlety and generality the simple arithmetic of the quantified heat of Black, Wilcke, Crawford and the rest. Leslie had, perhaps, shown the way into these ampler fields, and it is hardly surprising that later writers, such as Fourier and Poisson, came to regard the theories of heat-transmission as the sum total of all possible heat theories. Knowing the boundary and surface conditions, the specific heats, geometry and masses, one could, by applying the laws of conduction and radiation, obtain a complete description of the thermal state of a system of bodies at any time. Fourier, indeed, felt impelled to write that he had '. . . demonstrated all the principles of the theory of heat, and solved all the fundamental problems'.[44] But he, and those who thought like him, had overlooked the no less fundamental 'heat–mechanical power' nexus, which the development of the steam-engine was increasingly bringing to the fore and which was to lead to the science of thermodynamics.

In 1800 the great astronomer Sir William Herschel published a paper which was destined to exert considerable influence on the study of radiant heat.[45] Herschel was, at that time, interested in the telescopic study of the sun, and for this purpose he needed suitably coloured filters to reduce the intense glare. The obvious first step was to find what parts of the solar spectrum, what particular colours that is, produced the most heat, so that an appropriate filter could be made to eliminate those colours. When Herschel applied a thermometer to different parts of the spectrum he found that the indicated temperature rose as he moved it from the violet towards the red and, astonishingly, continued to rise, reaching a maximum in the invisible region beyond the red. Clearly, there must be two distinct components in the solar spectrum, a luminous and a calorific one, the latter being less refrangible than the former. Herschel further claimed to have shown that the calorific rays obeyed the same law of refraction (Snel's) as did the luminous ones, and that the heat focus of a burning-glass was, accordingly, just beyond the luminous focus.

All this was in flat contradiction to the pulsatory theory of radiant heat, which Leslie was then busy developing. Accordingly Leslie intemperately denounced Herschel and ascribed his results to experimental error[46]—a truly egregious criticism to make of the great scientist whose observational and experimental skill had led him to the discovery, in 1781, of the planet Uranus; the first such discovery known to recorded history.

Leslie's challenge was answered by Sir Henry Englefield, who found experimental support for Herschel,[47] and in due course other scientists reported results which varied between confirmation of Herschel and location of the point of maximum heat in the red, or even in the yellow part of the spectrum.[48] Now the plausibility of Leslie's theory rested on the truth of the long-established principle that heat from a dark body, or a fire, *cannot* pass through a sheet of glass, while the heat from the sun can do so. Effects to the contrary could usually be shown to be due to the glass itself heating up—becoming 'saturated' as Robison put it—and so acting as a re-transmitter of the heat that had been conducted through it from the surface exposed to the radiant heat. But some doubt was thrown on the principle when Prévost claimed to have shown that heat transmission did take place even when movable screens were used so that the glass was always cold. And the matter was effectively settled when, in 1811, François Delaroche described a series of masterly experiments which proved (1) that radiant heat from a dark source can pass through glass; (2) that the amount of heat so transmitted is greater in proportion to the temperature of the source; (3) that radiant heat which has passed through one glass screen is much less attenuated on passing through a second one; (4) that the rays emitted by a hot body differ from one another in the facility with which they pass through glass and that heat rays, like light rays, differ from one another; (5) that thicker glass transmits less heat, but that this effect diminishes with the temperature of the source; and, lastly (6), that the heat radiated from a hot to a cold body increases in greater ratio than the temperature difference between them. Delaroche's general conclusion was that heat rays are of the same nature as light, and merge in family continuity with it.[49]

These splendid observations all pointed to a refutation of Leslie's idea of the nature of radiant heat and tended to confirm Herschel's discovery. At much the same time Wollaston and, independently, Ritter and Beckmann, again following Herschel's work, had extended his discovery by showing that the ultra-violet end of the spectrum could produce definite chemical effects (Scheele had first noticed the chemical action of light) so that there was, presumably, a third kind of light-type radiation; a chemical one. (A fourth variety was subsequently 'discovered'; Moricini, Mrs Somerville and others claimed to have detected a magnetising effect at the ultra-violet end of the spectrum. But this proved illusory.)

To clear the matter up Claude-Louis Berthollet commissioned Malus and Bérard to investigate the problem systematically. Unfortunately Malus, who had discovered the polarisation of light in 1808,[50] died prematurely in February 1812, leaving Bérard to complete the work alone. The most important thing in Bérard's subsequent memoir is his claim to have shown that radiant heat not only obeys the familiar optical laws of reflection and refraction but is also, like light, polarisable.[51] He was unable to confirm Herschel's observations beyond the visible red, but this was not regarded as in any way decisive. The important thing was that Leslie's idea had certainly been refuted; and in any case the last detail was resolved to almost everyone's satisfaction (Leslie was, presumably, an exception) when, in 1818–19, T. J. Seebeck showed that the position of the point of maximum heat varied with the material of the prism, being in the infra-red region for a flint-glass prism, in the red region for a crown glass prism and in the orange or even yellow for a hollow prism filled with various transparent liquids.[52] The conclusion was drawn with admirable clarity by J. B. Biot: 'These experiments of M. Bérard demonstrate that different parts of a solar ray, dispersed by a prism, possess very unequal energies [sic] to produce vision, heat and chemical combinations.'[53]

Does this mean that there are three independent species of rays, each capable of only one effect? Or, in other words, have we three superimposed spectra? But if this were so each of the three sets of molecules would have to be polarisable, like those of light; and they would have to obey all the other laws of optics as well. And so, observes Biot: 'In place of this complication of ideas let us conceive simply and conformably to the phenomena, that solar light be composed of an ensemble of rays, unequally refrangible . . . which supposes original differences in their (molecular) masses, speeds and affinities.'[54] This simple hypothesis, in accordance with the commonly accepted ideas about light, explains all the phenomena known to Biot. Our eyes, he suggests, are sensitive only to rays of a certain refrangibility; other 'eyes' could, conceivably, see radiant heat, or ultra-violet chemical rays.

By 1816 therefore the new facts about radiant heat had been brought into harmony with the known principles and theories of light, and radiation generally was interpreted in terms of the old corpuscular theory. What, then, of Thomas Young's undulatory theory of light and of his and Rumford's brilliant suggestion that radiant heat must be of the same nature? According to Whittaker: 'The wave theory at this time was still encumbered with difficulties. Diffraction was not satisfactorily explained; for polarisation no explanation of any kind was forthcoming.'[55] But these desiderata were soon to be met; in the same year that Biot published his *Traité de physique* (1816), Augustin Fresnel put forward an explanation for diffraction effects by postulating the destructive interference of secondary wavelets emitted by the original wave front, thus superseding Young's clumsy and unsatisfactory suggestion that diffraction was due to reflection from sharp edges. A considerable point in Fresnel's favour was that he predicted some unusual phenomena, which were experimentally confirmed; for example that there should be a spot of light in the centre of a shadow cast by a round, opaque object. Fresnel's work impressed Biot, Laplace and Poisson: but they remained corpuscularians. With two powerful but quite irreconcilable theories to explain the basic phenomena of radiation in competition with one another we can sympathise with the puzzled scholar who, with a singularly inapt choice of words, wrote: 'The nature of heat, like that of light, is still in darkness.'[56] The debates about the nature of light and about the nature of radiant heat indicate that a few experiments, no matter how judicious, are insufficient to falsify a well-established hypothesis; generally speaking a host of ancillary questions have to be answered satisfactorily before a new hypothesis can be accepted in place of the old one.

In 1816 D. J. F. Arago told Young that two pencils of light polarised at right angles to one another cannot produce mutual interference effects.[57] On 12 January 1817 Young wrote to Arago pointing out how this could happen if the undulations of light were at right angles to the direction of propagation; in other words if the light waves were transverse, and not longitudinal like sound waves. The theoretical development of the new hypothesis of transverse light waves was difficult, and to

E

discuss it would take us too far into the history of physical optics.[58] For our purposes it is enough to note that from 1821 onwards Fresnel developed the theory, showing how it could be applied to the optics of crystals and demonstrating how it could account for dispersion. In this way the picture was progressively transformed; from being on the defensive the undulatory theory was, in a comparatively few years, so refined that it could account for everything that the corpuscular theory could, and much more besides. It was a quantitative theory (dealing with wavelength and frequency) while the corpuscular theory was merely qualitative. By the mid-1820s the wave theory was in the ascendant and the accumulation of new knowledge tended to confirm this ascendancy. Thus two French physicists with high professional skills had saved Young's theory; had it been left to the English there is little doubt that it would have been submerged under the waves of Newtonian obscurantism.

If light is held to be propagated by waves travelling through an aether it follows that radiant heat, shown to be closely analogous to light, must also be undulatory in nature. But if this is so what is the status of 'radiant caloric', or indeed caloric in general? The problem was put very clearly by Philip Kelland: 'If, as there is every reason to believe, light is the result of vibratory motion, we are naturally driven to enquire into the possibility of applying the same consideration to heat.'[59] The matter was not entirely settled, however. For one thing the properties of the aether, through which the undulations must be propagated, were quite unknown; for another, the velocity of radiant heat had not been measured and it was difficult to see how it could be. Lastly, the Rev. Baden Powell tried to repeat Bérard's experiments but was unable to get an effect indicative of polarisation.[60] This in fact proved to be a red herring, for the splendid experiments of Melloni and, independently, Forbes soon confirmed the polarisation of radiant heat. All things considered, in the 1820s it was reasonable to hold that light was propagated by a transverse wave mechanism, that radiant heat was probably of the same nature, and that it would therefore be wise to reserve judgement on the ultimate nature of heat. In any case the basic axiom of the science of heat was the principle of conservation, and this was equally acceptable on either the caloric theory or the vibratory theory.

The analogy between radiant heat and light may well have been one of the factors that induced a number of scientists to adopt a rather more reserved attitude towards the nature of heat than had been apparent at the beginning of the century.[61] Another but less basic reason was that as the scientific study of heat was now very prosperous, both theoretically and experimentally, there was much less incentive to worry about fundamental questions. One man who took an agnostic attitude to the nature of heat and whose contributions were concerned with the theoretical development of the subject was Joseph Fourier (1768–1830). But before we consider Fourier's work we must mention that of William Charles Wells, whose *Essay on dew* was published in 1814.

Wells' *Essay*, which had been preceded by two years of careful experimental study, has been described as a paradigm of scientific method.[62] In essence he found that dew is precipitated on the coldest objects and that the phenomenon, under natural conditions, is determined by the radiation of heat from the earth. In daytime the earth radiates heat out into space but, on balance, it receives more radiant heat from the sun than it emits. At dusk and at night, when there is a net loss of heat by radiation, those objects which are the best radiators of heat—dark bodies with rough surfaces—cool most rapidly, while polished, shiny objects such as mirrors radiate much less heat and so remain warmer. The consequence is that the good emitters, the coldest bodies are the ones which collect dew first; the highly polished ones, the good reflectors, last. But Wells' brilliant combination of his own experimental results with the principles established by Leslie and Rumford and Prévost's theory of exchange did even more than this; it explained exactly why less dew is deposited on a cloudy night—the emission of radiant heat is cut down by the cloud cover—and why, as all but the crassly unobservant can confirm, clear, still winter nights are usually the coldest. Thus with Wells' *Essay on dew*, together with Prévost's theory of exchange, men were brought face to face, intellectually and almost literally, with the immense cold of outer space; a basic fact about the universe which only a developed science of heat could establish.

Although there was some overlap in time, Fourier's contributions were made, in the main, after the experimental work of Leslie, Rumford, Herschel and Wells. In contrast with his immediate predecessors Fourier was a master theorist, and it is tempting to cast him in the familiar role of the mathematical physicist who follows the experimentalist; as Kepler followed Brahé, Ampère followed Oersted and Maxwell followed Faraday. But Fourier does not quite fit into this pattern; it might be more appropriate to regard him as the Frenchman who brought the professional touch to the study of radiant heat. But this, too, is inadequate to describe the man whose salient characteristic is, perhaps, that he is *modern.*

His ambition was to do for the science of heat what the great geometers had done for rational mechanics. From a few basic principles they had developed a coherent group of theories, which explained the 'system of the world' so that the complex motions of the heavenly bodies were shown to result from the same laws as those of terrestrial dynamics. In short, Fourier, like his immediate predecessors, is interested not in the passive distribution of heat (as the chemists had been) but in the circumstances of its transmission. Radiant heat is a particularly intriguing mode of transmission; it raises the problem of reconciling the axiom that the faculty of receiving heat must always equal that of communicating it* with the experimental facts of the differential heating and cooling of dark and polished objects and the fall in temperature of a thermometer placed at the focus of a metallic mirror turned towards the sky.[63]

The amount of heat radiated from a surface is, as Leslie showed, proportional to the sine of the angle of emission; and this, Fourier points out, is a necessary consequence of the principle of the equilibrium of heat. If we imagine a number of different bodies all at the same temperature and within a closed space whose inside surface is also at the common temperature then common sense assures us and experiment will confirm that the bodies will remain in thermal equilibrium indefinitely. Were the sine law of radiation not true then thermal equilibrium would not be possible. As Fourier puts it:

> These consequences would not follow if all the heat transmitted, or if an appreciable portion of that heat was subject to a dif-

---

* This is, of course, Lavoisier and Laplace's axiom that a body which absorbs so much heat in rising a given number of degrees must give up exactly the same amount of heat in cooling down the same number of degrees.

ferent law of emission from the one we have enunciated. Bodies would change temperature in changing position. Liquids would acquire different densities in different parts, not remaining in equilibrium in a place of uniform temperature; they would be in perpetual motion.[64]

This remarkable argument is an early instance of thermodynamic reasoning. It was, as far as we know, the first occasion in the history of science when the acknowledged impossibility of perpetual motion was used as the *reductio ad absurdum* of an argument in the study of heat. As such, however, it does not seem to have aroused the interest of historians of science; possibly because Fourier, regrettably, did not develop the argument.

Three elementary properties of bodies, Fourier remarks, determine the activity of heat.[65] These are their heat capacities, their interior conductivities and their exterior conductivities, all of which must be determined by experiment. Heat capacity, or specific heat, has been known for some time and several methods are available for measuring it. The external conductivity Fourier defines as the amount of heat which flows in one minute from one square metre of the surface of the body in question maintained at 1°C into a stream of air passing over it at a uniform, known velocity and maintained at 0°C.[66] He does not explain how this property, which he denotes by 'h', should be measured, but no doubt the able French experimentalists of the time could have coped with it. For small temperature differences Newton's law of cooling is correct and the heat communicated to the passing air is proportional to the temperature difference. But generally speaking the external conductivity has two distinct components: the heat communicated by contact with the air and the heat radiated. The latter is small for low temperatures and only becomes appreciable as the temperature is increased.

The interior conductivity, denoted by 'K', is defined as:

... the quantity of heat which, in a homogeneous solid ... enclosed between two infinite parallel planes, flows during one minute across a surface of one square metre taken on a section parallel to the extreme planes supposing that these planes are maintained, one at the temperature of boiling water, the other at the temperature of melting ice and that all the intermediate planes have acquired and retain a permanent temperature.[67]

With this formal definition the ambiguities which had clouded the concept of heat capacity from the days of Black onwards were at last finally removed. It was, we should note, a rigorously quantitative definition; it was not qualitative in the way in which Ingen Housz, Rumford and others had understood the notion of conductivity. And although once again Fourier did not describe the experimental procedure for measuring conductivity as he defined it, there was no doubt that such measurements were entirely feasible.

Fourier was now in a position to develop the differential equations of the propagation of heat through bodies of different form (cylinders, cubes, spheres, rings, etc.), and then through any regular system of bodies at any given time. Moreover, he is able to solve these equations and so, for him, the science of heat is now complete and finished. An important example of a system of bodies whose thermal states can be examined by Fourier's methods is the solar system. The superficial temperatures of all the planets must represent a radiation balance between the heat they receive from the sun and the cold of outer space, slightly modified (in the case of the earth, at any rate) by such internal heat as still remains. The spin of the earth raises the problem of the alternate (diurnal) heating and cooling of a sphere together with the problem of the extent and nature of the propagation of the daily 'heat wave' beneath the surface. It is factors such as these, together with phenomena such as the absorption and radiation of heat by the atmosphere, that determine the great movements of air like the trade winds and the climates of the continents. In brief, Fourier sets up a new cosmology: the cosmology of heat:

> . . . a very extensive class of phenomena exists, not produced by mechanical forces, but resulting simply from the presence and accumulation of heat. This part of natural philosophy cannot be connected with dynamical theories, it has principles peculiar to itself, and is found on a method similar to that of the other exact sciences. . . . The dilatations which the repulsive force of heat produces . . . are in truth dynamical effects; but it is not these dilatations which we calculate when we investigate the laws of the propagation of heat.[68]

So, in setting up the new cosmology of heat Fourier at the same time explicitly separated it from the domain of mechanics and the mechanical cosmology of the seventeenth and eighteenth centuries. Fourier's contributions to physics were of great importance. He had a comprehensive understanding of the phenomena of heat and an insight into its cosmological significance. He succeeded in putting the science of heat on an analytical or mathematical basis and thus systematised a wide range of phenomena. Furthermore he made very important contributions to our understanding of the theory of dimensions and the nature of physical constants as well as to general mathematical physics.[69] But for all these impressive achievements his explicit separation of heat from the science of mechanics meant that his contribution to the establishment of a general mechanical theory of heat and to the elaboration of the concept of the heat engine could only be indirect.

Looking back over this chapter we see that between the times of Scheele and of Fourier the study of heat transfer and in particular of radiant heat had a number of important general consequences. In the first place, taken in conjunction with the progress made by the undulatory theory of light, it provided increasingly weighty reason for doubting the material or caloric theory of heat. In the second place, thanks to the works of Rumford, Leslie, Wells and Fourier, it led to a great extension of knowledge about the role of heat in geophysical, meteorological and even cosmological phenomena. Indeed a new cosmology—that of heat—began to emerge and through the efforts of Joseph Fourier it reached a stage of advanced mathematical theory: the first branch of theoretical physics to be established independently of classical mechanics.

The account we have given also epitomises appropriately the commanding role that France had come to play in science by the beginning of the nineteenth century. The Revolution seems to have had a double effect; it heightened the creative powers of a people who were already among the most scientifically talented in the world and, at the same time, it provided them with a new set of social institutions and values—of which the Ecole Polytechnique was one outward symbol and manifestation—which enabled them to bring their talents to bear rapidly and efficiently on important scientific problems. (In much the same way an efficient army can muster its forces rapidly and achieve the maximum effect with the minimum effort.) The result was that between about 1790 and 1825 France produced the brightest galaxy of scientific genius that the world has witnessed to date.* Revolutionary France also brought a new, professional touch to science; it became increasingly common for two or more men to collaborate in scientific research; the technical standards of papers and books were maintained at the highest level and science was systematically applied to industrial, military and agricultural problems.[70] In fact it seems clear that if national policies for science, its application to industry and for scientific and technological education had been the sole or even the main determinants then industrial revolution must have occurred in France as soon as or even before it did in England.

---

* The reader should beware of the spurious quantifications now so popular with politicians and others in their comments on modern science. How many of the type Joe Bloggs, Ph.D, are equivalent to a Fresnel, a Gay-Lussac, an Ampère, a Fourier?

Chapter 5

The Science of Heat Becomes Autonomous

THEORIES OF HEAT AND THE PHYSICS OF GASES

Viewed against the long perspectives of history, science in the years immediately following the Napoleonic wars appears to have advanced rapidly and confidently on all fronts. Men of science who were, judged by today's standards, few in number seem to have been unerring in their selection of the important problems and impeccable in the abilities they displayed in solving them. However, while hindsight can detect the great pattern of concept and theory that was being woven at that time, it would be wrong to assume that it was equally apparent to contemporary observers. A closer study of the period conveys a bewildering impression of scientific conjectures and refutations. Hypotheses that were later shown to be wrong were acclaimed while much sounder ones were ignored. More confusing still, the right ideas were sometimes put forward for the wrong reasons, as for example when a Dr Reade claimed to have found that merely agitating a half pint of water for a few minutes increased its temperature (which was of course correct) by about 8°F—which was incredible;[1] or when Marshall Hall asserted that the dynamical theory of heat was strengthened and the caloric theory correspondingly weakened by the discovery of the 'real' radiation of cold.[2] Confusions of this sort are no doubt inseparable from the advance of science in any age, and the historian must duly report them, without emphasising them. Again, the historian has to describe certain quite unrelated activities which were going on at the same time because in later years these activities merged in one great synthesis of knowledge. But he must be careful to avoid giving the impression that those who engaged in these unrelated researches envisaged the grand synthesis which he, the historian, knows took place much later. Contemporary expectations were usually quite different from the actual course of subsequent events; as, for example, in the case of the fairly widespread hope that heat, electricity and light would eventually be shown to be manifestations of one ultimate material substance.

During the eighteenth century comparatively little progress was made in the study of the physics of elastic fluids. As we saw, the idea of 'total heat' was put forward towards the end of the century; it had developed from Watt's researches on steam and implied that there is no such thing as a 'natural state' for any body: its state depends, *inter alia*, on its heat content. In fact 'total heat' was eventually transformed into one of the thermodynamic functions of state: enthalpy. But apart from 'total heat' and the work on compressive heating of air, very little was achieved in the century that had begun so promisingly with Amontons' study of the thermal expansion of air. In spite of the efforts of a number of experimenters,[3] and in spite of the fact that Martine had put his finger unerringly on the main source of trouble in such determinations—the presence of drops of moisture in the apparatus—there was, at the end of the century, still no agreement on the coefficient of expansion of air. This is the more surprising when we recall how fundamental this coefficient is for meteorological work. We can only suppose that, apart from the difficulties inseparable from experimental work with gases, 'Air' in the days before Lavoisier must have been no more uniform or tractable a substance than 'Earth'. With Lavoisier's recognition of the chemical individuality of gases the problem of the expansion of air—now understood to be a mixture of oxygen and nitrogen—became not only more important but also more manageable. Priestley, no disciple of Lavoisier, was actually the first to try to measure the expansion of different gases, but his results were, as he himself admitted, not very satisfactory. Nor for that matter were the results obtained by Guyton de Morveau and Duvernois,[4] but they had the entirely beneficial effect of stimulating two most able scientists—John Dalton and Joseph-Louis Gay-Lussac—to clear the matter up. This they did, independently and within a few months of each other.

John Dalton was, by any standards, a quite remarkable figure. He was born in 1766 at Eaglesfield in Cumberland, and it was while practising as a village schoolmaster in the Lake

District that his interest in meteorology was aroused. His work in meteorology led him, quite logically, to the physics of gases and from there he progressed to chemistry, to the great atomic theory with which his name is associated. In 1793 Dalton accepted an invitation to teach mathematics at a Dissenting Academy in Manchester, which was at that time a smallish town with no particular claims to historical or cultural distinction. Here, however, he found himself at one of the centres, perhaps *the* centre, of the industrial revolution; and while this did not directly concern him it expedited the formation of the local scientific group associated with the Manchester Literary and Philosophical Society (see pages 82–3). We may infer that men like Thomas and William Henry, Peter Ewart and Thomas Barnes were highly congenial to Dalton, and stimulating for him. As his recent biographer puts it: 'Within ten years he had made himself known throughout scientific Europe. In twenty years he was famous.'[5] But Dalton may be credited with something else besides his personal scientific achievements; he was perhaps the first of a line of Manchester scientists that was later to include such figures as Joule, Osborne Reynolds (engineer and scientist), Balfour Stewart, Schuster, Schorlemmer, Roscoe, Jevons (economist and philosopher of science) and Rutherford. In fact, beginning with Dalton a new and recognisable school of science—the Manchester School—joined the distinguished company of the Scottish Universities, Paris, Cambridge, London, Leyden and the Swiss cities as one of the major sources of world science.

As a scientist Dalton left his impress on his century as few other men have ever done. The atomic theory, as he set it out, provided nineteenth-century chemistry with its basic structure and proved to be of fundamental importance in a number of branches of physics. As a thinker Dalton combined singleness of mind with a strong preference for clear-cut explanations of physical processes; he liked his conceptual models to be clearly understandable, and when he was wrong—by the scientific standards of his time—he was clearly wrong; he did not try to confuse issues by sheltering behind face-saving clauses, ambiguous pronouncements and other sophistries. Quite possibly this massive intellectual integrity was the reason for his rapid success; and he seems to have bequeathed it to his successors in the Manchester School. Dalton died in 1844 and was buried amid scenes of civic mourning which have never been equalled in the annals of British science.*

Apart from his criticism of Rumford's assertion that liquids and gases are absolute non-conductors of heat (page 98), Dalton's first contribution to the physics of gases was his paper on the compressive heating and expansive cooling of a gas.[6] As Mr Greenaway points out, this was not in itself an important paper, for it adds little to Erasmus Darwin's contribution. But it foreshadowed later developments. Dalton argued that the actual rise in temperature may well be much more than the few degrees indicated by sluggish thermometers; it might even reach 50°F. The release of heat on compression shows that the specific heat, or heat capacity, of a gas must vary with its density: the more the gas is compressed, the smaller its specific heat. Such a conclusion is inescapable in the absence of the dynamical theory of heat, which postulates that the mechanical energy expended in compressing the fluid is actually converted into heat energy. But Dalton was a disciple of Lavoisier and thus a calorist. The consequences were interesting.

Dalton had long doubted the regularity of the mercury thermometer scale; the gas phenomena confirmed these doubts.[7]† If a pound of boiling water is mixed with a pound of water at freezing-point, the mercury thermometer indicates that the temperature of the mixture is mid-way between those of the boiling and freezing-points and can thus be marked as 122°F, or 50°C. This was always taken as proof that the mercury scale is regular. But to Dalton it proves just the opposite. As the pound of boiling water cools down by mixture so it contracts, and in contracting it must, on the 'adiabatic' principle, give out heat; similarly the cold water on heating up expands and must absorb heat. Now the 'compressive' heat given out must be greater than the 'expansive' heat absorbed, since the specific heat is

* Newton was buried with great pomp in Westminster Abbey. But Newton was a Cambridge scientist and his funeral was a national event. London scientists have never been particularly honoured by London. Why, for instance, is there no Michael Faraday Street in central London, comparable to John Dalton Street in Manchester?

† It is possible that John Gough's paper on the heating of rubber by rapid expansion also influenced Dalton; see *Manchester Memoirs* (1805), i, p. 288. I am grateful to Professor L. R. G. Treloar for calling my attention to this important paper: in effect the first observation of the 'adiabatic' effect in a non-gaseous material.

greater in the first case than in the second. From which it follows that the real temperature of a mixture of a pound of boiling water with a pound of freezing water must be *greater* than the arithmetic mean. If the mercury thermometer indicates, as it does, that the mixture temperature is the arithmetic mean then the mercury thermometer must be incorrect. To obtain a true arithmetic mean one should mix equal *volumes* of water, taking (in theory at any rate) precautions to prevent any subsequent

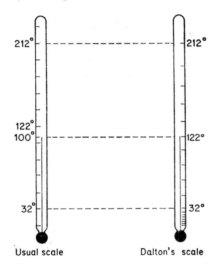

Usual scale   Dalton's scale

Figure 10

change in volume. Dalton's observations suggested to him that with the mixture at the true arithmetic mean a mercury thermometer would indicate only 110°F. Dalton says that he had long suspected that mercury and water both expand as the square of the true temperature measured from freezing-point, or from the point of maximum density. A thermometer marked with fixed points at 32° and 212° but so calibrated that the divisions on the stem are as the square roots of the expansions from the freezing-point of mercury will read 122° at the point corresponding to 110° on a conventionally calibrated but other-wise identical thermometer (Figure 10). Not only would this new scale be in accordance with the 'adiabatic' principle but it would re-establish Newton's law of cooling, which was then known to hold only for small temperature differences. Accord-ing to ordinary thermometers hot bodies cooled more rapidly than the law predicted, but on Dalton's scale this is exactly compensated for by the greater intervals between the marks as the temperature is higher. This, thought Dalton, was a strong argument in favour of the new scale. What we can now see is that Dalton was, in effect, trying to establish an absolute scale of temperature, common to all thermometric substances and consistent with Newton's law of cooling.

A further advantage of the new scale was made clear in the second of the four 'Experimental Essays' which Dalton read to the Manchester Society in the autumn of 1801.[8] In this essay he described his experiments on the pressure of saturated steam as a function of the temperature—the same problem that had interested the young Watt. Dalton asserted that if the temperature increases in arithmetic progression, the pressure increases *nearly* in geometric progression, but the ratio becomes exact if the new temperature scale is used. It was therefore a pity that, with all these advantages, Dalton's new scale proved to be unjustified in the end: over the range of temperatures he investigated the expansion of mercury is to all intents and purposes uniform and the 'adiabatic' effect can be neglected.

It was in the fourth of the 'Experimental Essays' that Dalton announced one of his most important contributions to this branch of science,[9] the discovery that all gases expand uniformly with the same increase in temperature. He had dried the gas and the apparatus as much as possible, using sulphuric acid for the purpose, and had thereby avoided the mistake that Guyton de Morveau and the other investigators had made. Dalton found that a volume of air, or any other gas, amounting to 1,000 units at 32°F expanded to 1,376 units when it was heated up to 212°F. This would suggest a coefficient of expansion of $\frac{1}{266}$ per degree centigrade, a reasonably accurate result. But Dalton did not try to determine the *law* of expansion experimentally; he was content with these two readings, and he did not believe that between the two temperatures gases expanded *linearly*. A gas, he believed, consists of a swarm of tiny, infinitely hard atoms each surrounded by its own atmosphere of caloric. The atmospheres touch and the atoms are maintained in static equilibrium. When heat (caloric) is added to a gas, the atmospheres are augmented, with the effect that the gas expands. The absolute zero of temperature is that temperature at which the atmospheres of caloric vanish.

Dalton was prepared to accept that the law of expansion was linear in the case of solids but not, as we have seen, in the cases of liquids, gases or vapours like saturated steam. He first believed (1802) that gases expand as the cube of the temperature measured from absolute zero (Vat^3). This would have the convenient consequence that, on Dalton's theory of the atomic structure of a gas, the cube root of the volume is the measure of the thickness of the caloric atmospheres; at absolute zero the atmospheres and the volume vanish. On this basis Dalton's experimental results give him an absolute zero at $-1{,}547°F$, in reasonable agreement with Adair Crawford's figure. Had he accepted the linear law of expansion for gases he would have obtained $-450°F$, or $-268°C$. But his fundamental doubts about the linearity of the expansion of liquids dissuaded him from such a simple solution, and he did not attempt to determine experimentally the law of expansion of gases between 55° and 212°, much less above and below these temperatures. As it was he was forced to abandon his t^3 law of expansion when he found that it was not consistent with the law of cooling. For it he substituted a law which asserted that the expansion of all gases is in geometrical progression for arithmetic increments in temperature (1808).

Having been forced to give up the basis for his calculation of absolute zero, Dalton returned to Irvine's method—which had been refuted by Lavoisier and Laplace some thirty years earlier —and concluded that the most probable value for absolute zero was $-6{,}150°F$; this was deduced from results that varied between $-11{,}000°F$ and $-4{,}000°F$!*

The mistakes which Dalton made in these difficult pioneering researches should not blind us to their importance. His discovery that all gases expand uniformly was a major achievement, and his intuition that a universal scale of temperature could—and should—be discovered was in accordance with the true needs of science at that time. He realised that the study of the thermal properties of gases offered the best hope for

---

* Dalton's idea of the nature of heat was unoriginal. He was a supporter of the ideas of William Irvine. For a clear account of Dalton's Irvinism see Robert Fox, 'Dalton's Caloric Theory', in *John Dalton and the progress of science*; ed. D. S. L. Cardwell (Manchester University Press 1968) also G. R. Talbot, *Origins and solutions of some problems in heat in the eighteenth century* (Manchester University Ph.D. thesis 1967).

progress in the science of heat. When we examine the behaviour of gases the obscuring factors are reduced, or eliminated, and we come closest to the true effects of heat—the raw essence of heat, as it were—unmodified by extraneous influences like chemical affinity, inter-molecular forces, etc. No doubt this insight was based, in the first instance, on the 'erroneous' caloric theory; particularly the belief that gases are composed almost entirely of caloric. Indeed, the law of equal expansion could be taken as striking confirmation of the caloric theory; it is what you would expect if *all* gases are substantially composed of caloric. But this does not detract from the value of Dalton's insight; some kind of working hypothesis is always necessary. As Dalton put it, 'It seems therefore that general laws respecting the absolute quantity and nature of heat are more likely to be derived from elastic fluids than from other substances . . .'[10] and, 'When we consider that all elastic fluids are equally expanded by temperature and that liquids and solids are not so, it should seem that a general law for the affection of elastic fluids for heat ought to be more easily deducible and more simple than one for liquids or solids.'[11]

In fact Dalton's insight, like all profound intuitions (Prévost's theory of exchange, for example) proved capable of transcending its metaphysical origins. It was adapted to the dynamical theory of heat, for as Clausius was later to write:

> . . . the case of the expansion of a permanent gas presents itself as particularly simple. We may conclude from certain properties of the gases that the mutual attraction of their molecules at their mean distances is very small and therefore that only a very slight resistance is offered to the expansion of the gas, so that the resistance of the sides of the containing vessel must maintain equilibrium with almost the whole effect of the heat. Accordingly the externally sensible pressure of a gas forms an approximate measure of the separative force of the heat contained in the gas; and hence, according to the foregoing law, this pressure must be nearly proportional to the absolute temperature. The internal probability of the truth of this result is indeed so great that many physicists since Gay-Lussac and Dalton have without hesitation presumed this proportionality, and have employed it for calculating the absolute temperature.[12]

The historical importance of the physics of gases was not only that it provided a direct insight into the nature of heat but

Plate VII. The search for economical power, i: masonry arch dam on the river Bollin built by Peter Ewart to provide power for Quarry Bank Mill.

Plate VIII (*below*). The search for economical power, ii: mill at Congleton, Cheshire, showing weir and breast-wheel.

Plate IX (*right*). The search for economical power, iii: mills at Congleton, Cheshire, showing weirs for breast-wheels.

Plate X. The search for economical power, iv: water-turbine installed about seventy years ago in the old waterwheel house at Quarry Bank Mill. The tapered axle of the waterwheel can still be seen. It is in the lower left-hand part of the photograph.

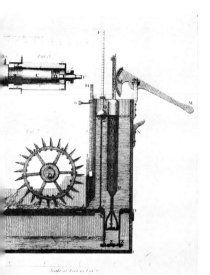

Plate XI. Diagram of John Smeaton's model for the experimental study of the efficiencies of waterwheels. The model can easily be adapted to be driven undershot, as shown, or overshot. The power developed is measured by the weight which the wheel can lift one foot in one minute; the 'effort' expended, or the energy of the stream, is proportional to the number of strokes of the pump, M, in the same time, and the head of the water indicated by the scale FG. (*Philosophical Transactions of the Royal Society*, 1759)

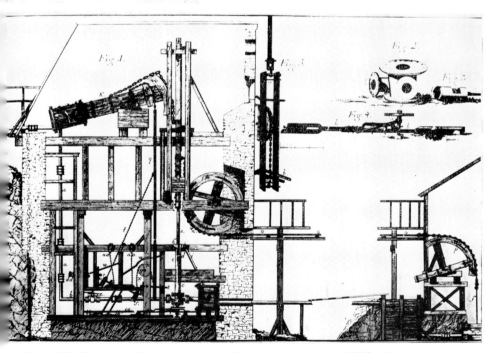

Plate XII. Column-of-water engine at Schemnitz in about 1770. The water is led in by the pipe *a* to the vertical cylinder *c*, fitted with the piston *d*. *A* is a large counter-balance of the sort that was common in early Newcomen engine practice. The oscillations of the piston are transmitted by the connecting rods *o* to the pump rod *f*. This engine should be compared with the much more advanced one illustrated by A. M. Héron de Villefosse (Plate XVI).

(After C. T. Delius, *Anleitung zu der Bergbaukunst . . .*)

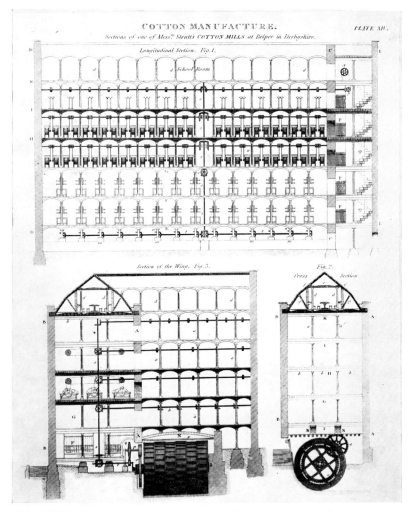

Plate XIII. A water-powered cotton mill of the industrial revolution. Power is generated by means of the large breast-wheel in the basement of the mill and is transmitted by vertical and horizontal shafting, pulleys and belts to the carding and spinning machines on different floors. The mill is as compact as possible to minimise power loss in transmission. Later in the nineteenth century the vertical shafting was replaced by rope and pulley drive. From Rees' *Cyclopaedia*, 1819. (*Photograph by Mr Harry Milligan*)

also that when heat was applied to gases and vapours its effects were more obvious and more easily quantifiable than when it was applied to solids and liquids. In other words, gases and vapours provided the means whereby heat could readily produce *mechanical* effects. Robison had understood this, for he had written:

> The laws which regulate the formation of elastic vapours, or the general phenomena which it exhibits, give us that link which connects chemistry with mechanical philosophy. Here we see chemical affinities and mechanical forces set in opposition to each other and the one made the indication, characteristic and measure of the other. We do not have the least doubt that they make one science, the Science of Universal Mechanics . . .[13]

Dalton must almost certainly have read this. But it should not be forgotten that Robison's idea of mechanics was fundamentally old-fashioned. He envisaged mechanics in terms of Newtonian central forces; not in terms of *vis viva*, work and energy. Dalton, on the other hand, regarded himself as a Newtonian, but he had almost certainly learned his Newton not from *Principia* and *Opticks* but from Desaguliers and the other popularisers of the eighteenth-century interpretation of Newtonian mechanics.

At much the same time that Dalton was developing his ideas and carrying out the experiments that he was to describe in the 'Experimental Essays', Joseph-Louis Gay-Lussac (1778–1850), a 'pupil engineer' and a protégé of Berthollet, was carrying out a similar series of experiments in France.[14] Provoked, like Dalton, by the failure of Guyton de Morveau and his predecessors to determine the expansion of air and other gases he saw where the earlier experimenters had gone wrong and, having corrected their mistake, was led to the same conclusion as Dalton, that all gases expand equally when heated through the same temperature range. But Gay-Lussac was more systematic than Dalton. He had the new, professional touch; he gave full details of his apparatus and the experiments he carried out. He showed that oxygen, nitrogen, hydrogen, carbon dioxide and air all expand by the same amount between the freezing- and boiling-points of water; and, if the law of expansion between these two temperatures is linear, the coefficient would be $\frac{1}{266}$ per degree centigrade. He then went on to measure the expansion of water-soluble gases and special cases like ammonia and sulphuric ether vapour. In all cases he found that the expansion was the same, no matter what the density and solubility of the gas. Finally he studied moist air and showed that, provided the water content remained constant it, too, had the same expansion as all other gases and vapours. It remained, observed Gay-Lussac, to determine the actual *law* of gaseous expansion; if it was linear then the coefficient $\frac{1}{266}$ would be correct at any temperature. Dalton of course would have denied this.

Gay-Lussac's work, although published a few months after Dalton's, was generally more comprehensive, more convincing and, indeed, more correct in that it was unencumbered by false assumptions about the law of expansion. Accordingly it became customary on the Continent to refer to Gay-Lussac's law. This was quite reasonable; just as it was understandable for English-men to refer to Dalton's law, or sometimes to Dalton and Gay-Lussac's law. But towards the end of the nineteenth century a curious thing happened. A Victorian 'improver' glancing at Gay-Lussac's original paper noticed a reference to a certain 'cit. Charles' who had, it seemed, discovered the law some fifteen years before Gay-Lussac and Dalton. The 'improver' in question (P. G. Tait) was a respected author of textbooks; he gave currency to 'Charles' law' and, in the English-speaking world at least, 'Charles' Law' it remains to this day.

But the facts do not in any way justify the ascription of this law to J. A. C. Charles. Gay-Lussac was an honourable and a meticulous man. He mentioned Charles' work in his paper, but his words make it perfectly clear that he had heard of Charles' efforts purely by chance, that Charles' apparatus could not have given reliable results and, in any case, that Charles had found only that oxygen, nitrogen, hydrogen, carbon dioxide and air expand equally; the water-soluble gases, he believed, all had different expansions. In other words, Charles had rejected 'Charles' law' in advance. Gay-Lussac's comments make this perfectly clear:

> Avant d'aller plus loin, je dois prévenir que quoique j'eusse reconnu un grand nombre de fois que les gaz oxigène, azote, hydrogène, acide carbonique et l'air atmosphérique se dilatent également depuis 0° jusqu'à 80° (Réaumur), le cit. Charles avait remarqué depuis 15 ans la même propriété dans ces gaz;

mais n'ayant jamais publié ses résultats, c'est par le plus grand hasard que je les ai connu. Il avait aussi cherché à déterminer la dilatation des gaz solubles dans l'eau, et il avait trouvé à chacun une dilatation particulière et différente de celle des autres gaz. A cet égard mes expériénces diffèrent beaucoup des siennes.[15]

These words indicate that Gay-Lussac was in no way relinquishing his discovery to the credit of Charles. What claim indeed does a man have if he makes no effort to publish his discoveries, does not announce them through his own 'grapevine' (as Black did through the medium of his lectures) and, in any case, gets things fundamentally wrong? For these reasons we deny the claim made on behalf of Charles and shall in future refer to the law as Gay-Lussac's law.

The 'adiabatic' heating and cooling of gases seems to have aroused less interest among French scientists than it did among foreigners. In Dr Robert Fox's opinion* this may have been due to the comparative neglect of meteorology in France, most of the notable meteorologists of the period being either Swiss or English. However, in the last few years of the eighteenth century some interesting observations were reported in the French journals. In 1799 the engineer Baillet, a professor at the Ecole des Mines, described and explained the phenomenon of the formation of ice and snow by the Heronic engine at Schemnitz;[16] although, in fact, he added nothing to Erasmus Darwin's observations of eleven years earlier. Later on, through the *Bibliothèque Britannique*, there came accounts of Dalton's work in Manchester and of books like Richard Kirwan's *Of the variations of the atmosphere*, which called attention to 'Watt's law'.[17]

The debut of 'adiabatic' studies in France has been described as spectacular.[18] In 1802 Biot explained that he had been commissioned by Laplace to study the effect on the velocity of sound of the temperature variations accompanying the compression and expansion of air.[19] It was notorious that calculations

---

* Robert Fox, in *The caloric theory of gases from Lavoisier to Regnault* (Oxford, The Clarendon Press 1970), discusses clearly and in detail French work on the adiabatic problem and the determination of the specific heats of gases. We shall therefore give a brief account of these problems here, and for further information refer the reader to Dr Fox's work.

of the velocity of sound through air of normal density led to results that were widely different from those found by experiment. The scandal was the greater as Newton's authority was behind the calculations, while the experimental results were equally beyond reproach. If, however, sound waves are longitudinal undulations of alternate compressions and rarefactions of the air then the resulting 'adiabatic' heating and cooling must, if significant, affect the velocity of propagation. By taking this into account Laplace and Biot hoped to redeem Newton's calculations. Biot knew that on compression atmospheric air 'loses some latent heat which becomes sensible' and conversely, when rarefied, some sensible heat becomes latent. But he was unaware that the heat needed to raise a gas 1° in temperature when its volume is kept constant is not the same as the heat required to raise the same gas 1° when it is free to expand; in other words, that the specific heat at constant volume is not the same as the specific heat at constant pressure. His calculation was therefore necessarily simple. He assumed that if, as Gay-Lussac had just shown, a gas expands by $\frac{1}{266}$ when it is heated through 1°, then compressing it by the same amount will warm it up by 1°. Amontons had shown that air heated through about 100° increases in pressure by roughly $\frac{1}{3}$, so that Biot can compute the additional pressure of the air due to its 'adiabatic' heating. The resulting calculation gave a velocity of sound that was too high, but this was an error on the right side, for 'classical' calculations had yielded a velocity that was too small. Evidently 'adiabatic' heating and cooling must be a significant factor in determining the velocity of sound in air.

It was, *à priori*, reasonable to assume that since all gases expand equally through the same temperature range, they might also have the same specific heats. Biot's assumption that compressing or expanding a gas by $\frac{1}{266}$ of its volume will alter its temperature by 1° implies the equality of specific heats, and Gay-Lussac found further evidence to support this in some experiments he carried out in collaboration with von Humboldt. Encouraged by Berthollet and Laplace, Gay-Lussac devised a series of experiments to measure the temperature changes caused by the expansion of gases and to find, thereby, what the relative specific heats should be.[20]

Leslie had devised a seemingly plausible method for measuring the relative specific heats of gases.[21] A large vessel is partly evacuated so that if air is re-admitted the temperature will rise through a measurable number of degrees. If the vessel is once again evacuated and, instead of air, a different gas (at atmospheric pressure) is allowed to enter, a different temperature rise should be obtained. A comparison of the two temperatures should therefore give the relative heat capacity of the gas—air being taken as the standard. Leslie claimed to have shown that air and hydrogen have the same specific heat. But the method was unsatisfactory, as the entering air and gas(es) should, if their specific heats are different, bring different quantities of heat in with them.

Gay-Lussac's apparatus was similar to Leslie's. Two twelve-litre glass flasks were connected by a tube fitted with a cock

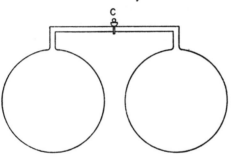

Figure 11

(Figure 11). Both were fitted with thermometers and contained a drying agent. One flask was filled with a gas, the other was evacuated; after a twelve-hour wait, to ensure temperature uniformity, the cock was opened and the temperature changes noted.

The temperature in the empty flask rose, that in the gas-filled one fell by the same amount. This was generally in accordance with the accepted doctrine that as a gas expands its specific heat increases. Several English disciples of Irvine and a number of members of the Swiss School had inferred from this that ultimately the specific heat of empty space, of a void, must be greater than that of any known gas.* But this was hardly logical. Specific heat, or heat capacity, is usually expressed as the thermal effect of a quantity of heat on a body of given mass

---

* This hypothesis was put forward explicitly by J. H. Lambert (see page 58). Its origins can be traced back, through him, to Boerhaave and ultimately to Descartes.

or weight and, of course, a void can have no mass or weight. If we are to talk meaningfully about the specific heat of a void we can only refer to the *volumetric* specific heat. And it by no means follows that the *volumetric* specific heat of a gas increases as the density diminishes. Gay-Lussac placed a thermometer in the Torricellian space of a mercury barometer and, by suddenly altering the height of the mercury column, 'compressed' and 'expanded' the void. No thermometric effect at all was detected.

When Gay-Lussac tried reducing the pressure and therefore the density of the air in the first flask he found that while the temperature changes were still equal in both flasks they were proportionately smaller. This suggested that as the gas became increasingly rarefied so the *volumetric* specific heat became progressively less. When he tried substituting different gases (at normal pressure) he found that hydrogen, the lightest gas, showed the biggest temperature changes while heavier gases produced steadily smaller changes. The implication was that for different gases the specific heat by *weight* is inversely proportional to the density. We can, therefore, briefly summarise Gay-Lussac's observations:

(1) The specific heat by *weight* of a gas increases as the density is reduced; i.e. as the gas becomes rarefied. We can write this analytically as $S.H._w = f(1/\rho)$.

(2) The *volumetric* specific heat of a gas increases as the density increases; i.e. as the gas is compressed. We can write this analytically as $S.H._v = f(\rho)$.

(3) Empty space, or a void, has no assignable *volumetric* specific heat.

(4) For different gases the specific heat by *weight* varies inversely as the density; this could imply that the *volumetric* specific heats of all gases are the same, under the same conditions of pressure and temperature.*

---

* Dalton, no less than Gay-Lussac, had been impressed by the implications of the discovery that all gases are equally expanded by heat. He discussed the matter at some length in the first volume of his *New system of chemical philosophy* and concluded that, other things being equal, the same amount of heat surrounds the *atoms* of all gases. This led him to the corollary that the specific heats by *weight* of gases are inversely proportional to their atomic weights. Hydrogen should therefore have a significantly greater specific heat than any other gas; as Crawford and Gay-Lussac had asserted.[23]

Four years later Gay-Lussac attempted to measure the specific heats of gases using a method of mixtures.[22] A steady stream of gas was heated to a temperature about 20°C above zero while an equal flow of another gas was cooled to the same extent *below* zero. The two streams were mixed and the final temperature noted. If one gas was taken as a standard then the relative *volumetric* specific heat of the other could be immediately computed from the initial and final temperatures. But the latter proved to be so close to the arithmetic mean (0°C) in all cases that Gay-Lussac concluded that all gases have, *ceteris paribus*, the same *volumetric* specific heats while their specific heats by *weight* increase as their densities are smaller. This confirmed his previous suggestion, but his method was inaccurate and results unsatisfactory. Later in the same year he changed his mind and expressed the belief that the *volumetric* specific heats of gases are different from one another.[24]

It was now quite evident that the accurate determination of the specific heats of gases was one of the most urgent tasks for science. In 1811, at Berthollet's instigation, the *Institut* decided to make it the subject of an official prize. In the resulting competition two important joint papers were submitted; one was by Clément and Desormes, two able and respected industrial chemists, while the other was by Delaroche and Bérard. The latter pair were declared the winners and, in 1813, their work was published.[25]

Delaroche and Bérard used a constant-flow method; a technique dictated not by theoretical considerations but by the practical requirement of getting a gas to impart a significantly measurable amount of heat to a given weight of liquid (to the calorimeter in fact). A calorimeter full of water contains a spiral copper tube through which a heated gas flows. The temperature of the gas is taken just before it enters the spiral tube and just after it leaves; the temperature of the water is also taken, and this can be kept constant by pouring in a steady stream of cold water. When a steady state has been reached (i.e. when the temperatures are no longer changing) the *volumetric* specific heat of the gas can be deduced by comparing, over a given time, the volume of gas passing through the spiral and its temperature-fall on the one hand with the amount of heat imparted to the flowing water on the other. The relative *volumetric* specific heats of gases could be found by comparing the times they took, at uniform flow, to heat up a given amount of water to a given temperature.

Delaroche and Bérard took great precautions to obtain accurate results, paying particular attention to the elimination of heat losses by radiation, conduction and convection. Accordingly their results were accurate; indeed they were the best obtained until the time of the great experimentalist Victor Regnault in the middle of the nineteenth century. A table of some of their results will confirm this:

SPECIFIC HEATS AT CONSTANT PRESSURE

| | Delaroche and Bérard | | Modern data |
	Volumetric	*By weight*	*By weight*
Air	1·000	1·000	1·000
Hydrogen	0·903	12·340	14·100
Oxygen	0·976	0·885	0·910
Nitrogen	1·000	1·032	1·030
Carbon dioxide	1·258	0·828	0·830

They had briefly considered trying to make measurements of specific heats at constant volume but had decided against it as a less convenient and accurate method. Evidently they did not appreciate the significance of the difference between the two forms of measurement. What seemed clear, however, was that the *volumetric* specific heats of gases were not the same and, even more important at that time, that Irvine's theory must be false since the assertion that heats of chemical combination are due to changes in specific heats when two gases combine to form a compound was not supported by their measurements of the actual specific heats before and after combination. So Lavoisier and Laplace were confirmed and Irvine refuted. This has been described as a crushing victory for the orthodox view in France.[26] Unfortunately Delaroche and Bérard went even further and confirmed yet another very plausible doctrine. After very careful measurements they asserted that: 'The specific heat of atmospheric air, in the ratio of its volume, increases with its density but at a slower rate; of course considered in the ratio of its mass, it diminishes as the density increases'; and, 'Everyone knows that when air is compressed heat is disengaged. This phenomenon has long been explained by the change supposed to take place in its specific heat; but the explanation was founded upon mere supposition, without any direct proof. The experiments which we have carried out seem to us sufficient to remove all doubts upon the subject.'

In other words, the erroneous but highly plausible theory that the specific heat by weight of a gas increases as the density is reduced now had full official support, as it was held to have been confirmed by rigorous experiment. This mistake was one of the most significant ones ever made in the history of science.[27]

THE LAWS OF EXPANSION

The problem of the specific heats of gases together with the phenomena of adiabatic heating and cooling having been settled, authoritatively it was believed, there remained the question of the true form of the laws of gaseous expansion: unsettled by Gay-Lussac; confused by Dalton. There was also the fundamental problem of a rational scale of temperature, which Dalton had raised in such a way that it could hardly be overlooked. This was an unacceptable, challenging situation for the highly professional French scientists of the period; accordingly two of them, P. L. Dulong and A. T. Petit, resolved to remedy the matter.

In their first paper, read to the *Institut* in May 1815, Dulong and Petit[28] established that the expansion of a gas and the expansion of mercury in a standard thermometer are compatible and sensibly uniform between about −40° and about 200°C. At about 260° the mercury scale is about 10° ahead of the gas scale, which suggests that over a wide range of temperature mercury expands slightly more rapidly than a gas. At about 300° mercury boils and is therefore unusable as a thermometric liquid. Nevertheless their results were accurate enough and extensive enough to falsify Dalton's version of the law of gaseous expansion and to confirm that of Gay-Lussac: that the thermal expansion of gases is, for each degree of the mercury scale, a constant fraction of the volume at a fixed temperature. After some observations on the expansion of glass and of metals (unless all results are to be relative, *one* part of an observational or experimental system must be fixed), Dulong and Petit conclude their paper by remarking: 'It is extremely probable that the expansion or the increase in pressure of gases is constantly proportional to temperature. . . . In accepting this we see that indications on the mercury thermometer will always be above the real temperature, the more so the higher the temperature.'

In their next paper,[29] Dulong and Petit refer to Dalton's work with great respect; he had tried to establish general laws for the measurement of temperature, for it had been a common fault in all works on heat that the thermometric scale was uncertain and the actual divergencies over even a small range were quite unknown. They define an ideal thermometric substance as one which expands equally for equal increments of heat. But if, as everyone believed, the specific heats of gases and liquids increase with their volumes, and if the regularity of their thermal expansions is in doubt, it is impossible to see how an ideal thermometric substance could be identified.

To underline the point they then proceed to a detailed study of the thermal expansions of liquids and solids. They describe their famous apparatus for the measurement of the absolute expansion of liquids (that apparatus known to many generations of would-be physicists) and remark that once the absolute expansion of mercury is known it is easy to deduce the expansion of glass from measurements of the expansion of mercury-in-glass. In this way they find that while mercury expands slightly more rapidly than air, or any other gas, glass expands more rapidly still: and it is this compensatory expansion of glass that makes the mercury thermometer reasonably accurate over a wide range of temperatures. Judged by the standard common to all gases, liquids expand more rapidly at higher temperatures, solids more rapidly than liquids, and iron most rapidly of all. Indeed, all substances, according to an iron thermometer, would show a decreasing rate of expansion at progressively higher temperatures. According to a gas thermometer all solids and liquids expand more rapidly at higher temperatures. Logically there is no reason why one should prefer one substance to another for the construction of a thermometer. But the uniformity of expansion of all gases is, as it were, nature's hint. They finally decide on the gas thermometer. The fact that all gases expand equally:

> . . . renders it very probable that, in this class of bodies, perturbing causes do not have the same effect as in solids and liquids and consequently that changes in volume produced by the action of heat are thus dependent more immediately on the

producing forces. It is thus likely that most phenomena relating to heat will appear in simpler form when we measure temperatures with an air thermometer.

This opinion was very similar to the view expressed by Dalton (page 128). To confirm the point made by Gay-Lussac they go on:

> ... in calculating temperatures with an air thermometer by volume increase under constant pressure, one consistently obtains the same results as are obtained from measuring the change of elasticity, the volume being constant. This confirms that Mariotte's [Boyle's] law is exactly true, whatever the temperature.

In 1814 A. M. Ampère had tried to derive a general equation relating the pressure, volume and temperature of a gas.[30] His model of a gas was similar to that of John Dalton; more or less static atoms were held apart from each other by the repulsion of their associated caloric, indeed such was the repulsive force of the caloric that any inter-atomic attractive forces were inappreciable by comparison. Making these assumptions Ampère showed that if H was a surface area bounding the gas, P the force upon it, n the number of atoms per unit volume and t the temperature, then:

$$P = nH \, F(t).$$

As the pressure, p, of the gas is given by P/H and as n is obviously proportional to the density, or to the reciprocal of the volume $(1/v)$, we have:

$$p = (1/v) \, \mathrm{F}(t), \text{ or, } pv = F(t).$$

As Biot was to put it in the following year, $pv = F(t)$ was the implicit equation, the general formula to be used with regard to all the circumstances.[31] Now, with the work of Dulong and Petit and the establishment of Gay-Lussac's law over a wide range of temperatures, it was possible to write $V = V_0(1 + \alpha t)$, and, since Dulong and Petit had confirmed the law of increase of pressure, the volume being constant, its analogue, $p = p_0(1 + \alpha t)$. From which it is very easy to deduce that $pV = k(1 + \alpha t)$.

Dulong and Petit then went on to study the laws of cooling. In spite of Martine's 'little known' dissertation of 1740 and of the much more recent work of Delaroche on radiant heat (see page 111 above), Newton's law of cooling was still widely accepted. Leslie and Fourier, both aware of its limitations, had made use of it, and Dalton had tried to eliminate divergencies from the law by ascribing them to inadequacies of the ordinary thermometric scale. Dulong and Petit therefore proposed to examine the law of cooling first *in vacuo* and then in gases at different pressures; by subtracting the first from the second they would be able to find the amount of cooling due to 'contact' (i.e. convection) alone.

As a cooling body they used a mercury thermometer, calibrated by an air thermometer, and suspended it in the middle of a copper sphere, which was blackened on the inside and immersed in a constant-temperature water-bath. As a result of their experiments they concluded that: 'When a body cools in an evacuated enclosure at constant temperature the cooling rate, for arithmetic increases in temperature will increase geometrically, less a constant number.' They express this law by means of a formula which, by extrapolation, led them to the conclusion that for null cooling rate the temperature of the surroundings must be at minus infinity.[32] The fact that, for Dulong and Petit, absolute zero is at minus infinity does not necessarily imply that bodies at normal temperatures must contain an infinite quantity of heat, for the specific heats of bodies is commonly believed to fall with their temperatures and may well vanish at the absolute zero of temperature. The total heat in bodies can therefore still be finite.

The laws of cooling by contact (convection) which can now be deduced are, they find, different from those which hold *in vacuo*. For one thing the nature of the surface of the body—whether it is light or dark, polished or matt—does not affect the rate of cooling by contact. But beyond this, the rate of cooling in gases varies as the (non-integral) powers of the excess temperature and of the pressure and, furthermore, is different for different gases. In short, the laws of cooling by contact turn out to be somewhat inelegant and, no doubt for that reason, are rarely accorded enthusiastic treatment in the textbooks. The law of 'total' cooling is therefore complex, and on this account attempts to discover it, up to the time of Dulong and Petit's researches, had met with little success.

Their next paper, published in 1819, was not directly relevant to our problems but as it was of great general importance we must briefly mention its main points.[33] Having mastered the laws of cooling, Dulong and Petit resolved to apply them to the determination of specific heats. It is possible that they had been stimulated to do this by a paper published some two years earlier by Despretz: 'Memoir on the Cooling of Some Metals; to Measure their Specific Heats and Exterior Conductivity'.[34] The advantage of the technique was that it offered a way of measuring the specific heats of substances that were only available in very small quantities and for which the method of mixtures was therefore unsuitable.

After they had determined the specific heats of some thirteen elements by this method of cooling they noticed—how, they do not tell us—that the products of the specific heats multiplied by the atomic weights were all constant. In other words, the atoms of all elements have the same heat capacity. The importance of this discovery in the history of chemistry can hardly be exaggerated. Not only did it confirm the importance of Dalton's doctrine of atomic weights but it also provided a way of avoiding ambiguities in their computation. Dalton's rules for the determination of atomic weights and his refusal to accept Avogadro's hypothesis led inevitably to uncertainties,[35] in many cases over which of several possible atomic weights were the ones to choose. Dulong and Petit's law eliminated these ambiguities, for it implied that those atomic weights that when multiplied by the corresponding specific heats yielded the common constant were the correct ones. But as Dr Fox points out,[36] the law also cast some doubt on the caloric doctrine. It was plainly inconsistent with Irvine's theory (in so far as anyone still accepted it) and, less decisively, with the received version of Lavoisier and Laplace. It therefore marked a stage in Dulong's* progressive abandonment of belief in caloric.

A curious sidelight on the standing of the caloric theory at that time was provided by certain experiments carried out in 1825. In that year Fresnel reported[37] that very small and very hot bodies appeared to repel one another. The effect was not due to the presence of air or vapours of any sort, for it was observed when the small bodies were placed in an evacuated vessel and heated by means of a strong lens. Further experiments convinced Fresnel that the effect was not due to

* Petit died prematurely in 1820.

electricity and was most unlikely to be magnetic in nature.[38] But that was all he was prepared to say about the mysterious 'calorific repulsions'. Twenty or thirty years earlier these observations would have been hailed as positive and final confirmation of the truth of the orthodox caloric theory; in the climate of opinion of the 1820s they were of little more than technical interest.[39]

Though the orthodox caloric theory was dying in France it would however be quite wrong for the historian to assume that the dynamical theory must, as the alternative, have been in the ascendant. In fact the situation was not one in which there was a simple choice of either one theory or its rival. A veritable spectrum of beliefs was available: one might believe in caloric as a real, highly elastic gaseous substance, physically as real as air; or one might reject caloric, together with all other theories about the ultimate nature of heat, and hold fast only to the basic principles of quantity of heat, specific heat, latent heat, the laws of radiation and conduction, and the basic axiom of the conservation of heat; or one might believe that heat is ultimately due to the vibratory or projectile motions of atoms, and still retain belief in the basic principles and laws together with the axiom of the conservation of heat. Or, finally, one might (in theory at any rate) reject the basic axiom of conservation and believe in the mutual convertibility of heat and mechanical energy. But in practice no one at that time was willing—or able—to entertain the last belief. John Herapath, for example, implicitly rejected it when he wrote that the 'adiabatic' heating of gas was a function of the speed of its compression: if one compressed a gas very slowly indeed then, he thought, no heat would be generated.[40]

Further illumination was to come from quite a different quarter. Biot, we recall, had given some intimation of Laplace's proposed solution of the problem of the velocity of sound in 1802; in 1807 Poisson had written further on the subject and had laid it down that if a gas is compressed 'adiabatically' by $\frac{1}{116}$ of its volume its temperature will be raised by 1°C. Then, in 1816, Laplace published a short paper[41] in which he gave, without proof or derivation, a formula equating the true velocity of sound to the 'Newtonian' velocity, multiplied by a correction factor to take account of the 'adiabatic' heating and cooling of the air. This factor was the square root of the ratio of the specific heat of air at constant pressure to the specific heat at constant volume: Cp/Cv. The paper ended with the comment that the direct experimental determination of the ratio of the two specific heats was very desirable.

This determination was actually achieved in a singularly roundabout way. In 1818 Gay-Lussac wrote that '. . . the determination of the absolute zero of heat appears to be a chimerical problem'.[42] If, he argued, 'adiabatic' compression could heat air so much that it ignited tinder, what possible limit could we place on the cooling of air by 'adiabatic' expansion? The simple example of the Schemnitz engine shows clearly enough what great cold can be produced by quite small expansions; if we really tried we could, no doubt, achieve such large expansions that cooling of the order of several thousand degrees would result.

It was partly in reply to these comments of Gay-Lussac's that Clément and Desormes published in 1819 the memoir which they had submitted, unsuccessfully, in the 1811 competition which had been won by Delaroche and Bérard. It is not difficult to see why they were unsuccessful; unlike the paper by the latter pair, which was workmanlike, conventional and authoritative, Clément and Desormes' paper was based on assumptions that were unconventional, while certain of the conclusions they reached were distinctly heterodox.[43] In fact they accepted the (by then) very unfashionable doctrine of caloric-of-space. When air is allowed to rush into an evacuated vessel then, as Lambert and many others had demonstrated, the temperature rises; for Clément and Desormes this 'adiabatic' heating is almost self-evident proof of the correctness of the caloric-of-space theory. The incoming air brings its own heat with it to add to that which, the thermometer assures us, must be present in the empty space; a rise in temperature must be expected and this is observed: what further proof is necessary? Gay-Lussac's demonstration that suddenly reducing the volume of a void space has no effect on a thermometer showed only that his thermometer was insensitive: he could not have had a *perfect* vacuum, and if his thermometer did not detect the 'adiabatic' heating of the residual air it could not have detected the 'adiabatic' heating of the void.

They therefore proceeded to measure the volumetric specific heat of empty space. A large spherical vessel is connected to an air-pump and fitted with a barometer. It is partially evacuated and then, after the rarefied air has returned to thermal equilibrium, the cock is opened for a fraction of a second so that outside air can rush in. When the 'adiabatically' heated air has returned again to equilibrium the valve is opened and the amount of air that must enter to restore atmospheric pressure is carefully measured. This enables them to calculate, by means of Gay-Lussac's law, the rise in temperature when the cock was opened. Now according to Clément and Desormes the partially evacuated vessel may be regarded as containing a certain volume of air at *atmospheric pressure* while the remaining space is void. When air enters to fill that void the rise in temperature is no more than the heating effect of the caloric-of-space upon the volume of air in the vessel. We know the total volume of air and we know the volume of the void; we can therefore easily calculate the specific heat of the void in terms of the specific heat of air. The whole problem is hardly more complex than the method of mixtures applied to air and the void. We can now repeat these experiments using different gases and so, by reversing the procedure, compute their specific heats in terms of that of the void and therefore that of air as well.

Clément and Desormes then went on to calculate the absolute zero of temperature first from specific heat considerations (Irvine's theory), and then using Gay-Lussac's law of gaseous expansion. Both methods gave, in good agreement, a figure of −266°C; and this, they claimed, was further confirmed by calculations from the known specific heats of ice and water and from the law of increase of gas pressure with temperature. The concurrence of the results obtained by the first and third methods with those obtained by the other two was quite fortuitous.

This unfashionable but competent and persuasive paper did not pass without criticism;[44] nevertheless it bore fruit when Gay-Lussac and Welter intimated how the experiment could be modified to measure the ratio of the two specific heats.[45] This modification is now, of course, described in the textbooks as 'Clément and Desormes' Experiment'. Shortly after this Laplace published the fifth and last volume of his *magnum opus*, the famous *Mécanique celeste*; therein he gave an authoritative account of the derivation of his formula for the velocity of sound,[46] and showed how the all-important correction factor, the ratio of the two specific heats, could be obtained by means of Clément and Desormes' experiment.

While Laplace was clearing up the problem of the velocity of sound and Gay-Lussac and Welter were modifying Clément and Desormes' experiment to measure the ratio of the two specific heats, Poisson was working on an important paper on the heat of gases and vapours.[47] Starting from the *explicit* formula $p = ap(1 + \alpha\theta)$, and assuming that the heat in a gas is proportional to the pressure and the density,* he derived an expression relating the pressure and volume (or density) of a gas when it is expanded or compressed without change of heat (i.e. adiabatically). This expression was $p' = p(\rho'/\rho)^k$, where k is the ratio of the two specific heats, Cp/Cv. (Today we should write this equation as $pv^\gamma = $ constant.) From this it was easy to deduce the relationships between the pressure and the temperature, and between the volume and the temperature.[48] Poisson then went on to derive formulae for the total heats of gases and vapours such as steam.

Commenting on this paper John Herapath noted the explicit formula $p = ap(1 + \alpha\theta)$ and transcribed it as $p = i\rho$ ($F + 448$), where i was a constant.[49] This was, of course, not far in form from the accepted expression relating the pressure, the volume and the absolute temperature of a perfect gas; or $pv = RT$.

JOHN HERAPATH

The intriguing example of John Herapath is a convenient one with which to close this chapter. Not because he was a particularly distinguished scientist—he was not—but because he provides an illuminating contrast with the eminent men whose works we have just discussed. In doing this he gives us some insight into the richness of scientific life during this very creative period of thought.

Set beside the publications of Fourier, Gay-Lussac and Ampere, Herapath's are not merely old-fashioned: they are

* The heat in a gas is also a function of the temperature, but if the pressure and density are known the temperature is determined by the equation $p = a\rho\ (1 + \alpha\theta)$, where a is a constant, α the coefficient of expansion of the gas.

F

archaic. All the familiar Newtonian paraphernalia are there: Propositions, Scholia, Lemmas; and mathematical demonstrations are expounded in the form of proportions, as were the arguments of Newton, Galileo and Archimedes. And yet with all this antique mumbo-jumbo there is a sense in which Herapath was ahead of all his contemporaries. He propounded the dynamical theory of heat, taking account of the recent discoveries of Gay-Lussac and Dalton, in a quantitative form.

In every way an amateur scientist, Herapath, properly speaking, belongs to the tradition of the English eccentrics. He was led to doubt the correctness of the material theory of heat through his reflections on the cause of gravity. This, he believed, was due to the rarefaction by heat of the subtle aether in the neighbourhood of celestial bodies (all of which, it was reasonable to suppose, had their own internal heat). But if this were so then what was heat? Some time in 1814 he came to the conclusion that the hypothesis of projectile atoms striking one another and rebounding from the walls of a containing vessel gave the best explanation of a number of the known thermal properties of gases. Among these were the famous phenomena of the 'adiabatic' heating of a gas and the diffusion of two different gases into one another.[50] The latter was, of course, the problem which had so puzzled Dalton and which he was never able to solve satisfactorily.

Herapath's early efforts brought him into touch with Humphry Davy, who as a member of the Rumford–Royal Institution School was naturally sympathetic towards kinetic theories of heat, and who had his own theory about it. This was that heat consisted in the vibratory or rotatory motion of constituent atoms. Herapath considered Davy's theory but rejected it in favour of the simpler projectile theory. Whatever the merits of Herapath's theories they were not well received. There were at least four reasons for this: his style was opaque, even to his contemporaries; he seemed to be unable to express himself without using his own clumsy jargon;* he challenged orthodox

---

* e.g. 'Numeratom', 'Voluminatom', 'Megethmerin' etc. The taste for inventing new words seems to be fairly common among scientists. Among those who were conspicuous in this respect were Clausius, only two of whose inventions survive—'entropy' and 'virial'—and G. Johnstone Stoney, only one of whose inventions survives—'electron'. Fortunately none of Herapath's jargon has survived.

opinions on a number of important points and, lastly, he made some serious mistakes.

Herapath's theory rested on the orthodox Newtonian idea that atoms are infinitely hard and therefore quite undeformable. Unfortunately such atoms cannot be postulated in a dynamical theory of gases, for in each collision there would be a loss of motion, with the result that before very long all motion would cease and the gas would lose its recognisable properties. Herapath had therefore to suppose that the atoms, although infinitely hard, were also perfectly elastic.[51] He managed to conceal this blatant contradiction by juggling with the dynamical properties of the mythical infinitely hard atoms.

Daniel Bernoulli had believed that the temperature of a gas was proportional to the square of the velocity of its atoms.[52] This may well have been an extrapolation of the *vis-viva* concept, with which he was, of course, very familiar. Herapath, on the other hand, was a Newtonian and therefore predisposed to accept *momentum* as the measure of temperature. As Dalton had recently demonstrated that the atoms of different gases have different weights and therefore different masses, Herapath could assume that the temperature was proportional to both the mass and the velocity (mv) rather than to the velocity only. The pressure exerted by a gas he took to be the total change of momentum of all the particles striking the sides of the container; as the momentum was mv and as the number of individual collisions was obviously proportional to v, it followed that the pressure was proportional to v^2. Herapath therefore concluded that the 'true' temperature must be proportional to the square root of the pressure; or, following Gay-Lussac's law, to the square root of the *absolute* Fahrenheit temperature.[53] The latter was measured from the absolute zero; the point at which a gas would have no pressure and occupy no space. This according to Gay-Lussac's law would be at $-448°$F. Accordingly Herapath defined 'true' temperatures by the relationship:

$$T' : T :: \sqrt{448 + F'} : \sqrt{448 + F}.$$

He described his idea of absolute zero to Humphry Davy:

You are aware that I conceive heat to consist in motion; and that the temperature of a body is the intensity of the intestine motion of its particles estimated, when you compare the temperature of different bodies, not by their velocity but by their momentum. The degree therefore of absolute cold is where the particles have no motion; and my object has been to ascertain this by determining the ratio of the intestine motions at two fixed points.[54]

Almost immediately a penetrating criticism of this argument was made by 'X', who wrote:[55]

Unless it can be proved that such a state of absolute cold really exists it cannot be admitted that these numbers represent the real temperatures, or absolute quantities of heat in the gases, or absolute intestine motions measured from a point of absolute rest of the particles. They will only measure the temperatures above a certain point, and not above real zero.

The trouble with the absolute zero, as defined by Herapath, was that at such a point a gas would *cease to exist*. It must spring into existence, a nonentity must become an entity, as the temperature rises above absolute zero. But how can the temperature of a body rise if that body does not exist? It follows that the whole notion of absolute zero, as defined by Herapath, was self-contradictory. This argument, worthy of Zeno, suggests that 'X' had had a classical education.

If we draw up a balance sheet of Herapath's achievements we find that, on the credit side, he had resolved the awkward problem of gaseous diffusion, reconciled the dynamical theory with Boyle's law, related the determination of absolute zero to Gay-Lussac's law of gaseous expansion, and introduced the first notion of a 'perfect gas'. On the debit side he had based his theories on the contradictory notion of infinitely hard yet perfectly elastic atoms, he had devised a curious 'true' temperature scale, his theory of latent heat was quite unacceptable, and he had implicitly rejected Avogadro's hypothesis. The verdict of his peers went against him. Of course his rejection of the Newtonian dogma of atomic repulsions did not help, but the main reasons were probably the obscurity of certain of his arguments—Davy could not understand his mathematics and Davies Gilbert found his work 'somewhat obscure'—together with unsuitable presentation. Had he concentrated on the kinetic theory and left out the extraneous material that cluttered up his work he might have won more support for his ideas.[56] As it was Herapath retired from the fight and took up the editorship of a railway journal. But the din he had kicked up

—and it was very considerable—resounded down the years, so that when Joule was looking for support for his theories he found Herapath ready to hand. Not perhaps an entirely reputable ally; but far better than none at all.

Chapter 6

The Developments of the
Power Technologies

THE TECHNOLOGY OF HEAT

In the two previous chapters we discussed the development of the most important branches of the science of heat in the first quarter of the nineteenth century. The most satisfactory progress was made in establishing an analytical theory of heat, in clarifying the modes of heat transfer, in elucidating the two specific heats of gases and in devising an experiment for determining their ratio. Although these advances were to play important, indeed vital parts in the eventual establishment of thermodynamics, the most successful application of the science of heat at that time was to the solution of the problem of the velocity of sound in air. Its connection with the problem of power and the concept of the heat-engine seems to have been exiguous.

One possible explanation for this apparent separation of science and technology was political. England, with the most advanced industrial technology, and France, the leading scientific nation, were at war during the first decade-and-a-half of the nineteenth century. Although 'the sciences were never at war' in the sense that books and periodicals were exchanged and very distinguished scientists became temporary honorary neutrals for the purpose of visiting enemy countries, the inevitable hindrances to the free communication of other ideas and to intercourse at lower social levels must have retarded the progress of science in relation to technology. Not that there was lack of interest or effort; on the contrary, with the possible exception of Fourier virtually every distinguished French physicist, theoretical or experimental, concerned himself to some extent with the problems of the steam-engine. But apart from peace between the nations, further advances were necessary in the technology of heat-engines, and then a wide diffusion of knowledge of these advances. Above all, science awaited the genius of one remarkable man who was to weld all these disparate elements into one great system of thought.

Perhaps the first practical application of the 'adiabatic' effect was the commercial manufacture of 'fire pistons', which began in France in 1806.[1] If the air in a small cylinder is rapidly compressed by means of a piston its temperature can rise high enough to ignite tinder (which incidentally confirms Dalton's insight in 1802) and one has, in effect, a perpetual match. Joseph Mollet suggested the industrial manufacture of these 'fire pistons' in 1804, but they did not prove a commercial success. It was not until the days of Rudolf Diesel* that the principle of adiabatic heating found important practical application.

The converse effect, 'adiabatic' cooling, seemed of even less practical significance. Although there had been considerable interest in freezing mixtures and although Rumford had discoursed on methods of keeping ice-cream cold, observations like those of Baillet on the Schemnitz engine or of Ziegler on the freezing produced by the escape of compressed air from a Papin's digester[2] struck no inventive sparks. In 1813 an anonymous writer suggested the use of adiabatic cooling as a means of refrigeration but nothing came of it. More promising as a source of artificial cold was the system suggested by Leslie in 1810.[3] If a vessel full of water is placed in the evacuated receiver of an air-pump the water will boil until the rise in vapour pressure stops it. But if a drying agent, such as a vessel full of sulphuric acid, is placed besides the water then the vapour will be absorbed and the boiling will continue until the drying agent is saturated. By this means the continued absorption of latent heat by the boiling water can produce great cooling.

We have already noted that during the eighteenth century comparatively little interest was shown in the possibilities of the air-engine. In fact between 1700 and 1775 barely a dozen United Kingdom patents were lodged in connection with it.[4]

* Diesel did not invent the application of compressive or 'adiabatic' heating of air to the internal combustion engine, but he did use it, in conjunction with 'isothermal' combustion, more systematically and with far greater insight than anyone had done before.

It was not necessary to have an exact knowledge of Gay-Lussac's law in order to appreciate the virtues of steam and the disadvantages of air. A cubic foot of air expands by about one-third between the freezing- and boiling-points of water while a cubic foot of steam expands 1,800-fold. It is, therefore, rather curious that towards the end of the century interest in the air-engine suddenly increased, no fewer than 23 patents being lodged in the last ten years of the century. In the main this must have been due to the growing power shortage at that time, but it may also have been a response to the rising level of technological expectations and a reflection of a growing realisation that as gases have a small capacity for heat they might offer important advantages as 'working substances'. In 1791 John Barber patented an engine in which a mixture of inflammable gas and air was ignited and the resulting expanded gases used to drive a Branca-type wheel.[5] In 1794 Robert Street invented an engine in which a mixture of turpentine vapour and air was ignited inside a cylinder fitted with a piston and, in 1801, Philippe Lebon devised an engine in which the inflammable mixture in the cylinder was ignited electrically.[6]

That the air-engine was worthy of respectful consideration was made quite clear when, in 1807, Lazare Carnot reported to the *Institut* on an engine invented by the Niepse cousins and called by them the 'Pyreolophorous'.[7] The motive for the invention was said to be: '. . . to discover a physical power equal to that of the steam-engine without consuming so much fuel'. Their solution was an internal-combustion engine whose cylinder was supercharged with air by means of a bellows. It was said to have propelled a barge up the river Saône, and Carnot reported favourably on it.

A few months later Sir George Cayley, the aeronautical pioneer, described his air-engine and thereby took the very first step in the application of heat-engines to heavier-than-air flight.[8] Finally, in 1811, Cagnard Latour, the discoverer of 'critical temperature', announced his buoyancy-engine, a brief reference to which was made on page 76, note 14.[9] In this machine a bucket wheel was immersed up to its axle in warm water and an ascending stream of air-bubbles, expanded by the warmth of the water, rose into the downwards-facing buckets and so caused the wheel to rotate. An Archimedean screw was used to overcome the pressure of the water and pump air into a tank below the wheel. Some power was, inevitably, lost this way, but the expansion of the air more than provided for this and at the same time yielded a useful surplus. It was obviously a clumsy and thermodynamically inefficient engine, but it could do what no other comparable engine could do: derive useful power from low-grade waste heat, which is an unavoidable concomitant of many industrial processes. In view of its economic, if not its theoretical virtues, it was commended to the *Institut* by an inspecting committee which included Riche de Prony, J. A. C. Charles, Montgolfier and Carnot.

The seal of scientific respectability was set on the air-engine when Delaroche and Bérard observed that the very low specific heat of air made it a very suitable 'working substance'. But whatever advantages, theoretical and practical, the air-engine might now be thought to have, in the second decade of the nineteenth century it still laboured under two very severe handicaps. One comprised the still-unsolved design problems such as those of ignition and compression, while the other was the astonishingly rapid improvement in the steam-engine, whose lead increased every year.

POWER TECHNOLOGY IN THE EARLY NINETEENTH CENTURY

At the end of the eighteenth century an English writer could remark: 'The most ample information respecting . . . improved steam engines, which is to be met with in print, is found in the *Nouvelle architecture hydraulique* of Prony'[10]—a judgement which the modern reader who cares to compare Prony's work (volume 2, 1796) with those of contemporary English writers* can only confirm. It could also be asserted, in the same article, that steam-engines which worked at a steam pressure above that of the atmosphere consumed more coal and were therefore less efficient, although of course they might have practical advantages to compensate for this.

The last opinion was, however, soon to be refuted by the achievements of the Cornish engineers. The metal mines of that county were very valuable but the water-power resources were limited, and to overcome the problem of flooding steam-engines

* e.g. William Emerson, John Banks, Olinthus Gregory and Robert Stuart.

had to be used in increasing numbers. The only possible fuel for these engines—coal—had to be imported by sea, mainly from South Wales. There was, therefore, every incentive to develop engines of the highest efficiency and the greatest reliability. It is therefore understandable that for the first three-quarters of the nineteenth century the finest steam-engines in the world, judged by the criteria of economy and reliability, were manufactured in Cornwall.[11] The standard—and it was one which rose steadily year by year—was maintained by a careful attention to detail: new and better boilers were introduced, attention was paid to thermal lagging of pipes, improved valve mechanisms were used, and the cold water used to feed the boiler was pre-heated by exhausted steam, thus directly economising heat and therefore fuel.

The technological community of West Cornwall, which must have the credit for this achievement, was at the height of its fame at the beginning of the nineteenth century. The doyen was Richard Trevithick, a man who was responsible not only for such detailed improvements as the use of a steam blast to improve combustion, but also for developing the high-pressure steam-engine and building the first railway with a steam locomotive. This was at Samuel Homfray's iron-works at Penydarren in South Wales (1804), and although of tremendous historical importance it is actually of less interest to us than Trevithick's work on high-pressure steam-engines. Trevithick, in fact, seems to have intuited very early on that high-pressure engines might be *inherently* more economical than low-pressure ones. There is evidence to show that in 1802 he carried out a series of tests at Coalbrookdale which proved that the efficiency of an engine steadily improved as the operating pressure was increased.[12] This, and similar tests, might only have meant that the engine in question was being run below its optimum pressure, and that if it had been run above that pressure its efficiency would have fallen. Common sense and the established principles of the analogous field of water-power supported this reservation, so it would have been rash to have formulated the unrestricted principle that 'the higher the pressure the greater the efficiency'. Nevertheless experience in the field soon showed that there was plenty of scope in high-pressure steam and that the upper limit of 'optimum pressure' was nowhere in sight. For all practical purposes, then, a high-pressure engine was more efficient than a low-pressure one, and soon even the staid and superb Cornish pumping-engines were being run at steadily higher pressures. Of Trevithick we may say that he was one of the main sources of British wealth and power during the nineteenth century. Although he was rather erratic in his personal relationships, few men have deserved better of their fellow-countrymen, or indeed of the world; he was left to die neglected and in poverty.

After Hornblower's unfortunate encounter with Boulton and Watt in 1792 the two-cylinder compound expansive engine had been forgotten.* It was revived again by the Cornish engineer Arthur Woolf who, in 1804, took out a patent for an engine of this sort.[13] Curiously enough, the principle upon which Woolf had based his engine was the quite erroneous 'scientific dis-covery', which he claimed to have made, that steam at an excess pressure of two pounds per square inch would, if its temperature were kept constant, double its volume in expand-ing to atmospheric pressure; at an excess of three pounds per square inch it would triple its volume, and so forth. Woolf gave no details of the experiments by which he established this 'law' of nature, so completely at variance with other established principles such as Boyle's law; he merely mentions it in his patent specification. Wrong though it was, Woolf used his law as the basis for the design of a new compound expansive engine. High-pressure steam drives down the piston in the small high-pressure cylinder, after which it is led to the larger, low-pressure cylinder, wherein it expands according to Woolf's 'law'. The basis for calculating the relative sizes of the cylinders was wrong, and Woolf's initial efforts were accordingly unsuccessful. But the principle of compound expansive operation was correct, so that after trial-and-error adjustments a ratio of cylinder sizes was found which gave a much better performance. Not for the first time in history a useful advance had been made on the basis of a defective theory.

The first Woolf compound expansive engine was set to work pumping the Wheal Abraham mine in 1814. After some initial teething troubles which were soon corrected it yielded, in the

* Sadler's engine of 1798 used two cylinders; steam drove down the piston in the first cylinder, after which it was condensed under the piston in the second. This combination of a steam-engine with a Newcomen engine was not successful.

early months of 1815, the astonishing duty of about 52 million ft. lb. per bushel of coal consumed. A year later, in May 1816, it reached a duty of 56 million ft. lb. per bushel. Now these results, which represented something like a 100 per cent improvement on the best performance of the Watt low-pressure engine, did not go unnoticed. Woolf had collaborated in his researches with Alexander Tilloch, the first editor of the *Philosophical Magazine*,[14] and Tilloch lost no time in publicising the achievements of his friend and collaborator.

If, in a sense, the success of the Woolf engine represented a triumph of the native, Cornish engineer—a vindication of Hornblower and Trevithick against Watt—the withdrawal of Watt's influence in Cornwall at the end of the eighteenth century seems to have led to a temporary fall in engineering standards and a decline in the average performance of mine engines. But the Cornish engineers were too good to let this pass unnoticed and uncorrected, so they quickly organised, under the leadership of Joel Lean, the publication of monthly reports of the performance of all the engines in the county. The publication of the monthly *Engine Reporter* seems to have been quite unprecedented, and in striking contrast to the furtive secrecy that had surrounded so many of the notable improvements to the steam-engine. It was a co-operative endeavour to raise the standards of all engines everywhere by publishing the details of the performance of each one, so that everybody could see which models were performing best and by how much. The publication of these monthly reports had been started in 1811, with full publicity, by Alexander Tilloch, who reproduced the data in *Philosophical Magazine*; the excellent and authenticated performance of the Woolf engine at Wheal Abraham was of course particularly noted.[15] And it was not only the British scientific press that commented on these things.

When the Napoleonic Wars finally ended in 1815 the Continent was no longer cut off from these developments. When the last of the armies had been demobilised and people were free to travel and to take stock of their surroundings, Europeans found themselves face to face with a new type of society, for England had industrialised herself. An interlocking system of iron foundries and heavy machine-tools, steam-engines and textile mills, canals and mines clearly showed that England was, as the boast went, the workshop of the world.

And, in the near future, men could easily foresee the development of railways and steamships. For all the brave efforts of the French State and of the Ecole Polytechnique, and for all the genius of French science and engineering, nothing like it had been achieved on the Continent. It was entirely natural therefore that Frenchmen should have taken an acute interest in the complex technology that the ending of the wars had revealed to them. The industrious 'intelligencer' Baron Dupin set off on the first of several study trips to England. His preliminary verdict was that British prosperity rested on the steam-engine and (oddly enough) the Bramah press. But there is no doubt that it was the steam-engine which aroused the greatest interest.

Biot remarked on the excellent performance of the Woolf engine in his *Traité de physique*,[16] while Sir Charles Blagden, then staying in France,* wrote to the *Institut*[17] about it. The significance of Lean's monthly reports was immediately appreciated by the editors of *Annales de Chimie et de Physique* and they were therefore reprinted regularly in that journal.[18] France, of course, lacked plentiful supplies of coal, and fuel economy was as important a factor in the successful application of steam in that country as it was in Cornwall. France had quickly adopted the Newcomen engine and, later on, the Watt engine, but these machines were both heavy consumers of coal, and their usefulness was therefore very limited. Other things being equal the Woolf engine promised great economy, so that when, just after the wars, Woolf's old partner Humphrey Edwards emigrated to France and set up as a manufacturer of the new engines it proved a very successful venture. Professor Landes has recently written that: '. . . on the Continent far more emphasis was placed on fuel economy than in Britain. From the very start the Woolf compound engine . . . which made use of high pressure to operate two cylinders alternately and offered a fuel economy over the Watt engine of about 50 per cent found its greatest market in France.'[19] It had been understood for a long time in England that where coal was cheap and plentiful it was better economy to use a cheap and inefficient engine than to spend a lot of money on an efficient and economical one.[20]

Edwards must have been a good engineer, and the merits of

---

* Sir Charles Blagden, who had assisted Cavendish in his scientific experiments, and who had himself contributed to hygrometry, died in France at the home of his friend Berthollet.

the compound expansive engine were quickly appreciated. French engineers soon reported that they, too, found that the engines were more economical the higher the pressure at which

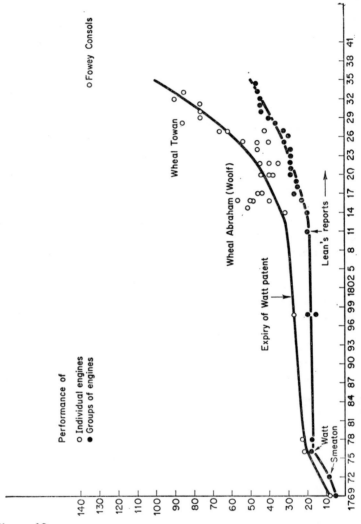

Figure 12

they were operated. This might have meant that these particular engines were being run below their optimum pressures. But if this were the case then the evidence being piled up month by month, year by year in the Lean reports must have meant that *all* engines everywhere were being run below their optimum pressures. Figure 12 (opposite) indicates the improvement in efficiency of (*a*) individual engines of particular interest or excellence, and, (*b*) groups of engines between 1769 (the date of Watt's patent) and 1835. We may interpret the long 'plateau' between 1780 and 1810 as due either to Watt's monopoly of the market and his refusal to sanction high-pressure operation, or else to the technical difficulties of assimilating the new steam-engine before a further leap forward was possible. In either case what is undeniable is that after the introduction of the high-pressure condensing engine the overall performance increased steadily and the performance of individual engines shot ahead. Correlated with these improvements was the steady increase in operating pressure. We must not forget that Lean's reports and the inferences that could be drawn from them related to engines that were universally acknowledged to be the best in the world.

The performance of the Woolf engine at Wheal Abraham fell off after its initial brilliant success. Nevertheless, false start as the Woolf engine may have been—the two-cylinder compound was soon dropped, to be re-invented for the third time when the McNaught design appeared in the mid-nineteenth century—the Cornish engineers had taken the point: low-pressure steam was out; the high-pressure expansive condensing engine was more economical. From this time onwards a succession of able Cornish engineers, men like West, Grose and Sims, steadily improved the performance of Cornish engines up to their mid-century pinnacle of excellence. In one or two cases the improvements they achieved were as dramatic as those of Woolf or even Trevithick. Thus Austen's engine at the Fowey Consols mine was said to have attained, in 1835, a duty of no less than 125 million ft. lb. for the consumption of a bushel of coal.[21] This was done under test conditions and the circumstances were therefore exceptional, but a sixfold increase over the best that Watt could envisage and a fivefold increase over what he thought was the theoretical maximum was a performance to conjure with.

From the point of view of the engineer, and still more of the scientist, the problem clearly was: why was the high-pressure condensing engine more economical than the low-pressure one?

It had been invented and developed because it offered possibilities of wider application than the low-pressure engine; it was much more portable and produced more power for smaller weight and bulk. The high fuel-economy it achieved was unforeseen and amounted to an unexpected bonus. There were three possible reasons why this should have been so. It might have reflected the detailed improvements, such as better boilers, new valve mechanisms, more efficient furnaces, to be expected from the normal process of evolutionary improvement; or it could have been the consequence of the smaller heat losses that must have characterised the relatively smaller high-pressure engines; or, finally, it might have resulted from some *inherent* superiority of high-pressure engines. In the last case some unknown law of nature was involved, and it is therefore necessary to consider what light contemporary science could throw on the problem.

It seems that the first person to publish a systematic and disinterested inquiry into the economy of high-pressure steam-engines was one John Sharpe, a Manchester solicitor and a close personal friend of John Dalton. In fact Sharpe was one of the two people to whom Dalton dedicated the second volume of his *New system of chemical philosophy* (1828), and it is quite possible that Sharpe was, besides being a friend, a scientific disciple of Dalton. After all, Dalton had an extraordinary flair for scientifically important problems, and we know that in 1800 he had read a paper to the Manchester Society entitled: 'Experimental Essays to Determine the Expansion of Gases by Heat and the Maximum of Steam or Aqueous Vapour which any Gas of a Given Temperature can Admit of; with Observations on the Common and Improved Steam Engines.' Unfortunately this essay was never printed and all trace of Dalton's ideas about 'Common and Improved Steam-Engines' has been lost.[22] Were the latter, by any chance, the very new high-pressure steam-engines?

Sharpe's paper was read in 1806 but was not published in *Manchester Memoirs* until 1813.[23] He was, he said, concerned to inquire into the reasons for the supposed high fuel-economy of the high-pressure steam-engine. The apparatus he used was simple: a high-pressure boiler, a still immersed in a measured amount of water, together with a thermometer and pressure gauge. Sharpe found that a given weight of steam always seemed to impart the same amount of heat to the still, irrespective of the temperature and pressure at which it had been generated; in other words, condensing high-pressure (and very hot) steam produced no more heat than condensing normal steam (at 212°F). Sharpe reasoned that the latent heat in steam certainly did not increase with the pressure and temperature. Unfortunately he did not know what *volume* of steam had been condensed in the still and he therefore had no idea of the densities at different pressures. As Dalton pointed out, the high pressures might have been due to the temperatures alone, the density being the same at all pressures. In this case there would be a distinct advantage in using high-pressure steam. But Dalton thought it more likely that steam obeyed Boyle's law, so that the pressure was proportional to the density. In this case, if the specific heat was constant, there was no evident advantage in using high-pressure steam since the weight of steam determined the pressure on the one hand and the amount of fuel burned on the other. In fact Sharpe did little more than confirm the general truth of 'Watt's law' at high pressure (he and Dalton were ignorant of 'Watt's law' at that time). But he did this in the context of a scientific publication, under the auspices of a very distinguished man of science, and with explicit reference to the economy of high-pressure steam-engines.

A few years earlier John Southern had carried out a similar series of experiments at Soho. He described these in a letter which he wrote to Watt eleven years later, in 1814; and they were finally made public in John Robison's *System of mechanical philosophy*.[24] Southern's views did not find much favour with his contemporaries and immediate successors,[25] but they are nevertheless of considerable interest. He pointed out that when air expands it must be supplied with heat if it is not to be cooled down; this heat, as he put it, goes latent. Perhaps the same thing is true of steam? If we boil water in the normal way, turning it into steam, then perhaps we have to supply a fixed latent heat to convert the water into steam, and a further latent heat to expand it 1,800 times at the same temperature? What we commonly call 'latent heat' should, on this hypothesis, consist of two parts: a latent heat of state, which is constant at all pressures and temperatures, and a latent heat of expansion, which obviously varies with the pressure and temperature. This would explain why the 'total' latent heat seems to increase as the volume of steam increases and the pressure falls. But it does not necessarily follow that the sensible heat must increase exactly as the total of the two latent heats decreases. It may well be that these two tendencies do not exactly compensate one another, so that the total heat is not constant. And, adds Southern, '. . . your experiments seem to bear this out'. As the addressee in this case was Watt himself it seems likely that he, at any rate, could hardly have accepted 'Watt's law' in the simple quantitative form it had come to have.

Perhaps the Manchester School provided a better environment for free scientific inquiry than did the somewhat secretive atmosphere of Soho. In any case Peter Ewart, who had been trained by Boulton and Watt and had been their agent in the Manchester area, felt no qualms about publishing his long paper, almost a book, 'On the Measure of Moving Force' in the *Manchester Memoirs*. Although this paper, as we have seen, deals mainly with problems in mechanics, he does make one important observation concerning the steam-engine. He was prompted to do this by Wollaston's Bakerian Lecture of 1805.

Wollaston had made the suggestion that a weight lifted a given distance—the engineer's common unit of 'duty'—should be taken as the measure of the power of, say, gunpowder. Wollaston went on to generalise this suggestion: '. . . similarly a distinct expression . . . is obtained for the quantity of mechanic force given to a steam engine by any quantity of coals . . . Although the rate at which mechanic force is generated may vary any quantity of work executed is the same, in whatever time it may have been performed.'[26] This perceptive assertion that there is an equivalence between a quantity of heat on the one hand and the amount of work that it can perform on the other was criticised by John Playfair in an *Edinburgh Review* article.[27] According to Playfair one could not equate work and heat since the performance of work by a steam-engine depends upon the way in which the heat is applied; a given amount of heat may be applied so quickly—or so slowly—that it yields no work at all.

Ewart's criticism of Playfair's comments was both convincing and penetrating. Obviously we can never hope to get the whole amount of work from a given quantity of heat; that would be asking too much of imperfect machines in an imperfect world. But this does not invalidate the principle of equivalence:

To these objections it may be replied that however slow or quick the combustion of the coals may be, if they be effectively burnt the full quantity of heat must be given out. If the heat be allowed to escape without being communicated to the water; or if after being communicated to the water the pressure of the steam be not wholly applied in producing the intended effect the loss must be owing to practical imperfections in the construction of the apparatus. Such imperfections must exist, more or less, in every apparatus and they will, no doubt, be greatest in extreme cases. But although the whole heat, or the whole force, can in practice never be completely transferred from one given object to another, yet there can be no doubt of the real existence of both the heat and the force in their full quantities, and we can form no idea of the portion of time being limited in which one must be evolved or the other transferred.[28]

This extremely important paragraph asserts the fundamental axiom of the *equivalence* of heat and work. It is not, of course, a statement of the mutual *convertibility* of heat and mechanical energy (Joule's doctrine), for Ewart was a supporter of the caloric theory, but it was an essential step in that direction. At this stage the future development of the dynamical theory of heat demanded as a prerequisite not so much the overthrow of the caloric theory as the clear recognition that to a given quantity of heat there corresponded a fixed quantity of work; once this had been acknowledged then the further recognition could be made that it was the *consumption* of this heat, not its mere passage from a hot body to a cold one, that actually produced the work.

Manchester, even with Dalton, William Henry and Ewart as residents, was not as important scientifically as Paris; nor could a paper in *Manchester Memoirs* carry the weight of one in a French journal, especially if the paper in question was pleading a cause long ago settled in France. But this was not really the point. Ewart's paper was a product of the tough-minded Manchester School, a significant contribution to that school of thought dominated by the formidable Dalton (it is significant, surely, that Dalton dedicated the second volume of the *New system* jointly to Sharpe and Ewart, making it quite clear that in Ewart's case the reason, apart from friendship, was admiration for the essay on the measure of moving force). It was, we infer, in the intellectual environment moulded by men like

Dalton and Ewart that the young Mancunian scientist James Prescott Joule grew up.

We have therefore found that contemporary science was in no position to explain the superior efficiency and economy of the high-pressure steam-engine. The one general principle which had by then emerged from the study of the steam-engine —that of 'total heat'—threw no light on the problem, even if one assumed 'Watt's law' to be correct. The knowledge that the same 'total heat' was required to generate high-pressure steam as was required to generate low-pressure steam was of little use, for the 'total heats' in question were by *weight*, and it was very probable that the density of steam increased with the pressure. Thus additional pressure would require a greater weight of steam, and this in turn would surely require more fuel. But in any case pressure is not the same thing as work, and the performance of an engine is measured by the work it does in terms of the heat it absorbs. What was required, then, for a scientific explanation was the generalisation of the engineer's practical unit of 'duty' into the scientific concept of work, and the discovery of general principles relating the heat absorbed to the work done.

In 1811 J. N. P. Hachette (1769–1834), Professor of Mathematics at the Ecole Polytechnique, published a notable work: the *Traité élémentaire des machines*. The second word in the title inevitably arouses the suspicions of the Anglo-Saxon reader: there was nothing 'elementary' about it. Indeed it dealt with the fundamental theory of machines, and in so doing made two important contributions to knowledge. In the first place Hachette, following the line pioneered by Lazare Carnot, succeeded in bringing the latter's abstract concepts into harmony with the practical engineer's measure of duty: a weight lifted a given distance. Owing to the general backwardness of French technology and industry at that time this measure had not achieved the general currency in France that it had attained in England. Nevertheless the acute French intelligence had, as we have seen, seized on the significance of *force vive* in both its active and latent forms. Hachette was to bridge the gulf between the practical and the theoretical. It is significant that he refers to John Smeaton's excellent works which '. . . have just been translated by M. Girard'.[29]

Hachette took as his basic concept the 'dynamic unit', or

'dynamode' whose numerical value was the raising of 1 kilogramme 1 metre high (from its size it was evidently an engineer's unit). The dynamic unit was the basic measure of *force vive* in either form, latent or active. But Hachette went even further. While men like Smeaton, Wollaston, Ewart and even Lazare Carnot had restricted their discussion of *force vive* to the grinding of corn, the duty of steam-engines, the effects of gunpowder—that is to the field of technics—Hachette extends the notion beyond mechanics to the phenomena of the physical world:[30]

> The engineer considers only four moving agents which can be applied to machines: animal power, water, wind and fuels. The physicist adds to these natural causes of motion electricity, magnetism, the force of affinity between air and water* and that force whose action on the atmosphere manifests itself in variations in the height of the column of mercury in a barometer.

We cannot overestimate the importance of this development. The generalisation of the full *vis-viva* doctrine and its application outside the confines of technology and mechanics represents the establishment of the science of physics in its modern connotation: not, that is, the old 'mathematics with natural philosophy', but a science concerned with energy and its manifestations and transformations.

Hachette explained that the weight of rain which falls on a given area in a given time multiplied by the height of the rain clouds ($mg \times h$) is the measure of the force of affinity between air and water; this 'force' (or potential energy as we should say) only becomes actual *force vive* as the raindrops fall. It needed but the rejection of the discredited theory that evaporation was the solution of water in air for this to become an important step towards the picture of the atmosphere as a heat-engine, together with the identification of the driving power of the hydrological cycle. Dalton had already taken this preliminary step.

Although the theory that evaporation was merely water

* Monge advocated the erroneous solution theory of evaporation. Hachette was Monge's protégé and dedicated his book to him. The persistence of this curious theory in a scientifically advanced country like France confirms Dr Fox's observation about French indifference to meteorology.

dissolving in air could be refuted by evaporating water *in vacuo*
this could, in turn, be countered by insisting that no vacuum
is ever perfect, and that the residual air is enough to dissolve
water. This answer was good enough to satisfy de Saussure but
it did not—perhaps could not—satisfy Dalton. He could no
more believe it than he could accept the idea that the main
gases of the atmosphere are held together in some sort of
chemical 'combination'. For Dalton oxygen and nitrogen were
physically distinct gases diffused among one another, and it
was a serious problem for him to decide what peculiar characteristics of their atoms prevented them from separating out into
layers with the heavy gas at the bottom.[31] Water-vapour was
also, as far as Dalton was concerned, an elastic fluid diffused
among the gases of the atmosphere, and the process of evaporation was a physico-mechanical one determined by heat and not
by chemical action or affinities. All this was excellent, but the
missing factor, which ran through most of Dalton's speculations
on gases, was the kinetic theory.[32] What was less contentious
but from our point of view no less important was the paper
which Dalton read to the Manchester Society in 1799.[33] In
this remarkable essay he carefully estimated the total amount
of rain, dew and other forms of precipitation falling on England
and compared it with the total volume of water carried off by
the rivers and lost by evaporation. This enabled him to show
that there was an approximate balance between the amount
precipitated and the amount run off or evaporated. The work
begun by Perrault, Mariotte and Halley had been taken a
step further.

Let us return to Hachette and the 'dynamode'. Leibnitz and
his followers had taken mv^2 as the *vis viva* so that the potential
vis viva, to use Daniel Bernoulli's expression, must be 2 mgh.
For Leibnitz mv^2 was as fundamental as mv ('motion', or
momentum) was to the Newtonians. But by the beginning of
the nineteenth century the engineers' common measure of duty,
weight multiplied by distance, was widely used and, generalised
by Hachette, became mgH; to have doubled it—2 mgh—
would have been pointless. It is probable that the engineers
were in the ascendant at that time: there had never been many
Leibnitzians in England and the prestige of the Ecole Polytechnique ensured that the opinions of engineers would be
heeded in France; accordingly the engineering unit became

fundamental. When the potential or latent *vis viva* becomes mgh, the *vis viva* must be $\frac{1}{2}mv^2$ and not mv^2 as before. Thus the relationship between the two forms of *energy* was determined by the progressive recognition that mgh was a basic and universally applicable measure in technology, mechanics and physics.[34] It was exactly these measures that Rumford lacked in his premature attempt to devise a mechanical theory of heat.

LATER STUDIES OF THE HEAT-ENGINE

One of the first attempts to compare the performances of heat-engines theoretically was made by A. T. Petit in 1818.[35] He computed the maximum possible performance of a steam-engine by comparing the amount of heat needed to warm up and vaporise a given weight of water with the work done by the water as it expands into steam, pushing back the atmosphere —the influence of Hachette may be detected at this point. He supposed that the water was contained in a very long cylinder fitted with a frictionless piston and of unit cross-section. It would take about 640 'units' of heat to warm up a gramme, or a cubic centimetre, of freezing water and then turn it all into steam. In so doing the steam would push the piston back some 1,700 centimetres against the weight or pressure of the air, which was equal to that of a column of water about 10 metres high. The total work done would therefore be, in dynamic units, equivalent to 17 kg. lifted one metre. But if we have only air in the cylinder and impart exactly the same amount of heat to it we find, using Delaroche and Bérard's value for the specific heat of air* and applying Gay-Lussac's law of expansion, that the work done against the atmosphere amounts to some 72 kg. metre, over four times as much as that yielded by steam.

This paper was soon criticised. Clément and Desormes, followed by Navier, pointed out that Petit had ignored the fact that the specific heats of all gases increase as they expand: the so-called 'latent heat of expansion', which was the inevitable

* If the specific heat by weight of water is taken as 1, Delaroche and Bérard found that 0·267 units of heat were required to raise the same weight of water by 1°C.

explanation of adiabatic phenomena in the absence of the dynamical theory of heat. The air-engine must therefore, they argued, absorb much more heat than Petit had allowed and its efficiency must be correspondingly less. Further, Petit had over-looked the enhanced efficiency which was obtained when steam-engines were operated expansively, he had paid no attention to the increased efficiency of high-pressure steam-engines, and we may add (with the wisdom of hindsight), he had failed to realise the necessity of considering the operation of heat-engines in terms of the closed cycle which brings the 'working substance' back to its original state. This last point may perhaps be over-looked. But the really cardinal defect in the paper was never actually made explicit. If we transform his measures into English engineers' units and if we assume, in accordance with Cornish experience, that one bushel of coal will vaporise some 14 cubic feet of water[36] then Petit's results indicate that the maximum conceivable duty of a perfect steam-engine would be some 46 million ft. lb. per bushel of coal consumed. *This duty had already been exceeded by Cornish engines on a number of occasions.* Petit's conclusion, therefore, could not possibly be correct, as Figure 12 clearly indicates.

In other words, the situation was exactly the same as was the case with the practice and theory of water-power in the decades that followed the publication of Parent's great paper: *the theoretical maximum had been exceeded in practice.*

In the light of the evident defects of Petit's paper, and in the light of Clément and Desormes's criticisms, Navier undertook a reconsideration of the problem, making the appropriate modifications to Petit's assumptions.[37] He used an empirical formula to enable him to compute the variation of the specific heat of air with volume—Delaroche and Bérard had proposed a formula of this sort—and then went on to compare the work done by air when it expanded in a cylinder, first at uniform pressure and then 'freely' according to Boyle's law (see Figure 6 above), with the heat that would be absorbed during the process. Having found an expression for the theoretical duty of a 'perfect' air-engine Navier next derived one for a 'perfect' steam-engine, also operated expansively. In all essentials Navier's calculation of the power derived by expansive opera-tion of a steam-engine was the same as that made by Davies Gilbert and Robison. It is worth noting that he rejected Watt's law when computing the heat absorbed by the steam-engine and he postulated that, in the case of both engines, the temperature remains constant during the expansive phase.[38]

Navier's perfect engine was not only mechanically perfect but was equipped with an infinitely long cylinder. The two expressions he derived showed that, contrary to what Petit had said, a steam-engine working at five atmospheres pressure would be five-and-a-half times as efficient as an air-engine. In fact, converting Navier's data into English engineers' units, the absolute maximum of work obtainable would be some 310 million ft. lb. per bushel of coal. If however we limit our considerations to a dimensionally feasible engine (i.e. one with a non-infinite cylinder) the figure is reduced to about 70 million ft. lb.* But even this comparatively modest figure is considerably greater than the confirmed performance of the best types of working engines, and it would, observes Navier, be useful to discover the exact cause of this substantial loss. As it happened, however, Navier's own conclusions were soon to be overtaken by events. Shortly after 1821 the Cornish engineers had pushed up the maximum performance to nearly 70 million ft. lb. and before the decade was out they had surpassed this figure (at Wheal Towan, 1828).

Navier was an able engineer, but like the others, he failed to realise the importance of considering the performance of a heat-engine in terms of a 'closed cycle', in which the working substance is brought back to its original condition, after it has done its work. And even in France he was hardly original in realising the advantage of expansive operations: as we shall see, he had been anticipated by Hachette, Bouvier and Clément and Desormes.

Apart from the merits of expansive operation there was little agreement at this time about the other causes of the advantages of high-pressure steam-engines. For his part Laplace argued that because the pressure of a gas increases as the square root of its caloric it follows that by doubling the caloric one quadruples the pressure, and this must account for the superior economy of the high-pressure engine.[37] But this conclusion was

---

* To express the duty of an engine in the familiar shorthand of the method of dimensions rather than in the cumbersome longhand of '70 million pounds one foot high' is no longer anachronistic: Navier himself expressed the duty of an engine in terms of 'kg. × m.'.

derived from highly speculative assumptions and, in any case, the static pressure exerted is not necessarily the measure of the work done; so Laplace threw very little light on the problem.

On the other hand a competent engineer like Charles Sylvester could deny that high-pressure engines were as efficient as had been claimed. If, he argued, the pressure of steam is proportional to its density (Boyle's law) and if, as everyone agrees, the total heat in a given weight of steam is constant (Watt's law), then making the reasonable assumption that the fuel consumed is proportional to the weight of steam produced it must follow that the fuel consumption will increase with the pressure; and there can therefore be no inherent advantage in high-pressure operation.[40] The apparent advantages must be due to exaggerated claims and, he wrote: 'The boasted advantage of the Cornish engines has chiefly arisen from their inventor (Woolf) assuming some erroneous data respecting the power of steam, and many others, even Mr Herapath, seem to have fallen into the same mistake.'

Sylvester had put his finger on Woolf's mistake (see page 155 above) and had assumed, wrongly, that all claims made on behalf of the Woolf engine must therefore be suspect. He was answered by an individual named Prideaux, who pointed out that the pressure of steam increases with its temperature (Gay-Lussac's law) as well as with its density.[41] Assuming that the expansion of steam is the same as that of gases (1/480 per degree Fahrenheit) then the pressure exerted by each pound of steam in a given volume will be doubled, trebled, quadrupled as the temperature is increased by multiples of 480°, and all the time the fuel needed to generate the steam will, by Watt's law, be the same. This extra pressure, over and above that due to the increase in density of the steam, is as Prideaux puts it, the 'profit' due to working at high *temperature*.

This short paper was duly noted in France.[42] The suggestion that the superior economy of high-pressure steam-engines was due essentially to the higher *temperature* of the steam was particularly interesting. Nevertheless, Prideaux was cautious about his argument, for he remarked: 'It is easy to illustrate this from the reports of working engines, but the effects in these cases are dependent on such mixed causes that no uniform conclusion can be drawn from them.' Among these other causes he instances expansive operation.

This particular problem had already been discussed by Hachette, who as early as June 1817 had read a memoir to the *Société Philomathique* on the comparison of the 'dynamic effects' of high- and low-pressure steam-engines.[43] Although he does not seem to have known of Watt's law he assumed that the sum of the sensible and latent heats of steam is approximately constant, and he demonstrated that the expansive use of steam must produce a greater effect the higher the starting pressure. There were, it seems, some objections to Hachette's unproved assertions about the total heat of steam, but these did not invalidate his point that expansive use progressively increased its 'dynamic effect' at higher pressures and that this explained the superior fuel-economy of high-pressure steam-engines.

At Hachette's suggestion the boiler of an Edwards steam-engine (Humphry Edwards was then Director of the Chaillot Foundry) was procured so that the exact amounts of water evaporated at different pressures by burning a given weight of fuel could be compared. These experiments were conducted by Clément and Desormes and they published their results in the form of an important memoir on the operation of the steam-engine. Unfortunately the bulk of this memoir has been lost; all that remains is a short paper which they published in 1819.[44]

Clément and Desormes begin by observing that 'One of the most interesting questions of natural philosophy is that of the mechanical power of fire . . .' They accept Watt's law, which they seem to have discovered for themselves following Hachette's approximation,* and they go on to point out that as the law of expansion and the variation of specific heats of gases with their volumes are both known it is now possible to determine the pressure of caloric in a gas, and so find the relationship between the quantity of caloric and the mechanical power it can exert. Unfortunately it is by no means easy to determine the specific heats of steam.

To find the maximum mechanical action of heat they propose an extremely ingenious 'ideal' engine: perhaps the simplest

---

* That Clément and Desormes were ignorant of Watt's law is partially confirmed by Thomas Thomson's comment that M. Clément knew nothing of Sharpe's work until he, Thomson, told him of it. Thomson, like so many others, mistook Sharpe's paper as establishing Watt's law. Thomson may have been instrumental in informing French scientists of the work of the Manchester School in this field. He was, after all, a close disciple of Dalton, the editor of *Annals of Philosophy*, and evidently had links with France.[45]

conceivable engine. The steam is introduced at the bottom of a
tall vessel full to the brim of water which can be assumed to be
at the same temperature as the steam. As the steam enters it
necessarily displaces water at the top, which overflows into the
reservoir, and this process continues as the steam forms a bigger
bubble. This stage corresponds to the first phase of the motion

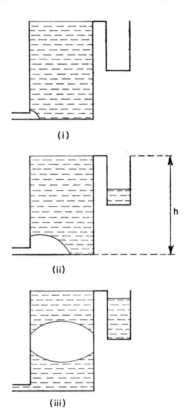

Figure 13

of a piston in a cylinder under the pressure of steam from the
boiler. As the bubble detaches itself from the bottom and rises
through the water it will expand as the hydrostatic pressure
falls, so displacing progressively more water until it finally
breaks surface. This, of course, corresponds to the expansive
phase when the steam in the cylinder is cut off from the boiler
and allowed to expand on its own.

The work done by the steam is the weight of water displaced multiplied by the height to which it has, in effect, been raised ('h' in Figure 13). With the data available they calculated that a kilogramme of steam should when used expansively* in such an engine generate some 115 dynamic units of work, the dynamic unit being taken in this case as 1,000 kilogrammes raised 1 metre. Using Davies Gilbert's figure for the amount of steam that can be generated by a given weight of coal and expressing Clément and Desormes's result in English engineers' units we find that, on their theory, a bushel of coal should produce 315 million ft. lb. of work: which agrees well enough with the duty that can be deduced from the calculations Navier made two years later. Incautiously Clément and Desormes argue that the best-working engines are only one-twelfth as efficient as their 'ideal' engine (i.e. realise a duty of 26 million ft. lb. per bushel); but they add, much more reasonably, that non-expansive engines must be even less efficient.

Implicit in the arguments of Petit, Navier, Clément and Desormes and others had been the notion of an abstract, ideal engine derived from experience of actual working engines, but purged of 'inessential' details and without such defects as heat and frictional losses. In Clément and Desormes's case we may hazard the guess that the Cagnard Latour buoyancy engine may have been in the back of their minds, although the appreciation of expansive operation comes from the Woolf engine. Now these men did not invent the idea of an abstract, perfect engine. The vast majority of engineers at this time were quite accustomed to dealing with both heat- and water-engines, and virtually all textbooks whether French or British dealt with both types.[46] Accordingly, engineers who had been accustomed to think in terms of idealised engines in hydraulic-power technology would almost automatically do so when they considered heat-engines. Déparcieux, after all, had envisaged a perfect reversible hydraulic engine as early as 1752, while Borda in his great paper of 1767 based his arguments on a generalised and perfect water-wheel.[47]

In the case of heat-engines one of the first, if not the first, to argue in terms of an ideal engine was the rather obscure engineer A. R. Bouvier. As early as 1816 we find him asserting that art, or inventive design, had brought the steam-engine to

* From a pressure of 5 atmospheres to one of $\frac{1}{68}$ atmosphere.

the point at which science was required for its perfection.[48] He was aware that high-pressure engines were more economical than low-pressure ones, and he ascribed this, naïvely enough, to the acknowledged fact that the specific heats of gases became smaller as their volumes were reduced and their densities increased. Bouvier then proceeded to describe expansive operation, which he appeared to think was an entirely original idea of his own, and to claim that it could lead to the perfection of the steam-engine. He derived the expression for the power generated by expansive operation as applied to a direct rotative engine (Figure 8 (6)), which again he appeared to think was an entirely original idea.

Whether or not Bouvier was a plagiarist must be in some doubt; the balance of probabilities would seem to be that he was not. He had evidently never heard of Watt's law or read Sharpe's paper, and the direct rotative steam-engine was not so well known in France, or anywhere else for that matter, that he must necessarily have heard of it. His claim to minor distinction rests not on the facts that he followed in Watt's footsteps and made the same calculations as Davies Gilbert and Robison, but he saw that the perfect heat-engine must be defined in terms of physics, and that such an engine must in principle be operated expansively. He did not know whether a direct rotative engine could actually be made; but if it could he was quite sure that its performance would fall considerably below the ideal.

Bouvier's claims to originality were quickly and authoritatively dismissed by Gengembre, who not only referred to Watt's direct rotative engine of 1769 but also gave a brief account of earlier work on expansively operated steam-engines in France.[49] Gengembre himself had made such an engine in 1808, and another one in 1812 which had achieved quite notable economies and had subsequently been exported to Germany. As far as direct rotative engines were concerned Watt himself had told Gengembre that in his experience performance had not matched expectations. After this crushing rejoinder nothing further was heard from Bouvier, which was, perhaps, rather a pity.

Possibly the most intriguing of all ideal engines was that which could be inferred from Manoury d'Ectot's machine as described in 1821 by Prony, Gay-Lussac and the inventor himself.[50] This machine was in fact an improved Savery engine which, rather like Kier's engine (see pages 75–6 above), used air as a cushion between the hot steam and the cold water; but it also included some additional improvements which justified the claim that it was an independent invention. Unfortunately the duty of this engine, quoted by the rapporteurs, indicated, when expressed in English engineers' units, an efficiency roughly equal to that of a Newcomen engine *before Smeaton's improvements*; that is, about 5 millions ft. lb. per bushel of coal. The rapporteurs claimed that the Manoury d'Ectot engine was more efficient than the Woolf engine recently installed at the

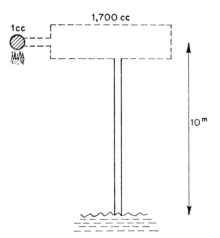

Figure 14

Quai des Ormes. Plainly there was some grievous error in their assessment of the performance of the Woolf engine, although their figure for the duty of the Manoury d'Ectot engine compares reasonably with that for the Kier engine quoted by John Farey.

The piquancy of the situation was that the wheel had come full circle. The latest high-efficiency, high-pressure engines were to be judged by the standard of a Savery engine! The irony was compounded when, in 1827, Davies Gilbert—of all people—formally enshrined the Savery principle as the abstract limit of power that could be developed by steam.[51] The ideal Savery engine is very easy to conceive. A given amount of heat, say 640 calories, will convert a gramme of freezing water into

1,700 c.c.s of steam. If this steam is now condensed it will leave a void space into which water can rise from a reservoir 10 metres at most beneath. The duty of this 'engine' will be, therefore, some 1·7 million c.m. gm for 640 calories, which, converted into English engineers' units, corresponds to roughly 40 million ft. lb. per bushel. The advantage of expansive operation coupled with the additional pressure 'due' to the higher temperature of high-pressure steam (cf. Prideaux) together account for the fact that the best-working engines considerably exceed this theoretical 'maximum'.[52]

We have now run through the spectrum of theories and concepts put forward to explain the economy of the steam-engine, and in particular the high-pressure condensing engine. Very broadly speaking two points have emerged. In the first place the superiority of the high-pressure condensing engine was best explained, before 1824, by the practice of expansive operation: a technique which became increasingly applicable and increasingly worth while the higher the pressure. In the second place engineers were able, to some extent, to think in terms of abstract or ideal engines in their pursuit of the utmost economy, and in their attempts to understand the factors determining the production of mechanical power from the generation and application of heat. In doing this they were hardly doing more than follow in the footsteps of the pioneers of water-power technology.

The success of the high-pressure steam-engine was now international. Its flexibility and economy were increasingly recognised in France and in Britain and if, in the latter country, it was on the point of making possible the widespread development of railways, in America it was being extensively and most successfully used in steamboats; a circumstance greatly favoured by the rivers of that continent. Under these conditions it is hardly surprising that there was little interest in alternative forms of heat-engine using different working substances. Air is abundant, safe and expansible but, as we have seen, it is difficult to heat and bulky to deal with. Furthermore, despite inventions like Stirling's engine that belong to this period, it did not seem to offer any advantage, in terms of duty, over steam. Thus M. I. Brunel was said to have carried out experiments on Joseph Montgolfier's behalf with a view to developing the power of heated air.[53] But these experiments

... demonstrated that the amount of mechanical power derived from the effect of any given amount of caloric on gaseous bodies was not greater than that produced by the expansion of water into steam and that practically it was not so generally applicable. The researches of Sir Humphry Davy and Mr Davies Gilbert confirmed this result.

Petit's brief but cogent criticism had (see above, page 76, note 15), exposed the fallacy in the plausible belief that great power might be obtained from the seemingly irresistible expansion of solids or liquids. But there still remained some possibility, in theory at any rate, that an advantage might be obtained by using the vapours of liquids other than water as working substances. In 1818 Andrew Ure published a paper in which he suggested that as alcohol has a latent heat of less than half that of water and boils at a temperature below 212° it should have considerable advantages as a working substance.[54] Indeed, as early as 1797 Cartwright's engine had been run on alcohol vapour, the condenser having served as a still. But, for very obvious reasons, such engines could only be feasible where alcohol was very cheap—on a West Indian plantation for example—and if any were ever put to serious test under these circumstances they cannot have been successful, for nothing came of the idea.

Finally, we must mention an interesting idea that was proposed by a Cambridge Don, the Rev. William Cecil. This was a gas-engine in which the combustion of hydrogen in a closed cylinder effectively creates a vacuum, into which a piston is driven by the weight of the atmosphere.[55] The invention was not successful at the time but it can be taken, not unfairly, to represent the first of the long line of gas-engines that were developed later in the nineteenth century.

In this way the principle of natural selection operated to ensure the survival of the fittest form of heat-engine. All manner of mutations and variations were tried out and found wanting so that, in the end, the high-pressure reciprocating steam-engine emerged as the undisputed inheritor of the earth; at least for the first half of the nineteenth century.

In one respect however the high-pressure steam-engine was admittedly inferior to the Watt machine. This was in the matter of safety. There had been some explosions, as Watt had gloomily anticipated, and there had even been a few fatalities.

G

It was as a consequence of one of these minor disasters—the explosion of a ship's boiler at Norwich—that Parliament resolved in 1817 to set up a Select Committee of the House of Commons to investigate the whole problem of the safety of high-pressure steam-engines.[56] This Committee, whose members included Sir Charles Pole and Davies Gilbert, interviewed some twenty-two leading engineers, and while there was some divergence of opinion in the evidence that was presented* they concluded that the high-pressure steam-engine could be safely used. They recommended that boilers should be made of wrought rather than cast iron and be fitted with adequate safety-valves. This approval was hardly surprising in view of the admission that:

> Your Committee entered on the task assigned them with a strong feeling of the inexpediency of legislative interference with the management of private concerns or property, further than the public safety should demand, and more especially with the exertions of that mechanical skill and ingenuity in which the artists of this country are so pre-eminent, by which the labour of man has been greatly abridged, the manufactures of the country carried to an unrivalled perfection and its commerce extended over the whole world.

And, in particular,

> . . . it is impossible for a moment to overlook the introduction of steam as a most powerful agent, of almost universal application, and of such utility that but for its assistance a very large portion of the workmen employed in an extensive mineral district of this kingdom would be deprived of their subsistence.

Six years later an investigation into exactly the same problem was initiated in France.[57] The Academy appointed Laplace, Gay-Lussac, Girard, Ampère and Baron Dupin to carry out the inquiry, and while the high professional qualifications of the French committee neatly underlined the different approaches to science and technology in Britain and in France the conclusions they reached were in line with those of the 1817 Select Committee.[58] High-pressure steam-engines were far too

---

* One or two engineers, like Galloway and Maudslay, doubted the economy of the high-pressure engines 'in the abstract' but agreed that Woolf engines were more economical in practice. Thomas Lean's detailed evidence on the performance of nearly all Cornish engines effectively settled the matter.

valuable to give up; they can be made safe enough by means of suitable safety-valves, and—a suggestion due to Girard—their boilers should be tested to four or five times their working pressure by means of an hydraulic press.[59] There was some point in these recommendations for, according to Girard, about three-quarters of the steam-engines in France at that time were of the Woolf type. A further consequence of this committee's work was the institution of long-term research into the law of the increase of steam-pressure with temperature. As engineers like Jacob Perkins were already experimenting with super-high-pressure engines,[60] knowledge of this law was evidently soon to become important.

With the seal of official approval set upon it in both Britain and France the future of the high-pressure engine was assured by the early 1820s. And yet in Cornwall the ten years that followed the brilliant success of Arthur Woolf's Wheal Abraham engine were, by comparison, rather pedestrian in tempo (see Figure 12). According to John Farey this was mainly due to the inadequate attention that was paid to thermal lagging and to the widespread practice of 'wire drawing' the steam.[61] 'Wire drawing'—the term is almost self-explanatory—meant throttling back the steam between the boiler and the cylinder instead of operating a clean cut-off to give proper expansive operation; the practice led to the build-up of a substantial pressure difference between the boiler and the cylinder, with a consequent serious loss of efficiency. Another, less tangible, factor would have been the unavoidable difficulties of incorporating improvements in the industry at large. The pace of advance generally is set not by the most brilliant and able engineers but by the capacity of the average individual—engineer or skilled mechanic—to master and use the improvements effectively as they come along. In 1827, however, Samuel Grose produced the Wheal Towan engine and the standards were once again raised. Finally, in 1835, came the Fowey Consols engine. The construction of these excellent machines marks a very convenient stage from which we can survey and summarise the main conclusions to be drawn from the engineering experiences of the period.

Although the theory of expansive operation was first put forward by Davies Gilbert in 1792 and publicised by Robison in 1797,[62] its importance could not be fully appreciated until the prejudice against high-pressure steam had been overcome and the economies due to high-pressure expansive operation fully confirmed. Bouvier, Hachette, Navier, and Clément and Desormes had all proved that expansive operation was essential if the maximum economy was to be obtained; but it did not follow that it was the only important factor. Poisson indeed had put his finger on another one when, in the course of a discussion on the theory of heat and the steam-engine, he remarked that '. . . it is in reduced heat losses and other constructional details that one must look for the reason for the superiority of such machines'.[63]

Our purpose is not to try to judge the merits of various explanations against the standards of subsequent knowledge but to select what would, in the context of the 1820s, have been the most reasonable conclusions to come to. The opinions of the Lean brothers come closest to this ideal for they were 'registrars and reporters of the duty of steam engines' in Cornwall, the place where the best and most economical engines were made. In their *Historical statement of the improvements made in the duty performed by the steam engines in Cornwall* the Lean brothers discuss the long saga of trial and error, frequent disappointment and occasional triumph, and they conclude that the progressive increase in efficiency of Cornish engines must be due to two main factors: the adoption of Trevithick's fire-tube boiler and the use of expansive operation pioneered, in effect, by Hornblower and established by Arthur Woolf. To these factors we must add the incorporation of minor improvements, the attention to such details as thermal lagging, and the general rise in industrial skills.

We can, in a broad sense, confirm these conclusions. Although it involves the wisdom of hindsight we are entitled to use modern knowledge to illuminate old scientific and technological problems; indeed it is important for our thesis that we do so. The Fowey Consols engine, under confirmed test conditions, was made to yield a duty of 125 million ft. lb. for a bushel of coal. During the test its steam-pressure varied between $50\frac{1}{2}$ p.s.i. and 59 p.s.i., which according to Figure 3 would correspond to temperatures of between 139°C and 148°C. Taking the highest temperature and applying the fundamental formula* we find

* i.e. $(T_1 - T_2)/T_1$, where T_1 and T_2 are the temperatures in degrees absolute of the source and sink respectively.

that for a perfect thermodynamic engine working between 148° and 15° (the temperature of the cold condensing-water) the efficiency should be 31·5 per cent while the same engine working between 100° and 15° would have an efficiency of 22·8 per cent. Now these efficiencies are of the same order and yet the Fowey Consols engine under test conditions yielded four or five times, and under normal working conditions about three times the duty of the best Watt-type engine working between 100° and 15°. Plainly and incontrovertibly the enhanced efficiency of this machine and the others like it must have been due to those factors which the engineers, mathematicians and physicists had recognised and which we have discussed above. The thermodynamic factor, by comparison, was so small as to be effectively masked by them.

We cannot close this survey of steam-power technology before the age of Carnot without some tribute to Arthur Woolf. He was said to have been a low, rough, quarrelsome fellow; he married a servant girl, and his brother had narrowly avoided execution as one of the naval mutineers at the Nore. Clearly Arthur Woolf did not find much favour with the scientific and business establishments and probably for this reason has not had the credit that is his due. He does not rank with Watt and Trevithick as one of the pioneers; yet it is apparent that at a very critical time his influence on the development of steam-power and, beyond that, of science, was decisive. The Lean brothers say of him:

> Besides being an experienced engineer, Woolf was a skilful workman; and the engines erected under his superintendence excelled in correctness of construction. After his example or by his instructions, other workmen also attained perfection in the art; and the engines made in Cornwall were found to yield in excellence to those of no manufactory however eminent.[64]

WATER-POWER TECHNOLOGY

The Woolf era in steam-power was paralleled by improvements in water-power technology, which although less dramatic were on a wider front. The use of cast iron for the construction of water-wheels enabled larger and stronger ones to be built, and the technique of taking the power from the moving rim of the

wheel rather than from the axle greatly reduced the stresses in
the spokes. The young William Fairbairn, who was responsible
for many of these improvements, also increased the efficiency
of power transmission very considerably by substituting light
and fast-spinning line-shafting for heavy and slow.

In 1812 Smeaton's writings on power were translated by
Girard. Their combination of meticulous experiment together
with precise argument caused them to be received with interest
and respect by French engineers. It may be that they played
some part in two important inventions that were announced in
1824 and 1825. The first of these was a machine which its

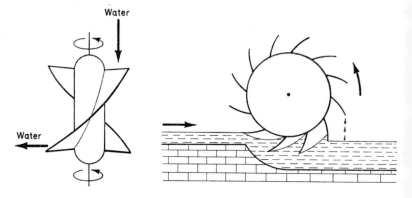

Burdin's turbine   *Poncelot's wheel*

Figure 15

inventor, Burdin, christened the 'turbine'.[65] It was a high-
speed water-engine in which the blades were, as Borda had long
ago suggested, curved in such a way that the water impinged
upon them without turbulent shock; the curve of the blades
then carried the stream of water smoothly through an angle of
90° so that it finally left the machine in the opposite direction
to that in which the shaft rotated (Figure 15). This design
meant that the machine, in theory at least, satisfied Lazare
Carnot's two conditions for the maximum generation of power:
that the water, or other motive agent, should enter the machine
without shock and leave it without appreciable velocity.

The specification for Burdin's turbine proved to be beyond
the technical competence of the engineering industries at that
time; it did not really become feasible until the middle of the

nineteenth century, and its potential was not fully realised until the development of hydro-electricity towards the end of the century. But the second of the two inventions—Poncelet's undershot water-wheel—was a very practicable proposition.[66] The overshot water-wheel is costly to make and often expensive to install; moreover if it is to be efficient it must be slow-moving. These considerations make it an inconvenient source of power in some circumstances, at least in comparison with the undershot wheel, which is both cheap to make and fast-running. The inherent defect of the latter is, as we saw, the loss of *vis viva* in the turbulent impact of the stream against the blades; but Poncelet claimed to have eliminated this by curving the blades so that they pointed upstream and the water impinged on them without turbulence (Figure 15). In fact Poncelet claimed that his water-wheel, although undershot, could achieve an efficiency of about 0·67, which was slightly better than the bucket- or breast-wheels could attain.

Towards the end of the second decade of the nineteenth century a number of excellent books on engineering were published in France. These included, besides Hachette's work, which we have already mentioned, Navier's revised and amplified edition of Belidor's *Architecture hydraulique* (1819), together with volumes by Borgnis,[67] Christian,[68] Lanz and Betancourt[69] and a superb book by A. M. Héron de Villefosse,[70] which, like Gabriel Jars' *Voyages metallurgiques*, was concerned with mining and its related technologies all over Europe. In all these books due attention was paid not only to water-wheels but to column-of-water engines as well. It is quite clear that the latter were being developed systematically, and they often included the latest design features of the new steam-engines. Furthermore, the rule seemed to be generally accepted that where the fall of water exceeded forty feet the column-of-water engine was to be preferred to any water-wheel. Some idea of the advances made in the design of column-of-water engines can be gained from Plate 3: (i) is a diagram of a double-acting engine taken from Christian; (ii) is taken from Héron de Villefosse and represents a later Schemnitz engine which is double-acting and incorporates Watt's parallel motion; (iii) is a rotative engine installed in an alum works at Whitby in 1812. This double-acting engine incorporated a device to provide a near-analogue of expansive operation and made use of Watt's parallel motion; also, following the latest steam-engine practice, it had a cast-iron working beam.

As we have seen (pages 87–8) the desirability of providing column-of-water engines with what amounted to an analogue of expansive operation was understood in France even *before the expansive operation of steam-engines had been heard of in that country.* Several different methods were said to have been available; one of them involved making use of the elastic properties of air.

The apotheosis of the column-of-water engine in France seems to have been the one installed at the Huelgoat mine in Brittany some time after 1816. The engineer responsible was a M. Juncker, who had the backing of Baillet, by then the Inspector General of Mines (see page 131). Juncker went to Bavaria to examine the machines constructed by Reichenbach, some of whose engines were reported to have attained an efficiency of 0·72. As a result of his visit Juncker was able to include the best German ideas in the Huelgoat engine, which was later said to have been designed and constructed with such precision and care that its operation was silent and smooth, there being no shocks or vibrations of any sort.[71]

For obvious reasons the column-of-water engine did not arouse so much interest in England. Its later applications in this country were mainly to those minor problems in the general technology of power which could not be conveniently resolved by the construction of special steam-engines. One such application was the provision of power for dockside cranes and varieties of lifts.[72] Another was to pump the bellows of the organs in the large churches that were built to serve the needs of the rapidly growing industrial cities in the north of England. I believe that a few column-of-water engines of this sort are still at work.

We conclude, therefore, that by the early 1820s engineers had in theory, and to a considerable extent in practice also, complete command over water-power. Whether the stream was large or small, slow or fast, the engineer knew how to derive whatever power was required and to do so with maximum efficiency: the exact limits having been defined by the theory evolved by men like Borda, Carnot, Hachette and Navier. And in practice, whether the object was to drive an English cotton mill, or a French factory, or pump out a German mine, an appropriate and highly efficient water-engine was available.

The art was, we may infer, regarded as finished, with only detailed, evolutionary improvement to be expected.

As we have already noted, practically every form of water-engine had its heat-engine analogue; and the development of technology—and of science—tended to emphasise the points of similarity.* The most successful form of heat-engine—the Cornish high-pressure expansive engine—had as its analogue the column-of-water engine, and the similarities between these two forms continued to impress engineers. Thus Poncelet, Navier and Arago compared the Huelgoat engine to a steam engine: '. . . (in the one) it is the pressure of the steam which causes the oscillations of the piston; (in the other) the same oscillations result from the action . . . of a long column of water whose pressure in atmospheres is obtained by dividing the vertical height by 10·4 metres'.[73] While Poncelet, writing somewhat later, pointed out that steam-engines, acting on the same principles as windmills and water-wheels, had been invented, but: '. . . as their application seems difficult in these last cases steam is usually applied to reciprocating engines. It is the same for great falls of water . . . Their pressures are applied to piston machines which are called column-of-water engines.'[74]

The significant thing about the column-of-water engine seemed to be that while it could, in practice, achieve efficiencies equal if not superior to those of any other water-machine, it was unquestionably the most efficient form of engine for harnessing the power of falls of water from forty feet up to the highest pressures that the pipes, valves and cylinder could stand. In this respect the column-of-water engine was eventually surpassed by the hydraulic turbine.

On the other hand the high-pressure steam-engine, so similar in other ways, had no theory behind it by which one could postulate the conditions for the generation of the maximum power; or determine the efficiency in terms of its capacity to restore the motive agent to the source. We may sum up the situation like this: water-power and steam-power had converged in practice by about 1820; was there any possibility that they could be made to converge in theory also?

* There are few points of resemblance between, for example, one of George Sorocold's water-wheels and a Newcomen engine of the same period.

Chapter 7

The Convergence of Technology and Science

SADI CARNOT AND THE BEGINNINGS OF THERMODYNAMICS

'Next to the watch the steam-engine is the highest development of mechanical art and the science of thermodynamics may be said to be the result of the study of the steam-engine.'—
OSBORNE REYNOLDS

The great achievements of science and technology in the first quarter of the nineteenth century certainly impressed contemporary observers; but probably puzzled them too, at any rate in their more private, reflective moments. Where, with so many new possibilities opening up, should one look for the main lines of scientific advance? Which of the many new theories and speculations would lead to those types of union, the linking of diverse fields of knowledge, which characterise the broad, secular development of science? May we expect to see electricity linked with heat, or, as Berzelius suggested, with atomism? Will light, heat and electricity be shown to be manifestations of the same fundamental substance, as Dalton and many others expected? What are the relationships between vital heat and living things? Will life itself be ultimately shown to be electrical in nature? With all our advantages of hindsight we can distinguish the main roads to the future from the blind alleys, although there were then no signposts for everyone to read. We can even hazard some guesses about the ground swell of scientific opinion which characterised this most fruitful epoch.

On the one hand there was the recognition of the general, or cosmological significance of heat. This was not the achievement of any one man, or even of a distinct group; no one won a Copley medal or was acclaimed by the *Académie* for it. It marked, in fact, a stage in the development of the science of heat which was brought about not only by internal advances

186

in that branch of science but also by progress in such collateral fields as meteorology and power technology. Over the greater part of the eighteenth century the study of heat had formed a part of the general science of chemistry; heat is a determinant and characteristic of chemical changes, therefore it is a proper object of study for chemists. Its roles in other natural phenomena were not fully appreciated. Heat, so regarded, was a provisional element in Lavoisier's system of chemistry, ranking in the series light, caloric, hydrogen, azote, oxygen, iron, etc. But while the chemists were studying heat firstly as a principle, then as an element or a subtle fluid, parallel developments in other sciences were paving the way for alternative theories of the nature and significance of heat.

Halley had realised that the origin of springs and rivers lay not in syphonic or capillary action but in rainfall, which was, in turn, due to the immense quantity of water evaporated from the oceans by the sun's heat. But the definitive solution to the problem of the hydrologic cycle was necessarily a rather slow business. It depended on such things as the accumulation of data about rainfall over a wide tract of country, the existence of accurate maps of the type that began to be available with the ordnance surveys of the eighteenth century and some means of assessing the amount of ground water lost by evaporation. When all these things were available it was possible to balance the equation, which had on one side the net precipitation and on the other the net amount of water carried off by rivers and streams. The motive power of the hydrologic cycle was understood to be that of heat. But while this was being established, during the eighteenth century other meteorological studies were delineating the great winds systems—the atmospheric circulations above the earth—and mapping the areas of rain, drought and evaporation.

Indeed, a kind of *cosmic machine* of a second type came to be recognised. In asserting this we are extending the seventeenth-century notion of a *cosmic machine* of the sort that was central to the thoughts of Galileo, Descartes and Newton: a strictly mechanical piece of clockwork. Fundamental to the new and complementary cosmic machine was *heat*. Heat is, in fact, the grand moving-agent of nature. Inertia, together with central forces, account for the local motion of planets and satellites; on the earth heat is the source of motion; without it the accidental factors of air resistance and friction would soon destroy all motion. There would be no rain, no winds, no variations in climate, no crops and indeed no life at all.

Such was the cosmological role of heat as revealed by the progress of science and discovery during the eighteenth century. Was it not entirely justified? Men of science had noted, and some had studied, the immensely impressive Newcomen engine: a unique achievement in so many ways, but one that was, unlike the works of Newton and Descartes, apparent to the eyes and to some extent the understanding as well of ordinary men. The radical improvements made by Watt, the rapid development of the steam-engine at the beginning of the nineteenth century and the construction of the first railways all combined to impress upon the consciousness of scientific men the immense *power* of heat; a power which they were only just beginning to exploit. It is difficult to recapture the general awareness of change, of progress, of revolution that all this added up to. It has recently been said that the last steam locomotives were the most impressive symbols ever created of power and of man's command over the forces of nature. Now if this is only partly true it does not require much historical imagination to realise how very impressive the first, crude steam locomotive must have seemed. As a German diplomat who visited Leeds in 1816 wrote.[1]

> Shortly after our arrival we went to the coal wharf to see the arrival of the coal waggons which are set in motion by steam machines and which bring coals from the mines at a distance of six English miles from the wharf. It is curious spectacle to see a number of columns of smoke winding their way through the countryside. As they approach we see them more and more distinctly until at length along with a column of smoke we also perceive the waggon from which it ascends, dragging a long train of similar waggons behind it, which gives it the appearance of a monstrous serpent.

To us the 'waggon' was Blenkinsop's quaint and grossly inefficient locomotive chugging along the Leeds–Middleton railway with its driver walking beside it. To the diplomat, a traveller and a man of the world, it was a remarkable and wholly novel experience.

Small wonder then that steam and its motive agent, heat, are put in the centre of the stage, for heat is the basic mover of the world. Harnessed by man it can pump hundreds of tons of water out of a mine, haul loaded coal-trucks up a gradient, drive the carding, spinning and weaving machines in a cotton mill; in nature it can raise many thousand tons of water from the sea to the height of the clouds.

In 1811 Hachette expressed the evaporation of water and its ascent to form clouds in terms of the 'dynamic effect', or work done; that is, the weight of the water multiplied by the height to which it is raised. This dynamic effect later reappears in the active form of the *vis viva* of the falling rain. Because he accepted the 'solution' theory of evaporation Hachette failed to relate the work (as Coriolis later termed it) done to the amount of heat absorbed. For him the work done, or the *vis viva* of the falling rain, is the measure of the affinity of air for water. Nevertheless he had applied the engineer's concept of work to the interpretation of a natural phenomenon, and we begin to have the two sides of our general equation of heat and work. In 1813 Peter Ewart asserted the relationship quite explicitly: to a given amount of heat there corresponds a fixed quantity of dynamic effect, although owing to the practical short-comings of steam-engines we cannot hope to extract the full quantity.

In this way the steam-engine represented a significant ex-tension of man's experience and knowledge of natural pro-cesses. For in it heat is used continuously to perform work, or dynamic effect. The same process takes place in nature—the expansion of rocks under the midday sun, the raising of water-vapour to form clouds—but in these cases it is rather elusive and not obvious to the senses. The steam-engine, on the other hand, manifests in dramatic and clear form the capacity of heat to perform work. This basic principle was, of course, very far from the Newtonian system of the laws of motion together with the concept of attractive and repulsive forces.

The idea of the cosmological significance of heat can be traced back to Francis Bacon or even, with suitable emenda-tions, to Heraclitus. But the credit for asserting it in a modern form seems to belong to Rumford. His appreciation of the significance of heat was deeper than that of his contemporaries, even if his scientific method was suspect. He called attention to the many subtle ways in which natural processes and pheno-mena are determined by the physical laws of heat. It is obvious that for Rumford the role of heat is as central as that of gravity in the Newtonian system.

Leslie, too, regarded the study of heat as forming an autonomous science and so, patently, did Fourier. The latter is particularly explicit; the study of heat is, he asserts, of great importance for physical science and for 'the civil economy'. Indeed, 'No subject has more extensive relations with the progress of industry and the natural sciences; for the action of heat is always present, it penetrates all bodies and spaces, it influences the processes of the arts and occurs in all the phenomena of the universe.'[2] One could hardly hope for a clearer brief definition of the scope of the science. And it is not surprising that, as Dr Herivel remarks, the analytical theory became '. . . the first branch of theoretical physics not to be based on Newton's laws of motion'.[3] Of course Fourier's interest in practical applications did not extend as far as the heat-engine. He had, in any case, formulated his basic ideas before the importance of the high-pressure steam-engine was fully appreciated. He can hardly be blamed for failing to realise that a close study of the operations of this machine would lead to the establishment of a science of heat even more general than the one whose principles he had announced in 1822.

The development of the general, or cosmological theory of heat took place during a period which was in several respects revolutionary. In France political revolution had let loose an abundance of creative talent while at the same time providing scientists with the organisation, the incentives and the moral stimulus to attack the widest range of problems. In England revolution was revealed in the rise of new forms of industry wherein manufacturing processes were analysed into their major components so that each stage could be mechanised. The factors which led to this new industrial technique have not yet been studied by historians,[4] but the consequences were clear enough. The fact of economic growth led, ineluctably, to a power shortage and this, in turn, put a premium on the efficiency of power generation and transmission. The invention of high-capacity machinery, such as Crompton's Mule, tended to heighten the power shortage, which thus became a built-in feature, an indicator, as it were, of rapid economic growth. Liebig may have been exaggerating, but he had some justification for his aphorism: 'Civilisation is the economy of power.'[5]

Lazare Carnot was by any standards a remarkable man. An able theoretical physicist who was at the same time a courageous and skilful soldier, a military organiser of genius, an active and sincere democrat and a man of integrity, his activities spanned the whole world of action and a good deal of the world of thought.[6] Any son growing up in the shadow of such a father, even a father ruined and exiled by the restored Bourbons, would, one feels, be hard put to it to make a mark in the world.

Sadi Carnot was born in 1796 when his father was just 43 years old. The young Sadi entered the Ecole Polytechnique in 1812 with the intention of becoming, like his father before him, an officer in the engineers, the 'Génie'. He saw action at Vincennes in 1814, but the long war was coming to an end and the years of peace that followed Waterloo afforded less opportunity for distinction and promotion to an army officer; especially to one whose father was a political exile. In 1820 Sadi Carnot went on half-pay and, Professor Mendoza tells us, devoted his energies to study, making the best use of the educational and research facilities of Paris.[7] He studied principally physics and economics and, according to Professor Mendoza, spent a good deal of time visiting factories and studying the organisation and economics of various industries, becoming an expert on the commerce and industries of different European countries.[8] The fruit of this devoted labour was a short book, some 118 pages long, which he published in 1824 under the title, *Réflexions sur la puissance motrice du feu*.[9]

The core of this book is an argument which, in Sir Joseph Larmor's words, '... is perhaps the most original in physical science',[10] and as its publication effectively inaugurated the science of thermodynamics we must devote some time to a consideration of its contents and of their relevance to contemporary knowledge. The book has been described by several writers as 'popular', and in the sense that it is very clearly written and that mathematical arguments are consigned to footnotes this is certainly true. But it is not a popular account of the steam-engine like contemporary works, such as those of Stuart, Partington and Dionysus Lardner.[11] In fact it presupposes considerable knowledge of the steam-engine as well as of physics; in so far as it can be briefly summarised, it is an attempt to set out, as clearly and directly as possible, the new cosmology of heat *in terms of the dynamical properties of heat—not* the dynamical *nature* of heat, for Carnot at this time still accepted the caloric theory. The philosophical foundations of this new science extend from the field of astronomy right through geophysics, chemistry, meteorology and the physics of gases to the advanced technology of heat-engines.

Carnot begins his book by making his position abundantly clear. Heat is the cause of all the motions, the kinetic phenomena, on the earth. It causes the winds and all atmospheric turbulences, the formation of clouds at different altitudes, rainfall and other forms of precipitation as well as the great oceanic currents and such geophysical phenomena as earthquakes and volcanoes. Indeed, Carnot overstates the cosmological significance of heat, for he leaves no room for the then totally unsuspected agency of radio-activity. But, as history confirmed, he did not overestimate the importance of the heat-engine when he observed that there seemed to be no limit to the practical applications of this immensely powerful agency, which he feels sure will one day replace all other sources of power. He is impressed by the widespread utilisation of steam-power in England, and he is unstinting in his praise for the engineers—most of them British—who have developed the steam-engine up to its latest, high-pressure form. With profound insight he remarks that the steam-engine is now more important for England's existence than is her Navy—evidently his readings in economics and his comparative studies of industry and technology had not been a waste of time.

Nevertheless, in spite of all this scientific knowledge and technological development there is no *general* theory of the heat-engine, applicable to all conceivable forms of heat-engine, whatever their working substances or their mechanical principles. He observes:

> Machines which are not driven by heat, those which are driven by the power of men or of animals, by a fall of water, by the movement of air etc., can be analysed down to their last details by mechanical theory. Every event is predictable, all possible movements are in accordance with established general principles which are applicable in all circumstances. This, of course, is characteristic of a complete theory. A similar theory is obviously required for heat engines.

From this point onwards the argument is developed with great clarity and assurance, characteristic of a work of genius.

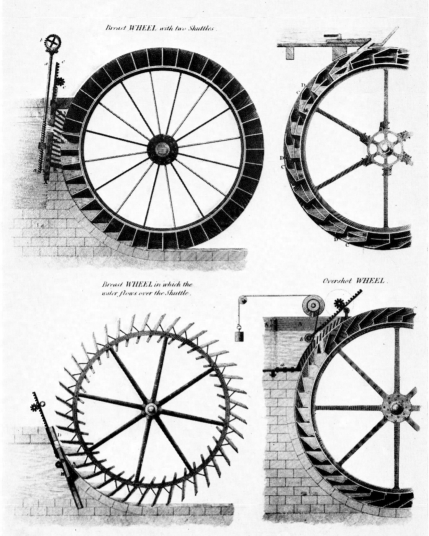

Breast WHEEL with two Shuttles.

Breast WHEEL in which the water flows over the Shuttle.

Overshot WHEEL.

Plate XIV. Early nineteenth-century industrial waterwheels designed to achieve the maximum possible efficiency. The top left-hand wheel is a 'suspension wheel' with very light spokes; it incorporates a 'feed-back' mechanism so that a governor coupled to the power output can control, via the two bevel wheels at the top, the flow of water through the 'shuttles'. These wheels should be compared with the Rossett wheel shown in Plate II. (From Rees' *Cyclopaedia*, 1819)

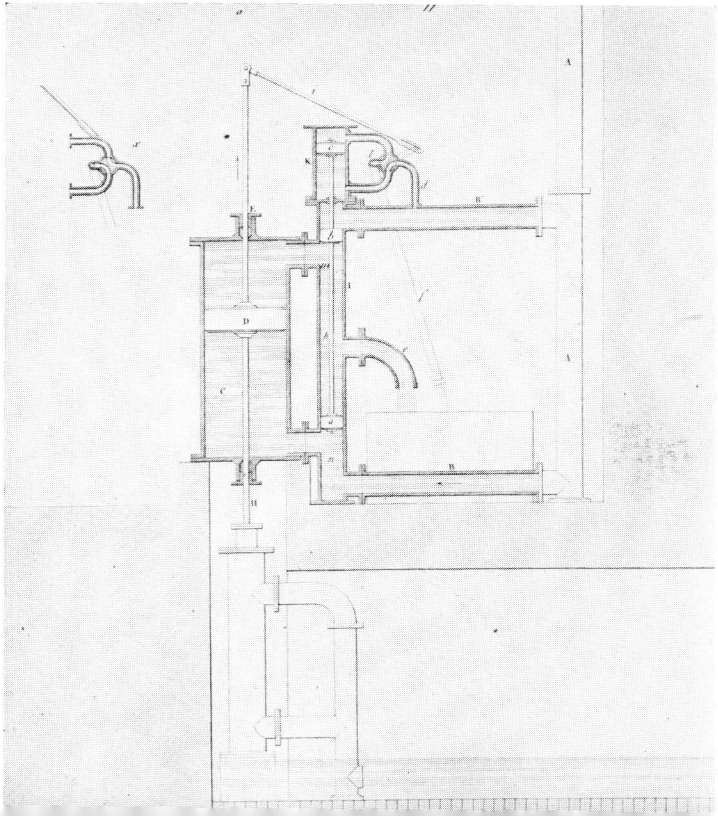

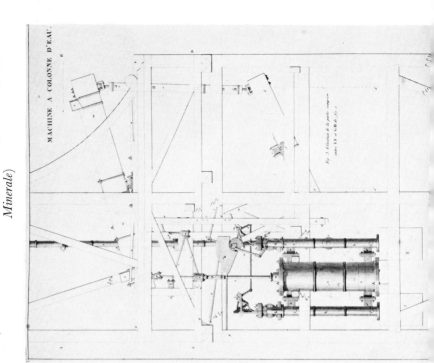

Plate XVI. A German double-acting column-of-water engine fitted with Watt's parallel motion. (From A. M. Héron de Villefosse, *De la Richesse Minerale*)

MACHINE A COLONNE D'EAU.

Plate XV. Diagram of a double-acting column-of-water engine. (From G. J. Christian, *Traité de Mécanique Industrielle*)

Plate XVII. George Manwaring's column-of-water engine of 1812.

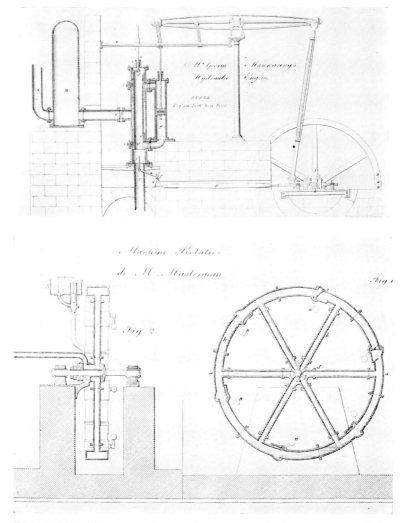

Plate XVIII. Masterman's rotative engine. A development of the engine invented by Amontons. (From G. J. Christian, *Traité de Mécanique Industrielle*)

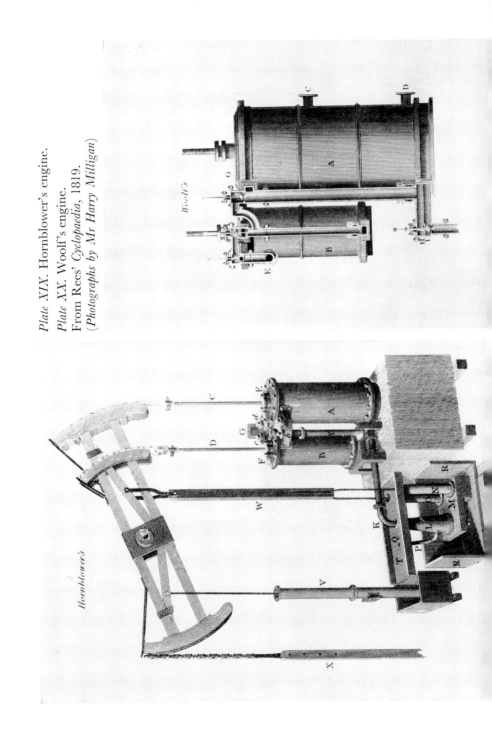

Plate XIX. Hornblower's engine.

Plate XX. Woolf's engine.
From Rees' *Cyclopaedia*, 1819.
(*Photographs by Mr Harry Milligan*)

In the annals of science it is almost unique, for it had no discernible predecessors and was built up from the assemblage of unordered opinions and problems, concepts, theories and measurements that were available at the time. Carnot, in short, was not standing on the shoulders of giants; he saw further than his contemporaries because he had a much clearer vision.

He begins by grasping and emphasising two essential points. The first one is that if you want to use heat to generate power you must have a cold body as well as a hot one. That is to say, a 'fall of caloric'—or a temperature difference—is absolutely essential. The water-power analogy is very close at this stage of the argument: a given quantity of water will only generate power if there is a lower level to which it can flow; a tail-race as well as a penstock. For Carnot, of course, the actual motive agent in a heat-engine, the analogue to the water in a hydraulic machine, is the highly elastic fluid, caloric.

The second essential point is that for the generation of the maximum amount of power there must at no point and at no time be a useless flow of heat. In other words, heat which could cause the working substance to expand and do work must not flow uselessly from the hot body to the cold one. This means that ideally there must be no contact between bodies at different temperatures during the operation of a heat-engine; a condition which, the logician will object, would make a perfect or ideal engine quite impossible. But this objection could, with equal force, be raised against Lazare Carnot's conditions for the maximum efficiency of a hydraulic engine; that the water enter the machine without shock or impact of any sort and leave it without relative velocity. Again the analogy between the working of an hydraulic engine and a heat-engine is a close one; the head or, since the one determines the other, the velocity of the water being the analogue of the temperature difference. In the case of the heat-engine Sadi Carnot answers the logician's objection by asserting that vanishingly small temperature differences are, for his purposes, sufficient to cause heat to flow without violating the prime condition that in a perfect heat-engine bodies in contact must be at the same temperature. Of course this removes the perfect heat-engine from the world of practical engineering to some Platonic heaven. But before we are led off into speculations about the metaphysics of infinitesimals let us see how the perfect but wholly impracticable Carnot engine works.

A perfectly smooth-bore cylinder is fitted with a frictionless piston. The sides of the cylinder are perfectly insulating while

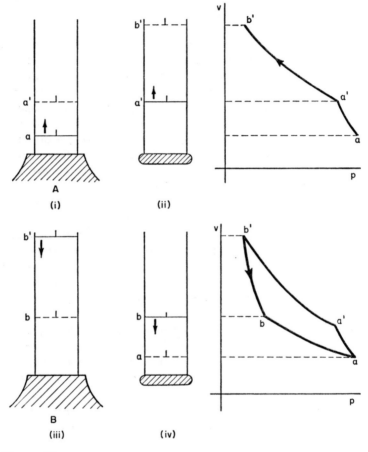

Figure 16

the base is a perfect conductor. We will suppose that the working substance in the cylinder is air, and that to begin with it is compressed and brought to the same temperature as the source of heat, A, with which the bottom of the cylinder is placed in contact (Figure 16). If the piston is now allowed to rise, so that the air does work as it expands, then so long as the cylinder is in contact with A it can draw the 'latent heat of expansion' from it so that the temperature of the air remains constant at t_a; this stage of the engine's *cycle* of operations is represented by (i) of Figure 16, and by the portion a–a' of the indicator diagram on the right-hand side.* When the piston has travelled a short distance to a' the source of heat A is removed and an insulating cap placed on the end of the cylinder. The air continues to expand, 'adiabatically' now since it is cut off from the supply of heat, and the pressure and temperature fall more steeply. This 'expansive' phase is allowed to go on until the piston reaches the point b', when the temperature of the air has fallen to that of the cold body or condenser, B; this is represented by (ii). The insulating cap is removed, the cylinder is placed on the cold body and pressure is applied to the piston. As the air is compressed it yields up heat to B: caloric is, as it were, squeezed out of it into the condenser. The temperature cannot rise—it remains at t_b—for B has, to all intents and purposes, an infinite heat capacity. The compression continues until the piston reaches the point b, by which stage as much heat has been squeezed out of the air as was absorbed during the initial phase, a–a'. The cold body is now removed, the insulating cap replaced and the compression continued, 'adiabatically' now, until the piston has returned to the point a when the air will be, as regards temperature, pressure and volume, back in its pristine state. Hence, Carnot infers, the total heat in the air will be exactly the same as it was at the very beginning, so that all the heat that was absorbed during the cycle will have been used to perform work, and to *perform work only*. The cycle having been completed, the operation may be repeated indefinitely if desired.

In the cycle of operations which we have just described the working substance, the air, was never at any time in contact with a body at a significantly different temperature. Even during the two 'adiabatic' phases the inside walls of the cylinder, being made of a perfectly non-conducting material, would need to absorb only a vanishingly small amount of heat in order to be at the same temperature as the expanding or contracting air. The condition for the maximum possible efficiency has evidently been satisfied.

* Carnot did not make use of the indicator diagram. In all probability he had never heard of it.

Apart from the boldness of the conception, the main features of this perfect engine operating on a strictly prescribed ideal cycle are the application of Watt's expansive principle (in itself a testimony to the penetration of Watt's original insight), the acceptance of temperature rather than pressure as the important parameter determining its efficiency and, lastly, the closeness of the analogy with hydraulic engines. Carnot leaves us in no doubt about the last point:[12]

> In accordance with the principles we have now established, we can reasonably compare the motive power of heat with that of a head of water: for both of them there is a maximum which cannot be exceeded, whatever the type of hydraulic machine and whatever the type of heat engine employed. The motive power of a head of water depends upon its height and the quantity of water; the motive power of heat depends also on the quantity of caloric and on what may be called—on what we shall call— the height of its fall, that is on the temperature difference of the bodies between which the caloric flows.

It is an interesting reflection that many writers interpret Carnot's hydraulic 'machine' as a reference to water-wheels, thus by implication oversimplifying and underestimating the water-power technology of that time. Carnot can hardly have avoided studying such machines as the column-of-water engine, and it is therefore reasonable to assume that since he thought in analogical terms when considering the operation of an ideal heat-engine, knowledge of hydraulic machines of this sort helped him to formulate his concepts.

So far, so good; but there are three points about the Carnot cycle, as he proposed it, that we must discuss. In the first place it is quite apparent that Carnot accepts the 'conservation of caloric'; indeed it is an axiom in his argument. This is not surprising, for calorimetry is based on the assumption that caloric, or at any rate heat, is always conserved: the validity of the method of mixtures for determining specific heats depends upon the principle of conservation being correct. Hence in the Carnot cycle the amount of heat absorbed by the air when expanding must equal the amount given up when it is compressed at the temperature of the cold body; in this way 'adiabatic' compression exactly restores the air to its original conditions as regards temperature, pressure, volume *and total heat*.

The second point concerns the two compression phases of the cycle. The unwary reader might wonder why it should be necessary actually to consume mechanical work in compressing the air; surely it would be more economical to put the air, at the temperature t_b, directly in contact with the hot body A and heat it up without doing work on it? But if we did this we should be violating the fundamental condition that the working substance must not be in contact with a body significantly hotter or colder than itself.[13] There would thus be a waste of power, a useless flow of caloric. If we assume the axiom of the conservation of caloric or heat we may imagine a small 'donkey' engine working between the hot body A and the cold air, and thus providing additional power, which would not be available if there were direct contact between A and the air.

There is nothing abstract or unrealistic about the compression phases. The starting point of the cycle, the high pressure and temperature, has to be 'paid for' in terms of work done on and/or heat imparted to the working substance; it is not a free gift of nature. The reader will be familiar with the 'compression' of the common internal-combustion engine, the motor-car engine, and will probably be well aware that the gas turbine, or jet-engine, has a compressor which is driven by the turbine. These are, of course, recent innovations, but even in Carnot's time the necessity of compression applied to the air-engine was well known; indeed air being so bulky it was, practically speaking, an essential requirement of such an engine. Thus in Philippe Lebon's engine (1801) the mixture of air and inflammable vapour was compressed before being ignited; in the 'Pyreolophorus' (1806) the combustion chamber is charged with air by means of a bellows before the inflammable material is ignited, while Cayley's engine (1807) made use of a cylinder and piston to compress the air (page 152, note 8).

In the case of the steam-engine, the most familiar form of heat-engine known to Carnot, the vapour is ejected into the atmosphere after it has done its duty and no compression is necessary. But of course heat was required to warm up and then vaporise the water that formed this steam so that, in the end, we find we have paid for the useful work obtained in the currency of heat rather than that of work (compression) plus heat. Indeed we have paid quite heavily, for one of the main defects of the steam-engine, recognised well before Carnot's time, was that much useful 'sensible' heat was wasted in the ejection of the exhaust steam into the atmosphere, or the condenser. A number of engineers, among them Trevithick, had sought to remedy this by using the exhaust steam to pre-heat the boiler feed-water.[14] This technique constituted, in effect, an implicit recognition that at the end of each cycle the original conditions must be restored as economically as possible; in Carnot's work the recognition is explicit and general, or common to all forms of heat-engine.

The third important point that we must discuss is the proof which Carnot derives from his highly abstract ideal cycle. As he specifies it, it is reversible. That is to say, if at any point the balance of pressures is slightly altered then the cycle of opera-tions is reversed; the engine will work backwards, absorbing heat from the cold body and imparting it to the hot body, and in doing this the engine consumes more work than it produces. The heat-pump and the refrigerator had not been invented in Carnot's time, although there had been one or two interesting suggestions (see page 151); but the conception of a reversible heat-engine is not very far-fetched if the hydraulic analogy is kept in mind. All engineers knew that virtually every kind of water-engine could be reversed in action, in which case the function became that of a pump. The column-of-water engine was no exception, indeed it was perhaps the most obvious example of reversibility imaginable, and many inventors— Trevithick among them—had designed engines of this sort which could, if desired, be run backwards as pumps. As long ago as 1752 Déparcieux had, as we saw, conceived an ideal reversible water-engine in the form of an overshot water-wheel and had used it to prove that no water-engine could be more efficient than such a machine.[15] From the time of Déparcieux and Smeaton, if not indeed from the time of Parent, it had become customary for engineers to denote the efficiency of a water-engine as the fraction or percentage which indicated the engine's capacity to restore or recover the initial situation. Thus 0·7 or 70 per cent efficiency meant that the engine delivered such power that, applied to a mechanically perfect pump, it could restore 70 per cent of the driving water to the source from which it had fallen; in other words some 70 per cent of the energy was recovered, and the remainder lost in friction and inelastic collisions. The ultimate limit was, of course, 1 or 100 per cent which implied perfection and complete recoverability. Beyond this figure one could not go, for perpetual motion (in the Galilean sense) would ensue, and such a thing was repugnant to common sense and to the very basis of the science of mechanics.

In much the same way Carnot has no difficulty in showing that a heat-engine working according to the cycle he prescribed, and of course mechanically perfect, is the most efficient possible. For let us suppose that there is an even more efficient engine; then if it is used to drive Carnot's engine backwards so that it acts as a heat-pump, all the heat 'let down' through the more efficient engine will be restored to the source. But the latter, by definition, will deliver more power than is required to drive the heat-pump. The consequence will therefore be the generation of a net amount of power without any net 'fall' of heat and this would, in fact, mean the production of perpetual motion. Thus no heat-engine can be more efficient than one which can be reversed in the way in which Carnot prescribes. It is obvious that the nature of the working substance cannot affect this conclusion; all such engines, no matter what the working substance —air, steam, alcohol-vapour or anything else—are necessarily of exactly equal efficiency.

We have emphasised Carnot's indebtedness to Watt for the invention of and the insight represented by expansive operation; he was also heavily indebted to Watt for the latter's invention of the separate condenser, which became the 'cold body' or 'heat sink' in Carnot's argument.* The need for and true function of the cold body in the operation of a heat-engine could not be clearly inferred from the workings of the Newcomen engine or of the subsequent Trevithick high-pressure, non-condensing engine. As we have already noted—it is important enough to bear repetition—Dr A. J. Pacey has pointed out that the invention of the condenser made it much easier to envisage the operation of the heat-engine in terms of a *flow* of heat from a hot to a cold body. But Carnot also owed much to the founders of the science of heat and to his contemporaries who were working in the field of the physics of gases. Finally, he was indebted to the hydraulic engineers for the concepts of efficiency and reversibility and for the relationship between

* See D. S. L. Cardwell, *Steam power in the Eighteenth Century* (Sheed and Ward, London 1963), p. 84.

them. It is, incidentally, an interesting speculation that the simple caloric theory, which Carnot accepted at this time, may well have made it easier for him to visualise the operation of a heat-engine as so closely analogous to that of a water-engine; both sorts of engine were, after all, driven by 'fluids' descending from a high 'level' to a lower one.

In short, virtually all the individual components used by Carnot in his postulation of the perfect engine working on an ideal cycle, and in the subsequent arguments he based on this invention, were available in one form or another at the time when he commenced his work. But they were unrelated, disordered, and it was his particular genius and judgement that enabled him to select the necessary components and assemble them into one coherent argument. In this he had no discernible predecessor. Bouvier had argued in terms of an ideal engine, expansively operated, and had seen that physics would have to be invoked to complete the specification of such an engine. But he was quite incapable of carrying his argument beyond this elementary stage and was apparently entirely unaware of the fundamental and general implications of the concept of an ideal heat-engine. Even more able men, such as Petit, Navier, Clément and Desormes, while they may have used the implicit notion of a perfect engine, were well off the mark. It was Carnot, and Carnot alone, who saw that only if the working substance is brought back, at the end of each cycle, *exactly* to its original condition can we be sure that all the heat which has passed into (and for Carnot, through) the engine has been used to perform work, and that none has been diverted to cause a net change in the condition of the working substance. Again, it was only Carnot who saw the importance of reversible changes: a cycle, each one of whose stages is strictly reversible, must represent the limit of efficiency of the performance of work by a given amount of heat; for if this were not so the absurdity of perpetual motion must follow.[16] But this is a limiting condition, one end of the spectrum represented by the dynamical properties of heat, as it were. The vast majority of thermal and thermo-mechanical changes are, as Carnot recognised, irreversible; the conduction of heat from a hot body to a cold body, the appearance (or as we should say, the generation) of heat by friction and percussion and many other phenomena are irreversible, for no slight change of the circumstances can cause the process to reverse itself. Finally, Carnot was unique in that he perceived that a cold body—or at any rate a temperature difference—was as necessary for the performance of work by heat as was a hot body, or a source of heat; and he noted, too, that the performance of work by a heat-engine is always accompanied by the flow of heat from the hot to the cold body. These last were profoundly important reflections, for in them, as re-interpreted by Clausius, lies the essence of the second law of thermodynamics.

Perhaps one of the truest indicators of Carnot's greatness is the unerring skill with which he abstracted, from the highly complicated mechanical contrivance that was the steam-engine (even as early as 1824), the essentials, and the essentials alone, of his argument. Nothing unnecessary is included and nothing essential is missed out. It is, in fact, very difficult to think of a more efficient piece of abstraction in the history of science since Galileo taught men the basis of the procedure.

CARNOT AND THE PHYSICS OF GASES

The very neat demonstration that between two reservoirs of heat at different temperatures there is a maximum amount of work that can be done by a given amount of heat, irrespective of the working substance through which it falls, had certain very important consequences for the physics of gases in particular and for the laws of heat in general.

Let us imagine a Carnot cycle in which the temperature difference between the hot and cold bodies is very small (Figure 17). We can thus ignore the two 'adiabatic' phases, so that the net work performed by the engine will be the difference between that done by expansion and consumed by compression. Now if we repeat the procedure using different gases as working substances, if the initial conditions at '*a*' are the same it must follow by the known gas laws that at every point on the cycle the gases will have the same pressure, volume and temperature, or in other words, they will all trace out the same indicator diagram shown on Figure 17. The net work done will therefore be the same, irrespective of the working substance, and since Carnot has shown that all reversible engines working between the same temperatures have identical efficiencies it follows that the amount of heat absorbed to perform this net work will in each case be the same. From this Carnot concludes that *all gases passing without change of temperature ('isothermally') from a given pressure and volume to another given pressure and volume (e.g. from a to a', or from b to b') absorb or release the same amount of heat.*

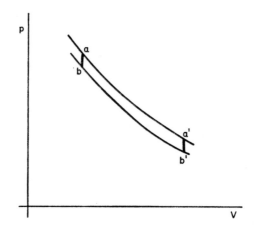

Figure 17

This extremely important law is very simply explained by the dynamical theory of heat: since all gases do exactly the same amount of work in expanding 'isothermally' from one given pressure and volume to another given pressure and volume, they must all therefore absorb exactly the same amount of energy *in the form of heat* in order to do this. But Carnot still accepted the caloric theory and was therefore in agreement with the vast majority who, whether they believed in material caloric or not, subscribed to the axiom of the conservation of heat. This axiom led directly to the common-sense view that the amount of heat required to bring a given weight of gas from one condition (pressure, volume, and temperature) to another depended solely on the temperature difference and the specific heat of the gas. That is to say, the amount of heat, or caloric, imparted to the gas was believed to be quite independent of the pressure–volume curve the gas followed as it heated up: it was the end-product that was important. As we saw, the 'adiabatic' heating or cooling of a gas, or conversely the emission or absorption of heat during an 'isothermal' change could, on this theory, be explained only by assuming that the specific heat by weight of all gases increases with their volume. A good deal of empirical evidence tended to support this view, and the authoritative experiments of Delaroche and Bérard were held to have established it beyond reasonable doubt.

Carnot's subsequent development of his thermo-mechanical ideas is therefore conditioned by his acceptance of the axiom of the conservation of heat and its experimentally confirmed entailments. Thus he deduced a simple numerical relationship between the two principal specific heats of air,[17] and his thermo-mechanical law enabled him to generalise it to cover all gases. He asserted that the difference between the volumetric specific heat at constant pressure and the volumetric specific heat at constant volume is the same for all gases, and he showed that the ratio of the two principal specific heats of air, calculated according to his theory, is very close to the experimental result found by Gay-Lussac and Welter. He went on to prove, un-exceptionably, that when a gas changes its volume 'iso-thermally' the amount of heat absorbed or released is pro-portional to the logarithm of the ratio of the volumes after and before the change.* Unfortunately he goes on to prove, consistent with the accepted theory of heat, that the specific heat of a gas changes in exactly the same way.

The 'adiabatic' heating of a gas by compression can, according to the axiom of conservation, be analysed into two equivalent operations: the release of heat by compressing the gas 'isothermally' to exactly the same extent followed by the restoration of this heat to warm up the gas at constant volume.

---

* i.e. $Q = \text{Const. Log} \dfrac{v_2}{v_1}$. The form of the expression for the specific heat is the same. Carnot's expression for the rise in temperature of a gas on 'adiabatic' compression is, accordingly, $t = \dfrac{A + B \log \dfrac{v_2}{v_1}}{A' + B' \log \dfrac{v_2}{v_1}}$

The resulting rise in temperature, he argues, is given by the quantity of heat released divided by the (reduced) specific heat of the compressed gas. The resulting expression is, of course, very different from that deduced by Poisson.

Thus while Carnot is able to indicate very clearly the relevance of his thermo-mechanical ideas to the physics of gases and to the general laws of heat, so that some of the basic principles of thermodynamics are foreshadowed, he is encumbered by the established doctrines of heat, so that his 'scientific' as opposed to his 'technological' insights and deductions had to be subsequently revised when the associated doctrines were superseded. We shall not therefore spend any more time discussing the former but, like Carnot himself, return to a consideration of the heat-engine.

Having shown that the motive power of heat was independent of the working substance used in the engine and that under ideal conditions it could depend only on the temperature difference between the hot and cold bodies, Carnot had now to consider the nature of this dependence. Will a fall of caloric, or heat, from (say) 100° to 99° produce the same amount of work as one from 1° to 0°? Common sense—and the hydraulic analogy—suggests that the answer is affirmative, but Carnot is more cautious. As the specific heat of gas increases with its volume the conclusion may well be unwarranted. Let us take a given mass of air at 1°C and with a volume v_1, and heat it up so that its temperature becomes 100° and its volume v_2. We can, in fact, ring an infinite number of changes on the way we choose to do this, but let us select for our discussion two ways which are both equally simple. We can either expand the air 'isothermally' at 1° until its volume is v_2 and then heat it up at constant volume until it reaches 100°, or we can begin by heating it up at constant volume and then expand it isothermally to v_2 (see Figure 18). According to the axiom of conservation the heat necessary to do this must be the same in both cases, for it is determined solely by the temperatures and specific heats of the air at v_1 (1°) and v_2 (100°). So we have, according to Figure 18, $q_1 + q_2 = q_3 + q_4$. But as v_2 is greater than v_1 it follows by the established doctrine that the specific heat at v_2 must be greater than it was at v_1, and therefore since q_2 must be greater than q_3, necessarily q_4 must be greater than q_1. Or in other words a gas absorbs more heat when it is ex-

panding 'isothermally' at 100° than when it is expanding by
exactly the same amount at 1°. But these unequal quantities of
heat will produce *equal* amounts of work for falls from 100° to
99° and from 1° to 0°. This can easily be proved by applying
the argument set out on page 202 (Figure 17). Let us suppose
that after the air has expanded 'isothermally' at 1° it is put in
contact with a very slightly cooler body at, say, $(1-h)°$, and
then compressed 'isothermally' back to v_1. The net work done
will be that produced by the expansion less that expended to
restore the original condition. If the air is now heated, at

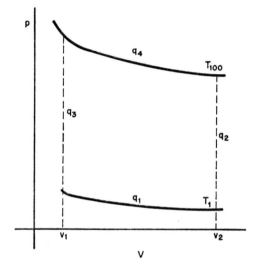

Figure 18

constant volume, up to 100° we can take it through an exactly
similar cycle by means of the hot body at 100° and a 'cold'
body at $(100-h)°$. It is very easy to show that the net work done
will be exactly the same as in the first case; but since more heat
will have passed through the working substance in the second
case we must conclude that the fall of heat, or caloric, produces
more work over the lower ranges of the temperature scale than
it does over the higher.[18] This conclusion was drawn from an
argument based on the false assumption of the conservation of
heat and from misleading experimental evidence that the

specific heat of a gas increases with its volume; nevertheless the conclusion happens to be correct.*

In the remainder of his short book Carnot estimates, from the data available, the work done by applying the same quantity of heat to air, steam and alcohol vapour, and he concludes that (within the limits of experimental error) it will be the same in all three cases: which is consistent with his abstract argument. He is unable to extend his discussion to a consideration of the mechanical effects of heat on solids, liquids and fusion, compared with vaporisation, but at least he says a few words about the rather obvious technique of harnessing the 'irresistible' force of expanding metals. An engine based on the last principle would, in its operation, violate the fundamental requirement that there must be no useless flow of heat: 'expansive' operation using a solid or liquid working substance is unthinkable. He then mentions the objections put forward by Petit, although he does not refer to the latter's short but conclusive demonstration of the fallacy involved in Pattu's engine.

Carnot finishes his book by recapitulating the requirements for the most efficient working of heat-engines: the maximum possible 'fall' of caloric and expansive operation, so that the temperature of the working substance is lowered 'adiabatically' by the amount of this fall and there is therefore no useless 'flow' of caloric, or heat. In practice these requirements can hardly be met—one can only approximate to them—but they do bring out, as nothing else at that time did, one of the cardinal problems of the steam-engine. The potential fall of heat from the temperature of the burning coal in the furnace to that of the cold water in the condenser amounts to about 1,000°C or more. But engines then in use working at the highest steam pressures—and therefore temperatures—could utilise a fall of heat of, at most, about 120°; that is to say, from about 160°, the temperature of steam at about six atmospheres pressure, to

---

* As Carnot observed, a given quantity of heat will do more work in falling from 1° to 0° than in falling from 100° to 99°. This should be compared with the modern formula: $W = J(Q_1 - Q_2) = J \cdot Q_1 (T_1 - T_2)/T_1$, which expresses the work done by the *conversion* of heat. This depends on the temperature difference $(T_1 - T_2)$ and diminishes as the (absolute) temperature of the hot body increases. Just as Carnot had proved from quite erroneous assumptions.

about 40°, the usual temperature of the condenser. Thus out of a potential fall of 1,000° we make use of only about 120° at best. And Carnot goes on to observe:

> It is easy to see now why so-called high-pressure steam engines are better than low-pressure ones: *their advantage lies essentially in their ability to utilise a greater fall of caloric.* Steam generated at higher pressure is also at a higher temperature and as the temperature of the condenser is nearly always the same, the fall of caloric is evidently higher.[19]

This was an extremely perceptive observation; whether he was strictly justified in making it is another matter, which we shall discuss in a moment. For the present we note that he goes on to describe, correctly, the discovery of the expansive principle in 1769 and its much more recent application in the Horn-blower and Woolf-type engines. He then goes on to point out that, contrary to established doctrine, contrary indeed to the promptings of superficial 'common sense', the fact that the pressure exerted by steam increases much more rapidly than the temperature is a *positive disadvantage.* Long before there is any chance of utilising the total fall of heat from the burning coal to the cold condenser the pressure of the steam has become dangerously great. On the other hand the pressure exerted by air does not increase so drastically with the temperature, and if the practical disadvantages of air as a working substance—we have mentioned them already—can be overcome then it has a significant advantage, thermodynamically, over steam in that it can be used over a much wider range of temperatures without blowing the engine to bits. This astonishing insight of Carnot's was the direct source of such high-efficiency internal-combustion engines as that of Rudolf Diesel, invented much later in the nineteenth century.

We can, in fact, assert that without an appreciation of Carnot's dictum the history of the heat-engine must be *un-intelligible.* The tremendous expansion, or the great increase in pressure, which follows the conversion of water into steam meant quite simply that here was a working substance which could generate a usable amount of power without, at the same time, making unreasonable demands on the skills of engineers

and metallurgists. The fact that, between the times of Newcomen and Woolf, the process was thermodynamically 'inefficient' was unimportant, indeed meaningless, for there was no
yardstick, economic or scientific, against which its economy
could be measured.

After Carnot's analysis the virtue of steam as a working
substance—the easy generation of high and therefore immediately usable pressures—was seen to be a drawback in that it
made it much more difficult to exploit the full temperature
range available. Everything that has happened since 1824 has
confirmed Carnot's judgement, and we must therefore divide
the history of the heat-engine into two very distinct periods:
before and after Carnot.

Let us now return to Carnot's explanation of the superior
economy of the high-pressure condensing steam-engine. The
fact of superior economy had been proved beyond reasonable
doubt by the Cornish engineers; sound practical reasons for it
had been proposed by authorities as diverse as Poisson and the
Lean brothers; thermodynamic factors, as we demonstrated in
the case of the Fowey Consols engine, were of minor importance
compared with the evolutionary, 'Smeatonian', ones: more
efficient boilers and furnaces, superior valve designs and generally higher engineering standards. Carnot, then, was not
strictly justified in ascribing the superior performance of high-
pressure steam-engines to the fact that they could utilise greater
falls of heat. His grounds must therefore have been on the one
hand his intuition, and on the other the analogy of the hydraulic
engine. In the case of the latter it was self-evident that the
greater the proportion of the 'head' that the engine could
utilise the more efficient it would be; this was especially true
of the high-pressure column-of-water engine, which was in fact
often called a 'water-pressure engine'. In the case of the former,
we must remember that the scientist usually has to frame his
most important hypotheses in the light of his intuition; if he
waits until he has absolute certainty on all sides then either he
will wait for ever or, more likely, he will find that some one has
jumped in before him.

Réflexions was the only work that Sadi Carnot published.
After his death from cholera in 1832 a small collection of his
scientific notes was found and, much more recently, a few other

unpublished writings have turned up.[20] These go some way to explain his failure to publish anything after 1824. It is quite clear from the notes found after his death that the accumulation of evidence against the caloric theory, coupled no doubt with his own deep insight into the processes of the heat-engine, convinced him not merely that the caloric theory must be false but that the whole axiom of the conservation of heat must go.[21] In flat contradiction of the doctrine set out by Lavoisier and Laplace in 1783 (see pages 62–3) and subsequently accepted by virtually everyone, whether calorist, agnostic or kineticist, he wrote:[22] 'Heat is nothing more than motive power that has changed its form ... Wherever motive power is destroyed ... heat is produced in a quantity exactly proportional to the motive power lost; conversely, wherever there is destruction of heat, there is production of motive power.'* He then goes on to state the principle that is now known as that of the conservation of energy: 'We can therefore lay down the axiom that the quantity of motive power in nature is invariable and that it is, strictly speaking, never created and never destroyed. In fact it changes its form, sometimes manifesting itself as one kind of motion, sometimes as another, but it is never annihilated.'

In his notes Carnot sketched out a series of researches to determine the exchange rate at which heat is converted into motive power and vice versa. He proposed to repeat Rumford's metal-boring experiment using a variety of different metals and—something Rumford never thought of—to measure in each case both the amount of work done and the quantity of heat generated. He then intended to measure the thermo-mechanical exchange-rate by agitating different liquids by means of a perforated piston moving in a calorimeter, and also by experiments on the 'adiabatic' compression and expansion of gases. He was evidently confident that these comprehensive experiments on very different substances would all yield exactly the same figure for the thermo-mechanical exchange-rate. Indeed, he can compute a rough value for it—he does not say how—from the data already available: 'According to my ideas about the theory of heat, the production of one unit of motive power requires the destruction of 2·70 units of heat.'[23]

* This is correct, strictly speaking. But when heat flows from a hot to a cold body by conduction there is no production of motive power. This was to be an important issue in the later development of thermodynamics.

H

We can easily envisage the difficulties that these revolutionary and far-sighted ideas must have caused Carnot. If he discussed them with his friends, as in all probability he did,[24] they would surely have been shocked by the suggestion that heat can be *destroyed*;* this would be turning the clock back with a vengeance and calling in question everything that had been achieved since Lavoisier and Laplace had first put the science on a systematic basis. The subsequent experiences of men like Mayer and Joule were to prove that scientists were very reluctant indeed to face the possibility that heat might not be conserved but might be convertible into other forms of energy.

For Carnot there must have been the additional and probably agonising problem of the validity of his one published work. On the new theory of heat certain arguments used in *Réflexions* were plainly wrong: what parts, if any, of the work could be saved from the wreckage of the old theory of heat? Carnot was, we infer, far too good a scientist not to realise that certain important sections of the *Réflexions* were valid, and no less fundamental in terms of the new theory of heat. But to have reconstructed his theory on this basis would have been well beyond any one man's capabilities; in the actual course of events the task required the combined efforts of Clapeyron, Joule, Rankine, Clausius and Kelvin with the assistance of other, relatively minor scientists. If Carnot could have achieved all this it would have been much the same as if Galileo had rounded off the *Two new sciences* by writing *Principia mathematica* and *Opticks*.

Let us in conclusion recapitulate briefly the main sources of Carnot's ideas. These, we believe, were to be found in the advances made in the physics of gases, especially in the study of 'adiabatic' phenomena; in the *convergence*, as we have called it, of the technologies of water- and steam-power and, lastly, in the very dramatic improvements which the Cornish engineers —including such expatriates as Edwards—were then making in the performances of steam-engines.

Recently two important papers, which add considerably to our understanding, have been presented by M. Jacques Payen and Dr Robert Fox.[25] M. Payen has called attention to the

---

* The importance of the axiom of conservation at this stage of the history of the theory of heat is well brought out in T. S. Kuhn's important paper, 'The Caloric Theory of Adiabatic Compression', *Isis* (1958), xlix, p. 132.

course of instruction—ostensibly in applied chemistry—which was given at the Conservatoire by Carnot's friend Nicholas Clément. This course, which began in 1819, was concerned with the theory of heat (based of course, on the caloric doctrine) and the theory and operation of the steam-engine.* Dr Fox on the other hand has shown that Carnot was indebted to Clément and Desormes for the fundamental insight, propounded in their paper of 1819 (see pages 171–2), that ideally the *temperature* of the working substance must fall during the expansion to that of the condenser or cold body: that is, of course, that the expansion must be 'adiabatic'. Dr Fox points out that only Clément and Desormes, and following them Carnot had realised this.

We have now assembled the main sources of Carnot's ideas. Although further researches and the changing perspectives of the study of history in successive generations will no doubt modify our views, the picture is now tolerably complete and fairly convincing. The image of Carnot as a wholly isolated figure without predecessors or precedents of any sort has been dissolved. Nevertheless in doing this we have not explained the thought processes whereby he came to make his extraordinary synthesis. Indeed we cannot do so, for he was one of the most truly original thinkers in the whole history of science. It might not be wholly inappropriate to say that in the independence of his spirit he was English, and in the rigour and clarity of his mind he was French; but in the depth and scope of his insight he transcends all such classifications.

AFTER CARNOT

Réflexions received a long and favourable notice in the *Revue Encyclopaedique* and was mentioned in *Bulletin des Sciences Technologiques*, but for the rest it was simply ignored in France, Britain and elsewhere. When, many years later, the young and

---

* Lecture notes, taken by a long-forgotten student, J. M. Baudot, give us some idea of the contents of this course, which dealt with such relevant matters as the principles of expansive operation and the problems of boiler feed-water.

enthusiastic student William Thomson (Kelvin) tried to buy a copy of the book he found that the Paris booksellers had never heard of it, or its author; the only Carnot they knew of was the father, the 'Organiser of Victory', Lazare Carnot.

It is not difficult to understand why *Réflexions* was a failure. To the engineer, the postulation of the ideal engine and the reversible cycle must have seemed closer to the more imaginative passages in the *Timaeus* than to the needs of power-engineering. Christian, Poncelet and others had advised French engineers to keep practical requirements always in mind and to avoid being too abstract and theoretical. Always there was the example of the Cornish engines, whose extremely high standards of performance had been achieved by careful attention to details and whose designers seemed to have made little use of abstract theory. In fact, Carnot had very little to teach the engineers of his time: the superiority of the high-pressure condensing steam-engine had been established, so had the advantages of expansive operation, and by 1824 engineers knew well enough that, all things considered, there was no better working-substance than steam. We must conclude that Carnot's message was essentially for engineers of a later generation with the resources of a more advanced technology.

On the other hand Carnot's ideas were general or cosmological in their scope and therefore, in principle, relevant to physics, chemistry and meteorology. In these fields however his mode of reasoning would, one infers, be unfamiliar and thus perhaps unacceptable to those who had had little or no experience of power-engineering. In sum, the *Réflexions* fell—inevitably, when one recalls its scope—between a number of specialist stools: the recognition of its technological importance lay in the future, while for the physical sciences the clarification and demonstration of its relevance called for the further, and highly original, work of Clapeyron and Kelvin (in physics), Gibbs (in chemistry) and Napier Shaw (in meteorology).

The years that followed the publication of Carnot's book appear to have been dull, anticlimatic, by comparison with those that preceded it. But this valuation is based on the wisdom of hindsight. For people of that time the progress of science was, according to temperament, as satisfactory or as unsatisfactory after 1824 as it had been before. In fact, Carnot's book made hardly a ripple on the surface of the main stream of science.

In 1826 the quest for a definition of quantity of heat was effectively ended, when Nicholas Clément postulated the unit of heat as the 'calorie'; this he took to be the quantity of heat required to raise one kilogramme of water by 1°C and, as in the case of Hachette's 'dynamode', it was intended to be a practical unit for use by engineers.[26] But the invention of the 'calorie' was a comparatively minor matter; much more important were the continuing studies of the specific heats of gases, and the work on radiant heat carried out by Melloni and Forbes.

Gay-Lussac's method of mixtures for measuring the relative specific heats of gases (see page 135) was revived and used by W. T. Haycraft in a series of experiments carried out in 1823.[27] Haycraft reported that the *volumetric* specific heats at constant pressure of the gases he studied were all the same. However, the validity of his and indeed all determinations of specific heats at constant pressure—including those of Delaroche and Bérard—was soon challenged by the Genevese physicists de la Rive and Marcet.[28] They insisted that the 'latent heat of expansion' absorbed when a gas expands at constant pressure cannot be regarded as part of the 'real' specific heat, which by definition is the heat required to raise the temperature of the gas by one degree and not the heat required to change its physical dimensions. Therefore, according to de la Rive and Marcet the specific heat of gas must be determined at constant volume, and in their experiments the increase in pressure of the gas was used to measure the rise in temperature—the gas thus acting as its own thermometer.

De la Rive and Marcet claimed that all the fourteen gases they studied had exactly the same *volumetric* specific heat at constant volume, and they confirmed that the *volumetric* specific heat of gases fell off as the pressure was reduced.* They found also that all gases had different conductivities.[29]

Unfortunately whether or not the specific heat at constant

---

* But as they asserted that the *volumetric* specific heat diminished less rapidly than the pressure it followed that the specific heat by weight would increase as the pressure was reduced or the density increased; they thus confirmed Delaroche and Bérard's finding, and the accepted doctrine of virtually every scientist at that time.

volume is more 'fundamental' than that measured at constant pressure, it is certainly much more difficult to measure in practice. Indeed, accurate direct determinations of the specific heats of gases at constant volume were not made until Joly's experiments in 1890.[30] An able and very experienced scientist like Dulong therefore had a comparatively easy task when he applied himself to a critique of de la Rive and Marcet's experiments.[31] From his extensive experience he was able to show that de la Rive and Marcet had almost certainly measured not so much the specific heats of gases but some combination of their specific heats and conductivities.

Dulong's criticisms were not, however, merely destructive; they were an integral part of a paper which is of great historical importance. To understand this we must pause for a moment to consider the development of the atomic theory in chemistry. It is well known to historians of science that the publication of Gay-Lussac and Humboldt's law, which states that when gases combine chemically they do so in volumes which bear a simple ratio to one another and to the product if gaseous, led directly to the formulation of Avogadro's hypothesis.* It is equally familiar knowledge that although Avogadro's hypothesis was published in 1811 it was effectively ignored by chemists until 1858, when Cannizzaro convinced them of its validity and merits. But although Avogadro's hypothesis was ignored the implications of Gay-Lussac and Humboldt's law could hardly be overlooked. Accordingly we find that by the 1820s it had become usual for chemists to accept the proposition that equal volumes of *elementary* gases at the same pressure and temperature contain equal numbers of atoms. With this in mind we can conveniently summarise the knowledge available to Dulong in 1828 in the form of two propositions.

Firstly: since equal volumes of elementary gases contain equal numbers of atoms, *ceteris paribus*, it follows that the weight of a given volume—i.e. the density—of such a gas must be proportional to the atomic weight. But Dulong and Petit had already shown that the 'atomic heat', or the product of the atomic weight and the specific heat, is constant in the case of elements, including gases; it therefore follows that the products of the densities and specific heats by weight of elementary gases

---

* i.e. that under the same conditions of temperature and pressure equal volumes of all gases contain equal numbers of *molecules*.

must also be constant. But the latter are merely the *volumetric* specific heats.

Delaroche and Bérard's measurements had indicated that the *volumetric* specific heats of different gases were not the same (see page 136), but they had not distinguished between 'elementary' and compound gases. Their measurements indicated that the differences between the *volumetric* specific heats of 'elementary' gases were so small that they could be ascribed to experimental error. Dulong therefore concluded, reasonably enough, that the *volumetric* specific heats at constant pressure of 'elementary' gases are all the same.

Secondly: it could be shown by means of the modified Clément and Desormes' method that the ratio of the two specific heats, Cp/Cv, was constant in the case of the 'elementary' gases, the value being about 1·4, but became smaller, tending towards 1, with the compound gases.[32] Combining this with the first proposition it follows that the *volumetric* specific heats at constant volume must, in the case of the 'elementary' gases, all be the same.

It was universally accepted by this time that the difference between the two specific heats was simply accounted for by the 'latent heat of expansion', and the straightforward deduction from these two propositions is, therefore, that when equal volumes of 'elementary' gases are compressed to the same extent they must release exactly the same quantities of heat.

The argument does not apply in the case of compound gases but, observes Dulong, it is very probable that the conclusion still holds. For one thing the reduction in the ratio Cp/Cv suggests that compound gases have greater specific heats; and this, together with the fact that their temperatures rise less on 'adiabatic' compression, is consistent with Dulong's general conclusions:

(1) Equal volumes of all gases at the same temperature and pressure suddenly compressed to the same extent release the same absolute quantities of heat; and conversely in the case of expansion.

(2) The resulting temperature-changes are in inverse ratio to the specific heats at constant volume.

These conclusions, published widely and backed by Dulong's considerable authority, were strong and timely support for the rapidly fading arguments of Carnot who, we remember, had deduced the same laws from the first principles of his own science of thermodynamics. They were also, paradoxically enough, taken by Joule to be strong supporting evidence for his dynamical theory of heat: paradoxically, for in Carnot's case the arguments had been based on the caloric theory while Joule, on the other hand, invoked Dulong's laws in confirmation of his axiom that the mechanical energy expended (in compressing the gases) must be equal to the heat energy into which it is converted.

Further enlightenment on the nature of heat was to come, as so often in the past, by an indirect route. In 1826 T. J. Seebeck discovered that if one junction of a bi-metallic circuit was heated while the other was kept cold an electric current would flow so long as the temperature difference was maintained. The 'Seebeck Effect' constituted an extremely sensitive thermometric device; in 1830 the Italian physicist Nobili used it as the principle of his 'thermo-multiplier', which consisted of a series of antimony-bismuth junctions so arranged that while half of them were exposed to heat the other half were screened. The exposed junctions were blackened to improve absorption and the current flowing in the circuit indicated by means of a galvanometer.

The introduction of a super-sensitive heat-detector like the thermo-multiplier put the study of radiant heat on a much higher plane of experimental accuracy. This was made quite clear by the series of elegant researches carried out by Nobili's compatriot, Melloni, in the early 1830s.[33]

These skilful investigations confirmed, accounted for and extended the discoveries of Herschel, Delaroche and Seebeck. Melloni found that a prism of, say, crown glass would affect the radiant heat from the sun in two distinct ways: it refracted it into a heat spectrum roughly co-extensive with the luminous spectrum, but at the same time it absorbed the *less*-refracted rays disproportionately. In the case of a water prism the absorption of heat rays at the red end of the spectrum was even more pronounced, so that the point of maximum heat-intensity was shifted to the yellow region of the luminous spectrum. Melloni, in fact, had found that transparent substances affect radiant heat in much the same sort of way that coloured glasses affect light. Transparent substances will readily pass heat of the appropriate refractive index but will progressively absorb rays whose refractive indices differ increasingly from that of the calorific 'colour' of the substance. Indeed, of all the transparent substances only one—rock salt—was 'colourless' in that it transmitted all heat rays equally well. It was, as Melloni put it, perfectly 'diathermanous'. Using rock-salt prisms and lenses Melloni found that there was an exact analogy between the laws of propagation of radiant heat and those of geometrical optics. He went on to show that different sources of heat produced rays of different refrangibility; the hotter the source the more the rays were refracted. He concluded that the observed differences between the transmission properties of terrestrial and solar radiant heat were due only to the different proportions in which differently refrangible rays were mixed.

Melloni had failed to detect the polarisation of radiant heat. But in 1834 Forbes, who had learned of Nobili's thermo-multiplier from Quetelet some two years earlier, was able to detect the polarisation of radiant heat after transmission through tourmaline and after reflection by mica.[34] This disposed of Baden-Powell's red herring and re-established Bérard's original discovery. Melloni confirmed these results and was able to show that the laws of polarisation by reflection and by refraction were the same for radiant heat as they were for light.[35]

As the fourth decade of the nineteenth century progressed, the similarities between light and radiant heat became more widely recognised and more firmly established, thanks to the experimental work of physicists like Melloni and Forbes. At the same time, the Young–Fresnel undulatory theory of light received very strong confirmation when Rowan Hamilton used it to predict the conical refraction of light emerging from a crystal, and Humphrey Lloyd very soon afterwards (1833) confirmed experimentally that the effect did, in fact, occur.[36] The logic of the situation was apparent: if light was undulatory in nature, then so also was radiant heat; but if heat can be propagated by means of an undulatory mechanism, implying a highly elastic universal aether, the material theory of heat must be wrong and only a dynamical theory—or theories—can be correct. Ampère expressed the point very clearly indeed when he wrote, in 1835: 'Thanks to the work of Young, Arago and Fresnel it has been clearly shown that the phenomena of light originate from the vibrations of a fluid filling all space and called aether. Radiant heat, which obeys the same laws of propagation can be explained in the same way.'[37]

Ampère had developed, from Gay-Lussac and Humboldt's law, his own independent version of Avogadro's hypothesis. This provided him with a model with which to account for the nature of heat and the mode of its propagation through material bodies. The ultimate chemical particles are molecules, which consist of two or more atoms held in position by a balance of inter-atomic attractive and repulsive forces. The vibratory movement of the molecules, as wholes, accounts for the propagation of sound; the vibrations of atoms, so that the molecules themselves distend and contract, accounts for the heat in, and the propagation of heat through material bodies. These atomic vibrations can set up oscillations in the surrounding aether and these, in turn, constitute the radiant heat emitted by a hot body.

The difference between light and radiant heat, which are both oscillations in the aether, is, observes Ampère, merely that they have different frequencies. The radiation of heat by a hot body consists essentially in the imparting of *vis viva* by the oscillating atoms to the aether, and in giving up its *vis viva* in this way the body cools down: it radiates its heat.* The converse process takes place when radiant heat falling on a cold body causes it to warm up.

This magnificent paper foreshadowed some of the important points in the dynamical theory of heat, and it earns for Ampère a place as one of the founders of that theory. Yet there were still some difficulties in the way. Thus, the destructive interference of radiant heat still had to be demonstrated, and no one had devised a way of measuring the velocity of radiant heat. An ingenious attempt to determine the latter was made by von Wrede in 1842.[38] He argued that if the velocities of light and radiant heat were different then the luminous and calorific aberrations of the sun would be different; that is, the luminous and thermal images of the sun would not overlap but would separate along the ecliptic. Von Wrede claimed to have found such a separation and from it he deduced that the velocity of

---

* This, of course, was the answer to the dilemma that caused Rumford such great difficulties. Ampère's hypothesis assumes that light and radiant heat are, by nature, mechanical oscillations in an elastic aether; the energy dissipated by radiation would be therefore mechanical and not, as on Maxwell's theory, electromagnetic.

radiant heat was about 0·8 that of light. But this order of difference was too small to base any really firm conclusions on and within the limits of experimental error it could have been argued that von Wrede had shown that the velocities of light and radiant heat were about the same.

All things considered it was still possible for an ultra-cautious man to argue that the material theory of heat had not been *completely* refuted—as indeed Kelland argued—although the objections to it were now formidable in the extreme. But whatever men thought about the nature of heat they were very unlikely to question the axiom, considered to be basic, of the conservation of heat and still more unlikely to speculate about the consequences of that axiom being discarded.

There is a certain agreeable irony about this, for in England, at any rate, the steam-engine was now the main source of industrial power—the very life of the country depended upon it—and all over the industrial areas steam-engines were busily at work doing what all the philosophers denied was possible: converting heat into mechanical energy. The steam-engine had in fact settled down to a comfortable, profitable and not very exciting middle age. Jacob Perkins' premature invention of the uniflow principle had been forgotten, while the steam turbine and the internal-combustion engine were more than fifty years in the future; steady, evolutionary improvement was the order of the day.

In response to the new stability men sat down to write long treatises on the steam-engine, confident that their books would not be out of date as soon as they were published. Prominent among these early treatises were those by Thomas Tredgold,[39] whose book was competent, detailed and dull, and by John Farey, whose work was scholarly, exhaustive and a mine of information for later writers.[40] Both writers were, in terms of insight, immeasurably inferior to Sadi Carnot. But Farey did raise one very interesting problem.

He defined the mechanical power of steam reasonably enough as the product of the pressure and the space through which it acts. As both the pressure and the stroke can be sizeable it followed that great power can be developed easily enough. But the *energy* [sic], which Farey defines as the power communicated to a gas or vapour to produce its motion, is measured by the product of the weight of the elastic fluid and the square of its velocity: i.e. its *vis viva*, And this has the interesting consequence that: 'The weight of an elastic fluid is so small that the energy of a quantity which exerts a very great mechanical power is but trifling.'[41] This explicit denial of equality between energy on the one hand and work, or power, on the other indicates how essential the kinetic theory of gases was for the establishment of the principle of the conservation of energy. It also reveals the basic difference between the old engineering, of which Farey was an exponent, and that of the present day.[42]

THE CONSEQUENCES: AN OUTLINE OF THE LATER HISTORY
OF THERMODYNAMICS

EMILE CLAPEYRON

Emile Clapeyron (1799–1864) was an engineer, a graduate of the Ecole Polytechnique, whose interests seem to have been in the theory of structures and in civil engineering generally. His posthumous distinction however rests on one paper, published in 1834, in which with superb timing (as it transpired) he rescued Carnot's ideas from the oblivion into which they had fallen and re-stated them with such conviction and clarity that they aroused the interest of the perceptive minority that exists in every generation. More, he made some highly original contributions to the science—thermodynamics—which Carnot had founded, for all that he, like Carnot, subscribed to the axiom of the conservation of heat.

One of his most important contributions is made quite clear at the very beginning of the paper.[43] He represented the Carnot cycle in terms of a Watt indicator diagram: the familiar form that it assumes in the introductory pages of all modern textbooks on thermodynamics. How Clapeyron came to know about the indicator diagram is an intriguing and, at present, unsolved problem in the history of science. The secret was, as we have remarked, very closely guarded by Boulton and Watt; so well in fact that it was not until 1826 that a well-informed engineer like John Farey first saw an indicator diagram being taken in Russia* and in this way learned of the principle. Some

---

* The engineers taking the indicator diagram would not have been Russian; they would almost certainly have been employees of the firm of Boulton and Watt.

time during the ten years prior to 1834 the principle of the indicator diagram must have become known in France, or at any rate to Clapeyron, who then had the inspired idea of transposing Carnot's rhetorical argument into the graphical form of the indicator diagram (Figure 19). The 'isothermals' *CE* and *FK* follow Boyle's law, but the 'adiabatics' *CK* and *EF* follow some law that was unknown to Clapeyron, for he did not accept Poisson's now famous equation (see page 145 above). Presumably this was because it was not consistent with Carnot's expression for the temperature rise of a gas when it is 'adiabatically' compressed (page 203, note). Accordingly

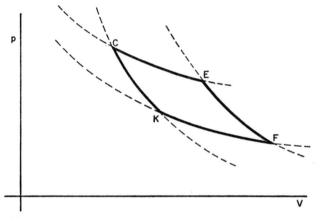

Figure 19

Clapeyron wrote: '. . . the unknown law, of how the pressure varies when the volume of the gas is reduced inside its impermeable envelope, is represented by the curve *CK* . . .' Ignorant of the law though he might have been, Clapeyron guessed at the form of the curves *CK* and *EF* confidently enough and later writers seem to have copied his diagram quite happily.

No less important than the graphical representation of the closed cycle was Clapeyron's other great contribution to Carnot's arguments: he put them in mathematical, or analytical, form. As a calorist and a conservationist he, like Carnot, accepted the simple doctrine of total heat: 'The different states in which a given mass of gas can exist are characterised by the volume, pressure and temperature and the absolute quantity of heat which it contains: if two of these four quantities are known the other two can be determined'. If, that is, a gas absorbs heat during an 'isothermal' expansion and gives it *all* up again during an 'isothermal' compression, then knowing the Boyle and Gay-Lussac gas laws in their combined form—which Clapeyron expresses as $pv = R(267 + t)$—together with the fact that the work done by the gas is shown by the area of the indicator diagram, we can calculate the maximum amount of work that the fall of a unit of heat can perform. When the ratio

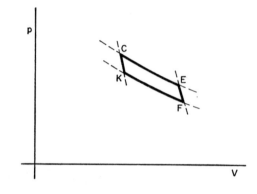

Figure 20

$$\frac{\text{Work performed}}{\text{Heat absorbed}}$$ is a maximum it must, by Carnot's axiom, be independent of the nature of the working substance and can only depend on the temperature. As Clapeyron does not know the form of the curves CK and EF he has to restrict his discussion to cycles with very small temperature differences so that he can assume that CK and EF are straight lines, and the indicator diagram, in effect, a small parallelogram (Figure 20). It is thus very easy for him to compute the area $CEFK$: the work done during the cycle. He is now able to deduce two interesting results;* the first being that the total heat Q in a gas is given by the equation $Q = R(B - C \log p)$, where B and C depend on the

---

* For very small 'falls' of temperature the heat actually converted into work is very small compared with that transmitted to the sink, or condenser, so that in this 'limiting' case, the caloric interpretation of Carnot's theory approximates to the later, dynamical interpretation. Accordingly most of Clapeyron's equations are, by modern standards, correct. See Mendoza, *op. cit.*[7] and Larmor, *loc. cit.*[10]

temperature. While the second result is that in a Carnot cycle in which the temperature difference between the 'isothermals' is the very small amount (dt), the ratio of the work done to the heat absorbed is given by $\dfrac{(dt)}{C}$. C is therefore very important, for its reciprocal, $1/C$, measures the maximum amount of work that a unit of heat falling through $1°$ can perform and its value is a function of the temperature. Clapeyron then went on to derive from his total-heat equation, Carnot's—or Dulong's—law relating the compression of a gas to the amount of heat released, and he deduced the relationship between the two principal specific heats.* His formulae for specific heats at constant pressure and at constant volume were, however, invalid, for consistently with Delaroche and Bérard's misleading result they implied that the specific heats must vary with the volume, or pressure, of the gas.

Having dealt with permanent gases Clapeyron goes on to apply his arguments to vapours in contact with their liquids (i.e. saturated) and, very generally, to all substances: solid, liquid or gaseous. The study of vapours gave him the famous 'Clapeyron's equation'. This he derived from the consideration that the maximum amount of work which a unit of heat can perform when used to vaporise a liquid (as for example in the case of a steam-engine) cannot, by Carnot's axiom, exceed the work which it could do if it were applied to any other working substance; it must therefore be equal to $\dfrac{(dt)}{C}$.† In the very general case, too, the upper limit is set by $\dfrac{(dt)}{C}$.

Clapeyron concluded his paper with an examination of the

* By differentiating the total heat equation with respect to temperature, first at constant pressure and then at constant volume, Clapeyron obtained expressions for the two specific heats. Their difference was:

$$Cp - Cv = \frac{RC}{(267 + t)}$$

† Clapeyron wrote: $L = (v_2 - v_1)\dfrac{dp}{dt}C$ and not, of course, the modern expression: $\qquad L = (v_2 - v_1)\dfrac{dp}{dt}T$ where T is the absolute (Kelvin) temperature (Clausius-Clapeyron).

fundamental factor C, or 'Carnot's function' as it came to be called. He had no direct way of measuring it, but special considerations enabled him to compute its value at $0°C$, while the application of 'Clapeyron's equation' allowed him to calculate it at certain given temperatures: specifically, those at which liquids like water, alcohol, turpentine and ether boil. These results confirmed that C increased slowly with the temperature, and accordingly $1/C$ diminished, just as Carnot had argued that it would. In other words, the work done by the fall of a unit of heat through $1°$ decreases as the temperature is increased (see pages 205–6 above).

So much for Clapeyron's all-important paper. Once again an engineer had made a decisive contribution not only to technology but, generally, to all physical science. Carnot's arguments had been put in mathematical, or analytical, form and their significance for the physics of gases and vapours abundantly demonstrated. But while Clapeyron had been doing all this the conventional engineers seemed to have been much more impressed by orthodox treatises. Roughly contemporary with Clapeyron's paper were the works of the Count de Pambours,[44] whose books received the accolade of translation into English: the language of the great steam-engineers. His *Theory of the steam engine* was a paradigm of exact, critical engineering study: an exhaustive investigation of the generation of steam and its progress from the boiler through the cylinder to the condenser or atmosphere. All types of engines were considered and the treatment was satisfactorily analytical. In fact de Pambours' meticulous attention to detail makes it easier for us to understand how difficult it must have been for engineers and physicists to appreciate the value of Carnot's very generalised notions.

De Pambours' main point was that the pressure of steam in the cylinder of an engine was equal to and determined by the resistance opposed to the piston. This meant that the faster the engine was running the less the pressure of steam in the cylinder until, when it was 'flat out' the force behind the piston would be virtually nil. There was nothing paradoxical about this; after all, a horse galloping at full speed can exert very little pull on the cart it is drawing, for if it could do more then the cart would go faster. But de Pambours' argument was a radical departure from the theory adopted by Tredgold and others that the pressure of steam in the cylinder must be the same as that in the boiler; or rather, since we are dealing with mechanically imperfect engines, it can be taken, for purposes of calculation, as a constant fraction of the boiler pressure. On the other hand it should, with the new theory, be possible to compute the correct proportions for steam-engines where previously, says de Pambours, this had to be done by trial and error. He mentions the indicator diagram, but it was obviously of little use to him as he was concerned with fast-running engines, such as railway locomotives, to which it could hardly be applied. He does not, of course, mention Carnot.[45]

An interesting minor figure who was influenced both by Clapeyron and by de Pambours was Carl Holtzmann, a schoolteacher and a competent mathematician, who in 1845 published a pamphlet on the heat of gases and vapours.[46] Holtzmann says that he had heard of Carnot's ideas through reading Clapeyron's paper but that he had been quite unable to get hold of a copy of *Réflexions*. He notes the importance of the Carnot function, C, and claims that he has found a way to determine it. What follows is very revealing.

Holtzmann points out that heat is known only by its effects and that of these the one that lends itself most readily to quantification is the amount of work it can do: a tribute indeed to the importance of the steam-engine. Work being the measure of heat he proposes to establish a 'work equivalent' for the unit of heat. This equivalent will be the work done when a gas in expanding 'isothermally' absorbs one unit of heat. This seems very promising: Holtzmann had realised that, according to the gas laws summarised at the beginning of Clapeyron's paper, the 'isothermal' absorption of a unit of heat by a gas will always produce the same amount of work irrespective of the pressure, volume and temperature.

Calling 'a' the work equivalent of a unit of heat,* Holtzmann

is able to show that the Carnot function C has the value $\dfrac{p}{a\varrho}$; this enables him to find a more explicit form for Clapeyron's

---

* To avoid confusion the reader should remember that Holtzmann's measure 'a' refers to an 'isothermal' absorption only and not to a 'fall' of heat through a closed cycle. Hence its value is independent of the temperature.

expression for the heat in a gas (see page 222 above)* and from this to derive new expressions for the two principal specific heats. The difference between the two specific heats turns out to be, very simply, $\dfrac{R}{a}$, and from the data available—Delaroche and Bérard's value for the specific heat of air at constant pressure and the known value for $\dfrac{Cp}{Cv}$ —it is easy to calculate the value of 'a'. This is 374 kg.m. the work equivalent of the heat that would raise the temperature of a kilogramme of water by one degree, and the order of accuracy is, Holtzmann estimates, between 343 and 429 kg.m. Holtzmann's method of calculating the work equivalent of heat is unexceptionable, and as he has made no assumptions about the nature of heat it is not open to the kind of criticism that was later made of J. R. Mayer's measure of the actual *conversion* of heat into work. Knowing the value of 'a' it is now possible to determine the values of $\dfrac{1}{C}$ at different temperatures by means of the expression

$\dfrac{a}{R(267 + t)}$ and these turn out, suggestively enough, to be in good agreement with those obtained by Clapeyron.[47]

It seems curious that neither Clapeyron nor Holtzmann took what seems to us to be the obvious step of using $(267 + t)$ or $(\dfrac{1}{\alpha} + t)$ to define a new temperature scale with a zero at $-267°C$. A possible explanation is that with the waning of belief in a real, material caloric, interest in a hypothetical absolute zero languished too. The notion of an absolute zero had, in any case, been kept alive largely by Scottish and English physicists.

Holtzmann, as we have seen, professed agnosticism about the nature of heat but even so he had to choose, if only by default, between the axiom of the conservation of heat and its then elusive rival, the axiom of the mutual convertibility of heat and mechanical energy. Holtzmann, eschewing metaphysical prejudices, chooses in effect the conservation axiom, with the result that his calculations are at one with those of the calorists. Indeed, the revealing thing about Holtzmann is how near a

* Holtzmann's expression is: $Q = F_t - \dfrac{R(267 + t)}{a} \log \dfrac{p}{p_0}$

competent physicist could get to a viable dynamical theory without making the final transition, or in other words, how close the two rival theories could get to one another. For Holtzmann's 'mechanical equivalent of heat' tells him nothing about the nature of heat; while for Mayer and for Joule their equivalents indicated the amount of work into which a unit of heat could actually be converted.

When Holtzmann came to consider the operation of the steam-engine—a procedure by then almost *de rigeur* for serious writers on heat—he accepted Watt's law, as de Pambours had stated it, and went on to argue that the decrease of latent heat with increase of temperature and pressure indicated that the great—and 'wasteful'—cohesive force of water became less at higher temperatures, so that high-pressure steam-engines should be more economical than low-pressure ones. We have been here before, but Holtzmann goes on relentlessly:

> The action of this quantity of heat is lost in all steam engines, which hitherto has been wholly overlooked; this loss amounts, as is evident, to much less at high temperatures than at low ones, whence it results that attention must be directed far more than hitherto to the construction of steam engines with very high pressures.

This was indeed an astonishingly unoriginal observation to make in 1845 and it seems, furthermore, to indicate that Holtzmann cannot have understood Clapeyron, still less Carnot, for they had demonstrated that the physical properties of the working substance cannot in principle affect the production of motive power, which is determined solely by temperature differences. Always provided, of course, that the individual operations that make up the cycle are strictly reversible.

If Holtzmann was not, therefore, in the first rank of contemporary scientists he was still of considerable interest; especially as he happened to be a German. Apart from the engineers like Höll and Reichenbach, who built the great column-of-water engines, and a few minor figures like Lambert, Haldat, F. T. Mayer, and Seebeck, Germans contributed remarkably little to the development of power technology and the science of heat and thermodynamics. This would appear to be just one aspect of the general backwardness of German science throughout the eighteenth century and during the first two decades of the nineteenth; a backwardness which was possibly the consequence of the fragmented political state and inferior economic and military situations of Germany. Certainly there were able Germans in each generation, but many of them emigrated in order to make names for themselves—in this way Herschel went to England and Humboldt to France— so that they enhanced the glory of their adopted countries while proportionately impoverishing the intellectual life of their homeland. With the political revival of Germany in the first half of the nineteenth century so the scientific situation improved too, and Germany came to play an increasingly commanding role in the advancement of science. Holtzmann was, in effect, an early swallow in the summer of German physics.

But he was certainly not the first.* In 1837 Friedrich Mohr had published a paper on the nature of heat which was later to win him considerable, if exaggerated, praise.[48] In fact he was not able to add very much to the excellent paper Ampère had written two years earlier, and his argument ran on very much the same lines: the discovery of the close analogies between radiant heat and light, and the establishment of the undulatory, non-material, nature of the latter, compels us to regard radiant heat as undulatory (see above, pages 217–18). Thus heat, being of a vibratory nature, is a manifestation of 'force' (*kraft*). This did not represent much of an advance, but Mohr went on to make a few interesting suggestions. He argued that the 'so-called absolute zero' of temperature must correspond to a state of absolute rest, and he queried the nature of latent heat. The latter, he said, amounts to nothing more than the destruction of the cohesive force of constituent atoms and the work done in separating them. When the body returns to its original state the process is reversed and the latent heat is *regenerated*. The term 'latent heat' is thus a misnomer, for the heat in question is actually converted into work as the atoms separate and is re-created when they come together again. Mohr also asserted that the difference between the two principal specific heats of a gas was accounted for, not by 'latent heat of expansion' as everyone had assumed, but by the work done when a gas expands at constant pressure.

* The great revival of German chemistry had, of course, begun even earlier; it may be conveniently dated from Liebig's appointment to the Chair of Chemistry at Giessen in 1824.

All this did not add very much to the stock of arguments available in support of the dynamical theory of heat, and Mohr did not rise to the conception of a numerical value of a unit of heat expressed in terms of mechanical work. Others indeed were at this time moving towards the idea of the dynamical nature of heat but the grounds for their acceptance of this belief were not necessarily the same as those of Ampère and of Mohr. Thus the Danish engineer L. A. Colding had been led to the dynamical theory of heat through his considerations of the working of steam-engines, while Marc Séguin, another engineer, was also favourably disposed to the theory and even tried to measure the difference between the amount of heat taken from the boiler and that delivered to the condenser; in this he was unsuccessful.[49] But the first man to realise the significance of, and actually to measure the mechanical value of a unit of heat was unquestionably Julius Robert Mayer.

J. R. MAYER AND J. P. JOULE

Julius Robert Mayer (1814–78) was born at Heilbronn and trained for the medical profession. It was while acting as ship's surgeon on a Dutch vessel trading to the East Indies that Mayer noticed that venous blood was of a much brighter red colour in the tropics than in higher latitudes. He ascribed this to the reduced rate of oxidation needed in hot climates, and this led him to further consideration of the problem of animal heat. He reasoned that the living organism's capacity to produce heat should be taken to include the external heat as well as the internal, animal heat that it can generate. A blacksmith can, for example, produce a good deal of external heat by percussion. This led him to conclude that the heat generated must be proportional to the work expended, for it was inconceivable that a mere variation in the means of production could lead to different quantities of heat being generated; there must, therefore, be a fixed relationship between heat and work.

In 1842 Mayer returned to Heilbronn and in May of that year published the first account of his theory.[50] He had generalised from the field of physiology to include the phenomena of inorganic nature, and he put forward as the basis of his theory an axiom: that of the indestructibility of *force*. The use of the ambiguous word 'force' and, indeed, the metaphysical undertones of his discussion were rather unfortunate; they may perhaps be traced to the relatively isolated state of German physics at that time as well as to Mayer's own intellectual background and training. But there can be no doubt about Mayer's meaning: energy is always conserved; it cannot be destroyed, it can only be converted into other forms. He then went on to discuss the now-familiar instances of the conversion of mechanical energy into heat, and vice versa, and set out to determine the numerical value, in mechanical terms, of a unit of heat.

His method of calculating this equivalent was very simple; outwardly, if not essentially, it was the same as the method used by Holtzmann. The difference between the two principal specific heats of a gas is interpreted as the amount of heat energy needed to expand that gas against the external pressure of the atmosphere. For Mayer this extra heat is therefore converted entirely into *external* work and the constant 'a', now a *conversion* factor, has, he computes, the value of 365 kg.m. per kilocalorie of heat.

There is, of course, an uncertainty, a gratuitous assumption, at the heart of Mayer's calculation. It is simply that while no one doubted that the difference between the two specific heats was accounted for by the heat required (consumed *or* made latent) to expand the gas, Mayer had not proved that *all* the extra heat was converted into work done against the external pressure. At the end of its expansion the gas is in a different state of aggregation—it occupies a bigger volume—than at the beginning and, on the face of it, it is a reasonable assumption that some heat at any rate may have been used up in altering the internal structure of the gas: overcoming the (hypothetical) mutual attraction of the gas molecules as well as the atmospheric pressure of the air. The only way to ensure that no heat is used up in this way is to bring the gas back to its original volume at the end of the operation.[51] Alternatively, one would have to prove that in the 'isothermal' expansion of a gas all the heat supplied is converted into external work so that the internal energy of the gas remains unchanged after the expansion. Mayer claimed, in his next memoir (1845), that Gay-Lussac's experiment (see above, pages 133–4) justified his assumption.

Having extended his theory to explain many forms of energy-transformation on earth he went still further in his memoirs of 1845 and 1848 and included the phenomena of the solar system. Meteorites, falling under gravity, build up a tremendous mechanical energy, which transformed by impact should yield considerable heat. Their constant bombardment may well account for the seemingly inexhaustible heat of the sun. So was born the meteoritic theory of solar heat, which became very popular in the days before the advent of explanations based on nuclear transformations.[52]

Mayer had considerable difficulties not merely in getting his ideas accepted but even his papers published. Indeed he was almost entirely ignored, and so unacceptable were his ideas, as far as editors were concerned, that he had to pay for the publication of his second memoir himself. A series of quite appalling personal disasters in 1849 coupled with the evident scorn of his peers led to a mental collapse, and his subsequent disappearance from the scientific scene for about ten years. During this time he is reputed to have cultivated his vineyard at Heilbronn.[53] It is a tragic irony that the great German physicists who contributed so much to the later development of thermodynamics —Helmholtz and Clausius—learned of the mechanical theory of heat and the significance of the numerical value of heat energy not from their distressed compatriot Mayer but from his English contemporary, Joule. Mayer was eventually rescued from his undeserved oblivion through the good offices and by the generous impulses of John Tyndall.

Mayer's significance in the history of science seems to have been threefold. Firstly, he took a cosmological view of the role of heat, just as Rumford, Fourier and Carnot had done, although of course he interpreted it as heat *energy*. Secondly, his vision was perhaps broader than theirs in that he included the world of living, organic things in his scope; in fact, he was the first physiologist since Adair Crawford to make a major contribution to the science of heat. Finally, and very disturbingly, he was outside the tradition of research and scholarship in the science and technology of heat; he stood apart from the then 'Establishment'.

This last point did not, on the face of it, apply to James Prescott Joule (1818–89). The son of a prosperous Salford brewer, he was wealthy enough to be able to devote his life entirely to science. And this he could do in a Manchester that was the home of the illustrious Dalton—Joule in fact studied under Dalton for a year—as well as the other distinguished members of the Literary and Philosophical Society: a group that formed a kind of informal university. We may infer that in those days, when textbooks and periodicals were few and when standardised degree courses and the other paraphernalia of organised science did not exist, local traditions and the *personal* influences of men like Dalton, the Henrys, Peter Ewart and Eaton Hodgkinson may have been much more powerful in moulding Joule's scientific character than could possibly be the case today. Certainly it is possible to detect the characteristics of a Manchester style of science: industrious, straightforward, and without philosophical subtleties. Equally these seem to be the dominant qualities of Joule's scientific career. All in all, his was one of the most satisfactory scientific lives that can be imagined. Profound insight, close reasoning, great experimental gifts and meticulous accuracy are all combined in his work. And his life was rounded off by the accomplishment—unambiguous and universally acknowledged—of a tremendous task: the establishment of the first law of thermodynamics.

Joule was led to consider the problems of energy conversion through an early interest in magneto-electric machinery. His first papers were contributions to William Sturgeon's* *Annals of Electricity* and they were of a practical, engineering nature: he looked forward to the day when electric motors would replace steam-engines for driving machinery. From electric motors he graduated to a study of the generation and the heating effects of electric currents. In 1841 he discovered the important law that the heating effect of an electric current is proportional to the resistance of the circuit and the square of the current flowing.[54] Two years later came the very important paper: 'On the Caloric Effects of Magneto-Electricity, and on the Mechanical Value of Heat.'[55] In this paper he demonstrated that the heat caused by the passage of an electric current is not *transferred* from another part of the circuit, which is correspondingly cooled (as the Peltier effect suggested might be the case), but is actually *generated*. In his own words, he showed that:

> . . . the mechanical power exerted in turning a magneto-electric machine is *converted into the heat* evolved by the passage

---

* William Sturgeon, the editor of *Annals of Electricity*, made some useful contributions to science, including the invention of the electro-magnet. He was self-educated and trained, having served as a private in the army.

of the currents of induction through its coils; and, on the other hand, that the motive power of the electro-magnetic engine is obtained at the expense of the heat due to the chemical reactions of the battery by which it is worked.

The mechanical power was provided—and measured—by the descent of a pair of weights placed, one each, in two scale pans; the latter were connected by means of fine strings passing over two pulleys to a vertical axle, round which a great length of each string was wound. As the weights descended a measured distance so the axle was rotated and drove the magneto-electric machine (dynamo).

As a result of a number of experiments Joule concluded that the mechanical value of a unit of heat was 838 ft. lbs of work expended to raise one pound of water by one degree (F).

These experiments, together with the conclusions drawn, were announced to the Chemical Section of the British Association meeting at Cork in 1843. They were received in silence; partly, no doubt, because some of his audience were unable to grasp the full import of what was at stake, partly because others were unwilling to throw overboard the grand principle of the conservation of heat, on which the entire science had been erected, and partly because still others felt that too much was being inferred from just one set of experiments. But Joule had already, in a postscript to this paper, transferred his attention from the electrical generation of heat to the purely frictional. By forcing water through tiny holes in a perforated cylinder, he had been able to evolve a detectable amount of heat, which compared with the work expended yielded a mechanical value for a unit of heat of 770 ft. lbs. This was in tolerable if not good agreement with the value obtained by means of the electrical experiments. But if expending mechanical power on water in this way can generate heat, what of the famous phenomenon of the 'adiabatic' heating of gases? Joule's next experiment therefore was to compress a gas in a strong copper vessel, which was placed in a large calorimeter.[56] The walls of the calorimeter were made as impermeable to heat as was possible, and it was filled with water. By comparing the heat imparted to the water with the work performed in compressing the gas Joule found a new value for the unit of heat. This was 823 ft. lbs.

At this point Joule faced the unresolved difficulty that had made Mayer's determination of the mechanical value of heat uncertain. What proof was there that *all* the work performed on the gas had been converted into heat; alternatively, that *all* the heat produced was due to the work performed? Might it not be that in compressing the gas—in reducing its volume— some of the heat released may have been due to 'latent' heat being made 'sensible'? These objections did not apply to the experiments made using water and a perforated piston, for water is virtually incompressible. But until they could be answered in the case of gases, determinations based on 'adia- batic' compression or expansion would be entirely uncertain.

Joule's answer to this dilemma was another and very famous experiment. Superficially it resembled Gay-Lussac's experi- ment (see page 133 above), but in principle it was very differ- ent. Two large copper vessels are connected by means of a pipe containing a stopcock (see Plate 4), and are placed in a large calorimeter full of water; the walls of the calorimeter are as impermeable to heat as is possible. One of the copper vessels is evacuated and the other is filled with dry air at a pressure of 22 atmospheres. The stopcock is opened so that the air rushes into the empty vessel, but no detectable change in the tempera- ture of the water in the calorimeter is noticed.

In other words, when air expands *without* doing work no heat is lost.

Joule now proceeded to analyse this experiment. He provided each of the two copper vessels with its own separate calorimeter and repeated the experiment. The results indicated that the calorimeter containing the evacuated vessel gained exactly the same amount of heat as the other one lost. The sum of the two quantities of heat imparted to and abstracted from the two identical calorimeters was, therefore, zero. The implication was plain: merely changing the volume of a gas does not produce or consume heat; this happens only when work is done on or by the gas. The 'latent heat of expansion (or compression)' is therefore an illusion: it is no more than the heat value of the work done by or on a gas, against or by an *external* force. Joule's calculation of the mechanical value of a unit of heat by measur- ing the amount produced when the gas is compressed to a given extent was therefore justified in principle; and so, incidentally, was that made by Mayer.

This splendid paper was rejected by the Royal Society. But it was printed in the *Philosophical Magazine*, and in this way the scientific world learned not only of Joule's fundamental insights and confirmatory experiments but also of his penetrating criticisms of Clapeyron and Carnot. For the first time the fundamental ideas of thermodynamics were subject to the (constructive) criticism of a first-class scientific intellect:

> I conceive that this theory . . . is opposed to the recognised principles of philosophy because it leads to the conclusion that *vis viva* (i.e. energy) may be destroyed by an improper disposition of the apparatus: thus Mr Clapeyron draws the inference that 'the temperature of the fire being from 1000°C to 2000°C higher than that of the boiler there is an enormous loss of *vis viva* in the passage of the heat from the furnace to the boiler'. Believing that the power to destroy belongs to the Creator alone I affirm . . . that any theory which, when carried out, demands the annihilation of force, is necessarily erroneous.

In the following year (1845) Joule read, to the British Association meeting at Cambridge, an account of a new method of determining the mechanical value of heat. A paddle-wheel, driven by the same mechanism of falling weights that he had used to provide power for his dynamo in 1843, rotated horizontally in a calorimeter full of water.[57] A comparison of the heat generated with the work performed yielded a mechanical value of 819 ft. lbs per heat unit. But no one seemed very interested in Joule's ideas or experiments. No one, that is, in Britain; abroad he was not without honour, for when in 1847 Helmholtz published his important paper *On the conservation of energy* he cited Joule's experiments in support of his arguments. In that year Joule described his improved paddle-wheel apparatus for the measurement of the mechanical value of heat and this time, at the British Association meeting at Oxford, his ideas aroused much more interest. This was largely due to the keen appreciation of the young William Thomson, later Lord Kelvin.*

The Oxford meeting of the British Association marked an historic stage in the history of thermodynamics: the real beginning of the acceptance of the first law. It is, of course, true that the measurements on which Joule had had to rely for his

---

* William Thomson was knighted in 1866 for his services in connection with the great Atlantic cable and was made Lord Kelvin of Largs in 1892. We shall refer to him hereafter as Kelvin, the name by which he is universally known.

determinations were small fractions of a degree, but he was a masterful experimentalist and he was supported by a most able instrument-maker.* The implication of Joule's ideas that one could heat up water *merely by shaking it*, seemed in the context of scientific—or common sense—thought in the 1840s quite extraordinary, and the confirmation that this was indeed the case (Dr Reade had long been forgotten) was taken as strong confirmation of his theory.

True to the British tradition Joule had inferred a value for the absolute zero of temperature from his theory. It worked out at 491°F below the freezing point of water (−273°C); this was the temperature at which the *vis viva* of the constituent atoms of matter would fall to zero and it was, of course, derived from Gay-Lussac's law of the increase of gas pressure with temperature. He accepts Herapath's hypothesis that the atoms of a gas behave like high-speed projectiles, '. . . constantly flying about in all directions with great velocity', and from it he infers that the atoms of hydrogen gas at 60°F and 30 in. pressure should have an average velocity of 6,225 ft./sec.

The way ahead was now, at last, clear for Joule. He had soldiered on and had won through to recognition and acceptance. In 1850 the Royal Society published his account of new and very accurate paddle-wheel experiments[58] which gave the figure of 772 ft. lbs for the mechanical value of a unit of heat. From this time onwards Joule could extend and consolidate his theory, devise even more accurate methods for determining his constant, and find further instances of the great principle of the convertibility of energy in nature.

Of Joule's epistemological position there can be no doubt at all: he was a straightforward Daltonian realist. If the dynamical theory postulates that heat is the energy—the *vis viva*—of the constituent atoms of material bodies then Dalton's doctrine that all discrete atoms possess determinate *weights*, or masses, makes the dynamical theory that much more plausible. The mechanism which Joule envisaged whereby mechanical energy is transformed into heat energy is extremely simple. Plate 5, taken from one of his laboratory notebooks,[59] shows how he thought that the energy of a descending weight could be trans-

---

* John Benjamin Dancer, of Cross Street in Manchester, the inventor of micro-photography. I can confirm, from personal experience, the superb quality of Dancer's instruments.

formed into heat: as in the case of the paddle-wheel experiments. The linkage between the weight and the moving, in this case rotating, atoms is direct.

Another important aspect of Joule's investigations is that he expresses the mechanical value of heat unambiguously in terms of *work*. There was no attempt to use that confusing word 'force' for, as Silvanus Thompson remarked, Joule was thinking as an engineer.[60] Here the source of his concept may, I think, be traced back to Peter Ewart's splendid paper of 1813, so much admired by Joule's mentor Dalton, and in particular, perhaps, to that passage in which Ewart related a given quantity of heat to a strictly determinate amount of work.* But we must not forget Hachette. He had suggested, after all, that the engineer's unit of work should be used as a measure of natural as well as of technological processes. It is unlikely that Joule had read Hachette's volume of 1811, but William Whewell certainly had, and he gave Hachette's suggestion currency through his numerous writings on mathematics and mechanics.[61] Few physical scientists of Joule's generation can have evaded the intellectual influence of William Whewell.

On one point, however, Joule did cause some confusion, which has continued to the present day. In his early efforts to win acceptance for his ideas he invoked the support of the published opinions of his illustrious predecessors. He claimed that Bacon, Newton, Locke, Rumford and Davy had all supported the dynamical theory of heat. This was true enough; but by a peculiarly gross distortion of history Joule contrived to give to Rumford's very meagre data such values that the figure of 1,034 ft. lbs could be extracted from them.[62] Now we have shown not only that Rumford had no concept of the mechanical value of heat but also that, in the context of his times, he could not have had one. Joule's reference to 'Count Rumford's equivalent' thus helped to propagate the still current myth that Rumford had actually measured Joule's constant. But there was nothing vicious in Joule's distortion of history: he deceived himself, for at the end of his own translation of Mayer's first paper he scribbled a pencil note to the effect that Rumford's

* Eaton Hodgkinson's obituary account of Ewart's work was one of the earlier complimentary offprints in Joule's private library (now in the possession of the Library, University of Manchester Institute of Science and Technology).

estimate of the mechanical value of heat had been more accurate than that of Mayer.[63] This note was obviously not intended for publication and we infer that Joule was merely trying to buttress his own hope that Mayer had not added anything significant to the work done by Rumford.

It is almost impossible to avoid comparing Joule with Mayer. But such a comparison is extremely difficult, if not impossible to make, for two more dissimilar characters, in terms of intellectual background, scientific attributes and personal destinies, can hardly be imagined. But one important thing they did have in common. Both had an extremely hard fight— in Mayer's case, an unsuccessful fight—to establish the dynamical theory of heat. They had, in fact, challenged the accepted orthodoxy, and if Joule at last succeeded in overthrowing that orthodoxy and establishing the first law of thermodynamics it was due to his great experimental skill, courageous persistence and great good luck in meeting Kelvin at the right time.

Chapter 8

The New Science

Lord Kelvin (1824–1907) was in some respects a typical Scottish philosopher; in others he was a rather more complex character than the simple biographical details would suggest. After a distinguished undergraduate career at Cambridge, where he was successively Second Wrangler and Smith's Prizeman, he spent a year in Paris working in Victor Regnault's laboratory. Almost inevitably, it seems, he met the leading French physicists and mathematicians of the day—Liouville, Biot, Cauchy and Foucault among them.[1] His own genius had earlier led him to the discovery of Fourier and the analytical theory of heat; now in Paris he was to come across Clapeyron's paper *On the Motive Power of Heat*: the impact was immediate, the conversion was complete.

Kelvin was, we suppose, predisposed by education, training and possibly the sentiments of the 'auld alliance' to adopt French views of the nature of heat and the laws of its action. His work with Regnault had familiarised him with the difficulties of thermometry and the desirability of some absolute scale of temperature, independent of the properties of any particular substance. Accordingly his own first insight into thermodynamics was the realisation that here was a means of establishing the long-sought-for absolute scale.* The efficiency of an ideal engine is necessarily independent of the nature of the working substance, and depends only on the temperature difference between the furnace and the condenser, or sink. The reciprocal of Carnot's function, as defined by Clapeyron, is the work done by a unit of heat 'falling through' one degree, and on the gas scale it decreases as the temperature increases.

---

* Kelland, an avowed admirer and disciple of John Dalton, may possibly have given Kelvin some insight into this problem, for he was the latter's tutor when he was an undergraduate at Cambridge.

Kelvin therefore suggested that $1/C$, rather than the increase in volume of a gas, liquid or solid, be taken as the measure of temperature; on his proposed scale each degree of temperature would be defined by the fixed amount of work done by a unit of heat falling through it. Thus from $T°$ to $(T - 1)°$ the work done is the same quite irrespective of the value of T, and Carnot's function therefore becomes a constant.[2]

The work done by the unit of heat must, by Carnot's axiom, be independent of the nature of the working substance, and an absolute scale is therefore achieved, although of course it must diverge from the gas scale. In practice the calculation of this new scale, in terms of the gas scale, required the application of Clapeyron's equation and hence knowledge of the latent heat of a volume of steam at different pressures, but fortunately Regnault was then engaged in providing this data. It is perhaps worth pointing out that in a footnote which is usually over-looked by commentators Carnot himself had remarked that, provided the specific heats of gases are independent of *temperature* there must be a relationship between motive power and thermometric degree; but as Kelvin had not seen a copy of *Réflexions* at that time he cannot have known of Carnot's strong hint.

In the following year, 1849, Kelvin, having obtained a copy of Carnot's rare book, followed up with a paper entitled: *Account of Carnot's Theory of the Motive Power of Heat; with Numerical Results derived from Regnault's Experiments on Steam.*[3] The full title is important, for it shows how the precise and extensive experimental data then being published by Regnault were providing a firm basis for the establishment of thermo-dynamics. Regnault thus appears to have played a part *vis-à-vis* the great theorists of thermodynamics very similar to the one played by Tycho Brahé *vis-à-vis* the great theorists of planetary motion from Johann Kepler onwards. Without Regnault's experimental data thermodynamics would have been impossible, just as Kepler's laws could not have been established without Brahé's extensive and accurate ephemerides. Regnault and Brahé, together with Kelvin and Kepler, exemplify the truth of T. H. Huxley's dictum that the advance of science depends on two contrasting types of men: the one bold, speculative and synthetic, the other positive, critical and analytic. It is worth remarking, too, that apart from his experimental gifts Regnault, like Brahé before him, was very conservative in his scientific views.[4]

Kelvin stresses the importance of closed-cycle operations whereby the working substance is brought back to the original condition so that the net effect of the flow of heat is the production of work and of nothing else, there being no net expansion or change of state of the working substance. Kelvin mentions Joule's ideas only to reject them: '. . . the conversion of heat, or *caloric*, into mechanical effect is probably impossible, certainly undiscovered'. But he then goes on to raise an important question; what happens when heat flows by conduction from a hot to a cold body?

> When thermal agency is thus spent in conducting heat through a solid what becomes of the mechanical effect which it might produce? Nothing can be lost in the operations of nature—no energy can be destroyed. What effect then is produced in place of the mechanical effect which is lost? A perfect theory of heat imperatively demands an answer to this question; yet no answer can be given in the present state of science.

This was, at one and the same time, an unoriginal and a very penetrating question. It was unoriginal in that the key phrase 'Nothing can be lost in the operations of nature—no energy can be destroyed' is hardly more than a rephrasing of Joule's remark: 'Believing that the power to destroy belongs to the Creator alone I affirm . . . that any theory which, when carried out, demands the annihilation of force is necessarily erroneous' (see page 235, above). Therefore Kelvin had picked up the gauntlet thrown down by Joule, even if he did not know what to do with it. On the other hand he had posed the question in the framework of thought established by Carnot: that relating to the efficiency of heat-engines. We can regard the mere conduction of heat from a hot to a cold body as the operation of a heat-engine of zero efficiency; in which case we must ask what happens to the work which a more efficient engine, in exactly the same situation, would produce? In this lay the key to the further development of thermodynamics. Science in fact is a matter of asking the right questions; having asked them someone, somewhere will sooner or later find the answers.

Kelvin then went on, quite inconsistently, to point out that Joule's experiments on the frictional heating of liquids explain why mechanical effects disappear when a fluid is set moving in a rigid closed vessel and then allowed to come to a stop by virtue of the internal friction. The position was indeed interesting. Kelvin had seen the force of Joule's arguments and had called attention to the desideratum in Carnot's theory; in fact in the general theory of heat as it then stood. But he did not regard the old theory as falsified—as perhaps he should have done according to the rules—for there was nothing adequate to put in its place. Science, after all, is not only a matter of destroying wrong theories. So, however logically vulnerable his position might have been he kept an open mind, and this was possibly the best thing he could have done.

He then went on to put forward a general expression for the work done by a quantity of heat falling through a given temperature* and to find appropriate formulae for Carnot's function, using air and then steam as working substances.† Basing his calculations on Regnault's extensive data he showed that $1/C$ decreased steadily as the temperature increased, and that his results were consistent with the four that Clapeyron had obtained using boiling water, ether, alcohol and turpentine. His expression for Carnot's function, using air as the working substance, enabled him to calculate the heat produced by compressing a gas 'isothermally', which consequently led him to the familiar result that all gases equally compressed at the same temperature yield the same quantity of heat. He computed also the work needed to compress the gas, and this gave him the important ratio of the work done to the heat released,

$$\frac{W}{Q} = \frac{1}{C}\frac{(1 + \alpha t)}{\alpha},$$

which is equal to $1/C(1/\alpha + t)$, or as Clapeyron might have

---

* i.e. $W = Q \displaystyle\int_{t_c}^{t_h} 1/Cdt$. The reader should note that Kelvin took the reciprocal of C as Carnot's function and denoted it by the letter μ. To simplify this account and make it consistent with Clapeyron and Clausius we have however altered Kelvin's notation.

† His expression for $1/C$ is, in the case of air, $\dfrac{\alpha p_o v_o}{v dq/dv}$. From this he deduces that the heat of compression would be expressed by $C\alpha p_o v_o \log V/V'$, while the work done, $\int p dv$ would be $\int p_o v_o (1 + \alpha t)\dfrac{dv}{v}$, which gives us $W = p_o v_o (1 + \alpha t) \log V/V'$ and the above ratio of W to Q follows.

put it, $1/C(267 + t)$, which is the work needed to produce one unit of heat. Kelvin pointed out that this yielded, when appropriate values for C and t had been put in, a figure for the amount of work needed to produce* one unit of heat that was in close agreement with one of the three sets of figures Joule had obtained in his paper 'On the Rarefaction and Condensation of Air' (760 ft. lbs per unit of heat at between 10° and 16°C).

Kelvin concluded this seminal paper, in which he coined the word 'thermodynamic', with a consideration of the performances of certain steam-engines, notably that of the still-supreme Fowey Consols engine.

It was at this time that another fundamental difficulty in Carnot's theory occurred to Kelvin. When water at 0°C turns into ice at exactly the same temperature (that is when it freezes), it notoriously expands with seemingly 'irresistible force'. Now although the practical difficulties might be insurmountable such a physical change could, in theory at any rate, provide useful work; it constitutes the scientific basis for a heat engine. But it would be a heat engine of a peculiar sort, for the latent heat that must regularly be given to and then abstracted from the working substance in order to complete the cycle does *not* flow from a hot to a cold body: it is always at the same temperature, which is 0°C. It is very easy to show that as work would be obtained *without* a flow of heat from hot to cold such an engine would violate the injunction against perpetual motion. Indeed it is almost intuitively obvious that such a machine would be self-acting. In this case, then, we have a fatal argument against Carnot. Or we would have had, had not salvation come from Kelvin's brother, James Thomson, Professor of Mechanical Engineering at Kelvin's own University of Glasgow. The latter suggested that under pressure the melting point of ice is actually lowered. In this case the *reductio ad absurdum* of perpetual motion cannot apply, for no work can be done if at the moment the freezing water starts to exert a pressure the freezing process is thereby stopped. Experiments showed that this was the case;[5] the melting-point of ice was lowered by pressure, and Carnot's theory was not merely saved,

* For Kelvin the production of heat by compression was, in accordance with the axiom of conservation, no more than the squeezing-out of heat already in the working substance; for Joule of course the heat was created by the conversion of mechanical energy into heat energy.

it had received something like a triumphant vindication. It had, in fact, led directly to a wholly unsuspected phenomenon being discovered, and there is not much more that one can ask of a scientific theory than that it leads us to new and unexpected knowledge.

Nevertheless Joule's work could hardly be ignored, and pressure from that quarter was mounting. Joule, too, could claim that his theory led to unexpected but confirmed results; for example, that one could heat water merely by shaking it. This phenomenon was hardly predictable on the caloric, or conservationist theory, and as we have observed, it appeared very paradoxical at the time. The arguments on both sides seemed very strong, and it is hardly surprising that in the words of Kelvin's biographer, Silvanus Thompson, 'The apparent conflict took possession of (Kelvin's) mind and dominated his thoughts.'

There can be little doubt that at the root of Kelvin's objections to Joule's theory lay his conviction that the whole science of heat rested on what since 1783 had been the basic axiom of *conservation*. If that were rejected what would happen to the impressive structure of experimental knowledge and theoretical development that had been built up by the labours of men like Delaroche and Bérard, Fourier, Dulong and Petit, Poisson, Victor Regnault and many others? Thus he quoted Carnot as saying, of the axiom of conservation, that: 'To deny it would be to overturn the whole theory of heat, in which it is the fundamental principle.' And he himself adds the comment that if the axiom of conservation is rejected: '. . . we meet with innumerable other difficulties—insuperable without further experimental investigation, and an entire reconstruction of the theory of heat from its foundation.'

While Kelvin was mulling over these problems he was being overtaken by two other men, R. J. E. Clausius (1822–88) and W. J. M. Rankine (1820–72). In some ways Clausius could almost be described as a disciple, at one remove, of Kelvin. He had not read the original Carnot volume and he knew of the theory only through the writings of Clapeyron and Kelvin. Possibly this was an advantage, for he could consider the theory dispassionately, without being unduly influenced by the persuasiveness of the whole work, and in the light of Kelvin's important observations about Joule's ideas. In any case, after comparing Kelvin's work very favourably with that done by Holtzmann, he takes as his starting point the basic problem to which Kelvin had drawn attention: what happens to the mechanical effect which is lost when heat flows from a hot to a cold body not by way of a heat engine but by simple conduction?[6] He sees that, in fact, the position is more open, there are rather more possibilities, than Kelvin had supposed. 'On a nearer view'—and how much is contained in that modest phrase!—the new (Joule) theory is opposed not to Carnot's theory but to the assertion that no heat is expended or lost in a cyclic operation. For it is quite possible that in the production of work *both* processes take place at the same time: '. . . a certain portion of heat may be consumed and a further portion transmitted from a hot body to a cold one; and both portions may stand in a definite relation to the quantity of work produced.' These words deserve to be pondered very carefully. In Clausius' view, Joule had established the consumption of heat in a cyclic process, Carnot the transmission of heat, and *both* phenomena are necessarily related to the work produced.

On the other hand, if we are now going to reject the axiom of the conservation of heat there are concepts which must, in the light of the new theory, be either dismissed or at least fundamentally reassessed. Prominent among these is the concept of 'latent heat' (still with us, alas) which implies that when a liquid is vaporised or a solid melted some heat, the 'latent heat', is in some way secreted among the particles of the vapour; wrapped around the atoms, perhaps, in the form of caloric atmospheres. Not so, argues Clausius, the heat does not go 'latent', it actually vanishes, being converted into the work done by expanding the volume of liquid into a gas or vapour, or by overcoming the inter-molecular forces in the case of a melting solid. Mohr's remarks were not referred to; presumably he had already been forgotten.

Again, the accepted idea of 'total heat' must be looked at once more, for it implies that if a gas is brought from a given temperature and volume to a higher temperature and an expanded volume, the amount of heat that has been added to it is quite independent of the way in which the change has come about. This belief, an immediate corollary of the axiom of the conservation of heat, was of course accepted without question by Carnot and by Clapeyron. Thus, if we turn back to page 205 and consider the indicator diagram (Figure 18), the working substance can be brought from volume v_1 and temperature T_1 to v_2 and T_{100} either by heating it up at constant volume (v_1) and allowing it to expand 'isothermally', or by allowing it to expand 'isothermally' at T_1 and then heating it up at constant volume (v_2). In both cases, so the doctrine ran, the amount of heat absorbed will be the same, so that $q_3 + q_4 = q_1 + q_2$. But on the new theory the amounts of heat absorbed will be very different, since the gas does more work in the first case, and so demands more heat, than it does in the second. Thus the concept of 'total heat' needs to be modified.

Nevertheless, on the new theory as on the old, the *net work done* at the end of a closed cycle, when the pressure, volume and temperature of a gas have been restored to their pristine values, can only be *external*; any interior work done in separating the constituent molecules of the gas is exactly compensated for when the gas is compressed.[7] Clausius closely follows Clapeyron in his subsequent analysis, save only that the heat *consumed* in the cycle is in direct proportion to the work done. From the equations which he sets up* he obtains the very famous relationship which is usually quoted as the analytical statement of the first law of thermodynamics:

$$dQ = dU + \frac{R}{J} \cdot \frac{a+t}{v} dv \dagger$$

where J is Joule's equivalent.

This implies, in general terms, that the heat imparted to a gas may be decomposed into two portions, U being that required for sensible heat and for internal work and the second

* i.e.
$$\frac{d}{dt}\left(\frac{dQ}{dv}\right) - \frac{d}{dv}\left(\frac{dQ}{dt}\right) = \frac{R}{vJ}$$

† This differential equation is not, of course, equivalent to
$$\delta Q = \delta U + p\delta v.$$

It does not mean that a small increment of heat gives small increments of internal energy and external work, for there is an infinite number of ways in which the body can be taken from one pressure, volume and temperature to another. And it certainly does not mean that Q is a function of p, v and U (the old total heat doctrine). Planck suggested that the equation be written $Q = dU + pdv$.[8] In fact the equation can be made exact and integrated *if the course of the change is prescribed.*

term being that which corresponds to the external work; at the end of a closed cycle U is the same, of course, as it was at the beginning. Clausius extends his analysis to vapours and shows that, Regnault having modified Watt's law, the experimental observation made by de Pambours that the steam exhausted from the cylinders of a locomotive is saturated—an observation previously taken as confirmation of Watt's law—is now consistent with an important deduction from the dynamical theory of heat: that the specific heat of saturated steam must be negative. He then derives the relationship between the principal specific heats of gases, showing that their difference must be equal to JR and therefore constant at all temperatures, and furthermore that equal volumes of all gases at the same temperature, equally compressed, release—or as we should now say, generate—the same amount of heat.

So far Clausius has based his discussion on the Joule theory of the convertibility of heat and work, the only use he has made of Carnot's theory has been the closed cycle. But even on the new theory there is transmission of heat from the hot to the cold body, as well as the conversion of heat into work during the cyclic process. We do not need a Joule-type equivalent, relating the heat transmitted to the work done, but there may still be a definite relationship between the two. The core of Carnot's argument is that there is just such a definite relationship: that the work done is determined by the heat transmitted and by the temperature difference, but not by the nature of the working substance. Is this argument still valid?

Clausius asserts that if there were some working substance, or engine, to provide more work for the fall of a given amount of heat than another engine; or, what amounts to the same thing, were it to provide the same work for the fall of a smaller amount of heat, then if the first engine drives the second one in reverse we would have an entirely anomalous result: no net work would be done and heat would, in effect, flow from a cold to a hot body. This, he postulates, is quite impossible; and thus he proves Carnot's fundamental principle. This, in fact, constitutes the first historical statement of the second law of thermodynamics. Carnot, we remember, went on to argue that a more efficient engine than a completely reversible one is impossible, since it would logically imply the feasibility of perpetual motion; and so indeed it would if the caloric theory, or the axiom of conservation, were true. The situation would be exactly the same as if a perfect column-of-water engine were to be driven in reverse, as a pump, by an even more perfect engine! But we are no longer dealing with fluids, subtle or

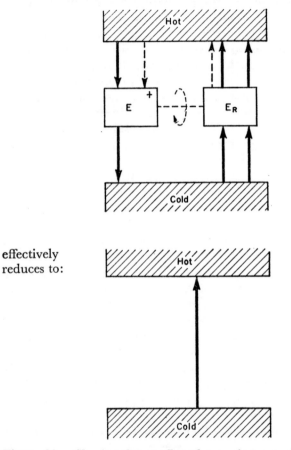

Figure 21.   *Showing the net effect of an engine more efficient than a perfectly reversible one. The vertical dotted lines denote the heat converted into work and vice versa.*

gravitating; we are dealing with energy, and the extension of the argument merely leads us to the conclusion that the flow of heat from a cold to a hot body would enable us to perform work by applying the heat to a third engine, which would thus be deriving its energy from that *stored in the cold body.* Now there is, *a priori*, nothing in such an arrangement to contradict the energy principle as enunciated by Joule and Mayer and, more recently, in a very general form, by Helmholtz. All that would be happening would be that the thermal energy in the 'cold' body would be being converted into the equivalent amount of mechanical energy—and there would be nothing absurd or self-contradictory about that. It certainly does not violate the great new principle of the conservation of energy. Clausius therefore rests his proof on the flat assertion that it is impossible for heat to flow of its own accord from a cold to a hot body. It is, of course, common experience that hot bodies always tend to cool down, that you cannot boil a kettle of water by putting it on a block of ice, and so on. Now the common experience of countless generations from time immemorial is to be elevated into a scientific axiom. We shall discuss the admissibility of this procedure later. For the present we can summarise the situation by pointing out that if the production of mechanical energy can only come about through a *fall* of heat then Carnot's demonstration is exact; but if the production of mechanical energy is the result only of the *conversion* of a quantity of heat then it has no force whatsoever.

When Kelvin came to accept the implications of the dynamical theory of heat he preferred to state the second law of thermodynamics in the form of an admission that one cannot obtain a continuous supply of work by cooling a body below the temperature of its surroundings; or as Ostwald later put it, a *primum mobile* of the second order is impossible. Long experience had convinced men that a *primum mobile* of the first order is impossible; a continuous supply of mechanical energy cannot be obtained from an arrangement of machines, no matter how ingenious; nature cannot be cheated, as the seventeenth-century mechanical philosophers had realised. This principle now received a significant extension: continuous mechanical energy cannot be obtained merely by exhausting the heat energy from a particular body. If this were possible then the Atlantic Ocean, which constitutes a vast heat reservoir, would provide an (effectively) inexhaustible supply of energy. This is regrettably not the case. The heat energy in the Atlantic Ocean is not available for practical purposes. An engine of the type represented in Figure 22 is a scientific impossibility.

Clausius then proceeded to study the implications of the

reformed Carnot principle that the *transmitted* heat bears a fixed relationship to the work performed and the temperature difference only. He divides the work performed in a closed cycle by the amount of heat *transmitted** and sets the ratio, the work done in proportion to the transmission of a unit of heat, equal to $1/C$ times the (small) temperature difference.

Applying this relationship to gases and vapours, he finds

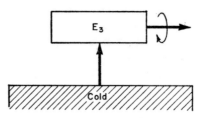

Figure 22

expressions which correspond to those obtained by his application of the Joule–Mayer theory to cyclic operations† and in the case of gases he is able to find a value for C; comparing the results of the Joule–Mayer procedure (work equal to heat converted) with those of the Carnot procedure (work proportional to the heat transmitted) he finds that $C = \dfrac{1}{J}\left(\dfrac{1}{\alpha} + t\right)$.

For vapours, like steam, he deduces the Clapeyron equation, now soundly based on the dynamical theory,‡ and goes on to

---

* By putting $\dfrac{\text{Work done}}{\text{Heat transmitted}} = F(t_1, t_2)$, where t_1 and t_2 are the temperatures of the hot and the cold bodies respectively, Clausius argues that when the temperature difference is very small, merely dt in fact, the function $F(t, t - dt)$ can be expanded and the higher powers of dt neglected so that we have, for $F(t_1, t_2)$ the function $f(t)dt$, where $f(t)$ is a function of t only. This he puts equal to $1/C$ so that the ratio of the work done to the heat transmitted becomes $1/Cdt$.

† i.e. on Carnot's principle we have $\left(\dfrac{dQ}{dv}\right) = \dfrac{RC}{v}$; on Joule–Mayer's,

$\left(\dfrac{dQ}{dv}\right) = \dfrac{R}{J}\left(\dfrac{1/\alpha + t}{v}\right)$.

‡ i.e. $L = \dfrac{1}{J}\left(\dfrac{1}{\alpha} + t\right)(v_2 - v_1)\dfrac{dp}{dt}$.

show that when C is calculated by means of his formula $\dfrac{1}{J}\left(\dfrac{1}{\alpha} + t\right)$ it increases in the same ratio as that found by Clapeyron and Kelvin. He concludes his paper with a detailed study of the behaviour of saturated vapours, showing that they depart widely from the general gas law, $pv = R\left(\dfrac{1}{\alpha} + t\right)$, and that the specific heats of such vapours must be negative. Finally he calculates the value of J from his gas and vapour equations and shows that the results obtained accord with Joule's experimental determinations.

Let us now consider the law on which he bases the re-established Carnot principle. That is, that heat cannot, of itself, flow from a cold to a hot body. What are we to make of this? As we remarked, it has always been common knowledge that hot bodies tend to cool down and that cold bodies placed on or near a fire tend to warm up. But 'common knowledge' is not necessarily a suitable basis for a scientific law. After all, it is common—and reasonably correct—knowledge that the heavier a body is the faster it falls; but two thousand years of misguided science had to elapse before Galileo put mechanics on a correct basis by explicitly denying this commonsense 'law', confirmed almost daily in human experience. It seemed to follow then either that Clausius' law was trivial or that it might well be wrong; so, at least, a number of his contemporaries reasoned. Let us consider the three following cases.

Wilcke had, in the course of his investigations into latent heat, searched for the cooling effect which melting ice would have, he thought, upon the ice-cold water in which it was floating.[9] Such cooling, 'necessitated' by the latent heat required to melt the ice, would amount to a flow of heat from a cold to a hot body (the melting ice). Whether or not Wilcke pondered the implications of this deeply we do not know, but evidently he did not feel that any fundamental issue was at stake.* In this simple case, then, it was not immediately obvious

---

* Dr Heathcote and the late Professor McKie suggested (*Discovery of specific and latent heats*) that Wilcke might, had he pondered the problem sufficiently, have arrived at a form of the second law of thermodynamics. This seems extremely improbable. The second law relates to the conditions under which various forms of energy are transformed into heat, or *vice versa*. Wilcke had no conception of such transformations, which were quite

that heat must *always* flow only from hot to cold bodies. We remember too that Rumford had argued that the heat necessary to melt a block of ice would, if applied to the same weight of gold, raise it to a bright red heat; implicitly this argument is not valid unless one assumes that in some way or another it is possible for heat to flow from a cold body (melting ice) to a hot body (hot gold) *without any other effect or change taking place*. Once again the second law of thermodynamics in the form Clausius gave it is not immediately obvious or trivial.

Secondly there is the very familiar case of the burning-glass. We may ignore the ridiculous fable of Archimedes and the legendary burning-glass which he is supposed to have used to incinerate the Roman fleet. But it was well established by the eighteenth century that a very large burning-glass, concentrating the sun's rays, could generate a temperature high enough to burn up a diamond. Such a high temperature was, of course, well beyond any contemporary thermometric device or computation; but *it was by no means obvious that it must be less than that of the surface of the sun*. Indeed the simple analogy of the speaking trumpet or megaphone, which can make a man's voice seem louder than it actually is, might suggest that a burning-glass can produce temperatures at its focus much higher than those of the source from which the radiant heat comes.

The third and last case was the subtle 'thought experiment' suggested by Hirn and discussed by Clausius.[10] Imagine two metal cylinders in which two frictionless pistons are linked together so that as one descends the other ascends by an equal amount (Figure 23). The cylinders are connected at the bases by a spiral tube which is immersed in a steam bath. Both cylinders contain air, the temperature of one being maintained at 0°C and the other at 100°C. If now a tiny, vanishingly small, weight be added to the piston in the cold cylinder it will start to descend, driving the air through the spiral into the hot cylinder. This air will heat up to 100° and its pressure will, accordingly, rise as the net volume is kept constant. The increase in pressure

---

outside the scope of eighteenth-century physics. At most he might have confirmed the commonsense impression that one cannot boil a kettle on a block of ice, and this is no more a statement of the second law than the observation that apples tend to fall *downwards* from apple trees is a statement of the Newtonian principle of universal gravitation.

will therefore heat the air additionally, above 100°, but the process will continue as the pressures in the two interconnected cylinders must be the same. In the end all the air will be in the hot cylinder, and its temperature will be about 120°, the additional 20° being due to the compressive heating. Virtually no work will have been done on the air—the small weight can be considered evanescent—yet the net effect is that a source of heat at 100° has heated air up to 120°; accordingly heat may

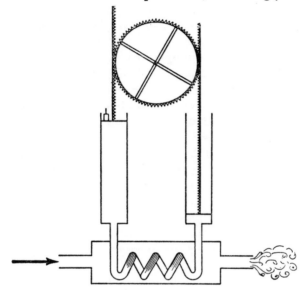

Figure 23

be considered to have flowed from a cold to a hot body. The basic axiom is therefore apparently wrong.

These three examples show us that in 1851, whatever 'common sense' might suggest, it was not firmly established that heat in all circumstances and under all conditions *must* flow from a hot body to a cold body and never in the reverse direction. Clausius had therefore to derive his axiom from general experience and then to postulate that in those cases where there might appear to be ambiguity or doubt—such as the three discussed above—it would on closer examination and in the light of further knowledge be always found to be true.

The second law of thermodynamics, expressed in the form of

the proposition that heat cannot flow from a cold to a hot body without some other change taking place, is not therefore a statement of the scientifically trite but an extremely fruitful law from which important conclusions flow. For the present we note that the Hirn 'thought experiment' can be simply explained if we regard the engine as in fact comprising two simpler engines: one is a heat-engine performing work by the expansion of air and absorbing heat from the hot body to do it, while the other is a compressor which absorbs work to produce high temperatures by compressing air. In much the same way a Watt engine working with steam at 100°C could produce very high temperatures by simple friction, percussion or compression. But a better insight is provided by Clausius' reply to Hirn: if the air in the cold cylinder was at 100° and not at 0° then there *would* have been a violation of his law if the final temperature of the heated and compressed gas was 120°. For in this eventuality heat would *everywhere* have flowed from a 'cold' to a hotter body and so the increase in temperature would have taken place of its own accord; or, as we might say, *without* a compensating flow of heat *somewhere* from a hot body to a cold one.

Kelvin formally accepted the dynamical theory in his paper of March 1851.[11] He had been anticipated not only by Clausius but by Rankine, who in a paper of great ingenuity had reached a number of Clausius' conclusions from a rather unconvincing hypothesis that heat phenomena resulted from certain specified molecular motions.[12] It was this metaphysical assumption, together with a certain obscurity of style, that deprived Rankine of the full measure of credit that would otherwise be his as one of the two founders of the modern theory of thermodynamics. Although Kelvin and Rankine were undoubtedly in friendly contact at this time—it would be interesting to know more about the relations between these two very distinguished men —it was Clausius rather than Rankine whom Kelvin followed. But we shall return to this question later on.

Kelvin was fair in ascribing the credit for the two main laws of thermodynamics to Joule and to Carnot and Clausius. But he does not accept Clausius' axiom on which the second law rests. According to Silvanus Thompson this was because he felt that it contradicted Prévost's law of exchange.[13] But it is very difficult to see the force of such an objection, or indeed its relevance, and it is surely more likely that Kelvin wanted to express his own ideas in his own way. Hence he put forward the second law of thermodynamics in the form of an assertion that it is impossible to derive a continuous supply of work by cooling a body below the temperature of all its surroundings. At this point he feels he has an answer to that fruitful query, prompted by Joule's challenge, of 1848: what happens to the mechanical effect, which does not appear in the case of straightforward thermal conduction? It is, he now replies, '. . . irrevocably lost to man and therefore "wasted" although not *annihilated*'.[14] This answer, although quantitatively—and perhaps emotion- ally—satisfying would hardly please a logician: if the energy is 'irrevocably lost' how can one possibly know that it is *not* 'annihilated'?

Re-interpreting the work done by a quantity of heat absorbed at a given temperature Kelvin finds a new expression, based on the principle of dynamical conversion, which yields smaller values than the expression based on the conservation principle, but which approximates to the latter as the temperature dif- ference is reduced.* When the temperature difference is infinitely small the two formulae are the same (see first footnote, page 242 above). At the same time he finds an expression for the heat transmitted to the sink so that the sum of the heat transmitted and that converted into work must equal that absorbed.

As early as December 1848 Joule had written to Kelvin sug- gesting that Carnot's function might be simply as the tempera- ture on the gas scale from the zero of expansion.† The values of

---

* If Q_1 is the heat absorbed and t_h and t_c are the temperatures of the source and the sink respectively, Kelvin finds that, on the dynamical theory, the work performed by an engine working a reversible cycle between t_h and t_c is:

$$W = JQ_1 \left[1 - \exp\left(-1/J \int_{t_c}^{t_h} \frac{1}{C} \, dt \right) \right] \text{ in place of } Q_1 \int_{t_c}^{t_h} \frac{1}{C} \, dt.$$

These expressions approximate as the difference between t_h and t_c is reduced. The expression for the heat transmitted to the sink now becomes:

$$Q_2 = Q_1 \exp\left(-1/J \int_{t_c}^{t_h} \frac{1}{C} \, dt \right)$$

† Joule actually followed Kelvin's designation of Carnot's function as the reciprocal of C; we have therefore departed from Joule's words in order to

C might therefore be calculated by means of the formula:

$$C = \frac{1}{J}\left(\frac{1}{\alpha} + t\right).$$

As we saw, Kelvin himself had derived this formula, implicitly, in 1849, and Clausius in his first memoir had used it to calculate C. If we substitute this value for C in our equations for the work done and the heat transmitted to the sink we get the modern formulae:*

$$W = J\frac{t_h - t_c}{(1\alpha + t_h)}Q_1$$

and

$$Q_2 = Q_1\frac{(1/\alpha + t_c)}{(1/\alpha + t_h)}$$

Kelvin then went on to repeat the steps of deriving the relationship between the two principal specific heats, showing that their difference is equal to R; to obtain the Clausius–Clapeyron equation and to confirm that the specific heat of saturated steam is negative.

In the following December Kelvin introduced the notion of the 'total mechanical energy' of a body: the total mechanical effect it would produce if cooled to the uttermost. But as we are ignorant of 'perfect cold' and of the intermolecular forces in a body, we must take the mechanical energy at a certain given state of the body as the datum from which to measure changes in energy.[15] He goes on to remark that the value of the

---

make the expression of his views consistent with the writings of Clapeyron and Clausius.

* Substituting for C in the expression $\exp\left(-1/J\int_{t_c}^{t_h}\frac{1}{C}\,dt\right)$ we get

$\exp\left[-\int_{t_c}^{t_h}\frac{dt}{(1/\alpha + t)}\right]$ and this gives us, $\exp\left[-\text{Log}\,(1/\alpha + t)\right]_{t_c}^{t_h}$ which,

of course, is equal to, $\dfrac{(1/\alpha + t_c)}{(1/\alpha + t_h)}$ and the above two formulae follow imme-

diately. In their modern form they are usually written $W = J\cdot\dfrac{T_1 - T_2}{T_1}Q_1$

and $Q_2 = Q_1\cdot\dfrac{T_2}{T_1}$, where the capital letter T denotes absolute temperature.

Kelvin referred several times to Joule's letter, to the annoyance of Clausius who, not unnaturally, assumed that Kelvin was trying, indirectly, to claim priority for Joule.

Plate XXI. James Prescott Joule as a young man. (*By courtesy of the Manchester Central Reference Library*)

Plate XXII. Peter Ewart. (*By courtesy of the Manchester Central Reference Library*)

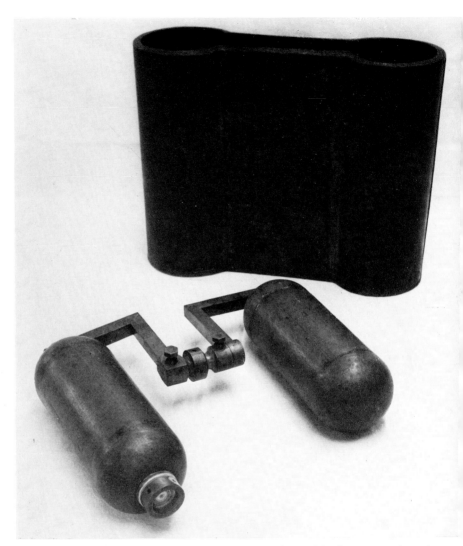

Plate XXIII. Apparatus used by Joule in his demonstration that when a 'perfect' gas expands without doing work its 'internal energy' remains unchanged. In the foreground are the strong copper vessels; in the background is the calorimeter designed to accommodate the copper vessels; it has double walls to minimise heat losses. (*Photograph by Mr Harry Milligan*)

Plate XXIV. Joule's idea of the transformation of mechanical energy into heat energy. (From one of his laboratory notebooks)

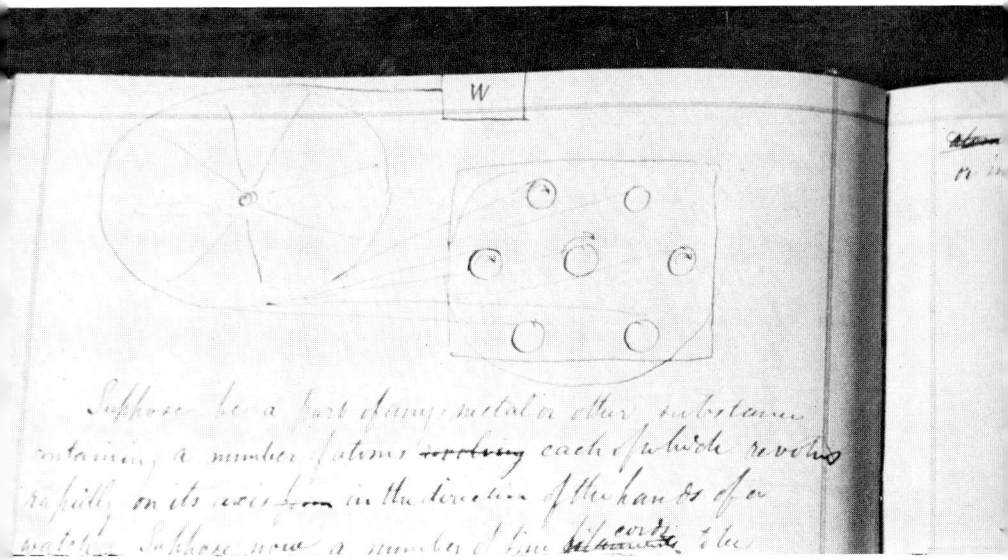

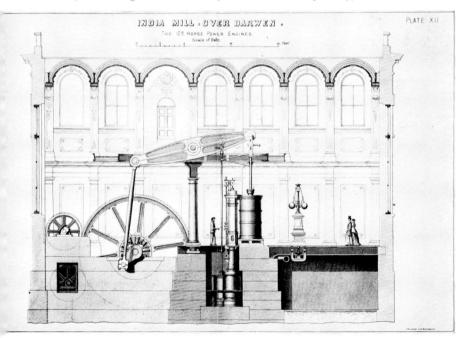

Plate XXV. The ultimate refinement of the beam engine: a rotative engine installed in a cotton mill near Darwen, Lancs., *circa* 1880. (Evan Leigh, *The Science of Modern Cotton Spinning*)

Plate XXVI. The last of the line. This steam engine, built by Galloway's, was one of the last to be installed in a cotton mill (Messrs Melland's, Elm Street, Burnley). It had previously been displayed, as an example of engineering skill, at the Empire Exhibition, Wembley, in 1924.

Steam is admitted to the high-pressure cylinder, on the left, at about 165 p.s.i. The low-pressure cylinder, on the right, works on the uniflow principle. Power for the textile machines is taken by means of rope-drive from the large flywheel in the background.

heat energy added to a body 'must be diminished by the work done . . . in expanding against resistance [in order to find] the actual increase in mechanical energy which the body acquires'.

Three months later, in February 1852, he points out that the heat radiated by the sun is the principal source of mechanical effect available to man; and then in the following year he produces one of his most perceptive papers: 'On a Universal Tendency in Nature to the Dissipation of Mechanical Energy.'[16] He observes that when heat flows from a hot to a cold body the energy cannot be annihilated but must undergo *transformation*. He continues;

> When heat is created by a reversible process (so that the mechanical energy thus spent may be restored to its primitive condition) there is also a transference from a cold body to a hot body of a quantity of heat bearing to the quantity created a definite proportion depending on the temperatures of the two bodies.

In all irreversible processes—friction, conduction, radiation, chemical change—there is dissipation of energy and full restoration is impossible. The concluding words of this brief paper are illuminating. Kelvin lays down that there is in the material world a universal tendency to the dissipation of mechanical energy; that any restoration of energy without more than an 'equivalent' dissipation is, and probably always will be, impossible; and that the world will become at a finite period in the future, unfit for man to live on, at least according to the laws of nature as now understood.

In this cosmic generalisation we note, clearly enough, the continuation of the philosophy which recognised heat as the grand moving agent of the Universe. One man who was not prepared to accept this generalisation immediately was Rankine. He wrote a short paper, hardly more than a letter,[17] to the effect that it was just conceivable that the Universe possessed certain quasi-optical properties so that the energy, bein dissipated in one place—our quarter of the cosmos—might be in process of reconcentration, or refocusing, in another. Thus the net effect over the whole would be that there was no dissipation of energy. We shall return to this point shortly. In the meantime the words 'transformation' and 'equivalent' should be remembered, as should the contexts in which Kelvin used them.

Kelvin's next important paper, on thermo-electric currents, appeared in May 1854.[18] He begins by restating the two laws which govern the conversion of heat into mechanical work and *vice versa* either in a continuous, uniform process or in a closed cycle of operations:

Law I. The system must give out exactly as much energy as it takes in, either in heat or in mechanical work.

Law II. If the process be perfectly reversible and if the heat is either absorbed or transmitted at one or other of two temperatures, then the amount of heat absorbed or transmitted at the higher temperature must exceed that transmitted or absorbed at the lower, always in the same ratio when the temperatures are the same, whatever be the nature of the working substance or of the operations concerned.

This is a general statement of the relationship

$$\frac{Q_1}{Q_2} = \frac{1/\alpha + t_h}{1/\alpha + t_c}$$

But he has now accepted Joule's suggestion that the Carnot function is simply as the value of the temperature measured from $-273°$ (specifically, $C = T/J$, where J is Joule's equivalent). If, now, we write $T = (1/\alpha + t)$, where T is the new, absolute temperature as defined by Carnot's function, we obtain the simple relationship:

$$\frac{Q_1}{Q_2} = \frac{T_1}{T_2} \text{ or, } \frac{Q_1}{T_1} = \frac{Q_2}{T_2} \text{ or, } Q_1 = \frac{T_1}{T_2}Q_2$$

As Kelvin put it: '. . . the absolute values of two temperatures are to one another in the proportion of the heat taken in to the heat rejected in a perfect thermo-dynamic engine working with a source and refrigerator at the higher and lower of the temperatures respectively.'

The new thermodynamic scale of temperature, unlike the first one, was naturally in agreement with the gas scale. And then, in a moment of penetrating insight, Kelvin saw that these simple relationships amounted to a quantitative, or mathematical statement of the second law of thermodynamics. If we consider the heat absorbed as positive and that transmitted as negative then the relationship becomes:

$$\frac{Q_1}{T_1} + \frac{Q_2}{T_2} = 0.$$

He is able to generalise this to include complex reversible cycles—we have come a long way from the simple expansive operation of steam-engines—and to lay down that in such complex cycles the quantities of heat are related to the temperatures (absolute) at which they are absorbed or transmitted by the linear equation:

$$Q_1 \cdot \left(\frac{1}{T_1}\right) + Q_2 \cdot \left(\frac{1}{T_2}\right) + Q_3 \cdot \left(\frac{1}{T_3}\right) + \ldots Q_n \cdot \left(\frac{1}{T_n}\right) = 0.$$

This, Kelvin, asserts, can be taken as '. . . the mathematical expression of the second fundamental law of the dynamical theory of heat'. The corresponding expression for the first law he writes as:

$$W + J \cdot (Q_1 + Q_2 + Q_3 + \ldots Q_n) = 0.$$

Thus the first law of thermodynamics expresses the equivalence between heat energy and mechanical energy, while the second law imposes a definite restriction on the ways in which the conversion from one form of energy to the other can take place. In the case of completely reversible changes the second law can be expressed mathematically by Kelvin's linear equation.

Kelvin had been rather slow to accept the dynamical theory of heat, together with its full implications. This may have been due either to an innate caution, or to his admiration for and understanding of Carnot as well as of other French writers,* or to the surprising confirmation of Carnot's theory when the lowering of the melting point of ice under pressure was detected, or, lastly, to some combination of all or some of these factors. On the other hand Kelvin had a close relationship with Joule, and it must remain a puzzle that while men like Clausius and Rankine accepted Joule's results easily enough Kelvin could not do so for some years.

Once Kelvin had accepted the new theory, however, he produced in the course of a few years an abundance of new ideas, which were to prove immensely important: many of them in the hands of other men, such as Clausius. Prominent

---

* It is ironic that Duhem, in the course of his attack on 'English' science, should have selected Kelvin as exemplifying some of the worst 'English' traits, while praising Rankine for his supposed conventionalist views. Duhem's judgement seems to have been wrong on both counts.

among these ideas were the concept of the 'energy' of a body, the insight into the process of dissipation of energy, the establishment of an absolute scale of temperature, the awareness of the importance of 'transformation' and 'equivalence' and, finally, the expression of the second law of thermodynamics in a mathematical form:

$$\frac{Q_1}{T_1} + \ldots \frac{Q_n}{T_n} = 0.$$

Finally, we note that the function C was originally proposed as the measure of the work done per degree temperature difference by a unit of heat at a given temperature. It was first visualised by Carnot, refined by Clapeyron, who measured its value at five different temperatures, and by Kelvin who, using Regnault's results, extended the range of measurement and established that it diminished steadily with the temperature. It was brought into the context of the dynamical theory of heat by Clausius and finally related by Joule and Kelvin to the reciprocal of the absolute temperature on the thermodynamic scale. If we turn back now to the earlier work and substitute 'T' for 'C' we can appreciate how close writers like Clapeyron were to the truth. Thus, with this substitution, his expression for the difference between the principal specific heats becomes correct, and so does the latent heat equation.

THE CONCEPT OF ENTROPY

Clausius saw clearly that the Carnot cycle represents a limiting, or ideal case. It is one end of a wide spectrum of thermal transformations, which go on in nature and indeed under man's control as part of his developed technology. In this respect the Carnot cycle is similar in kind to Galileo's intuition of the laws of falling bodies and of (circular) inertia: eliminate all imperfections—air resistance, friction, etc.—and under these ideal conditions all bodies fall with equal acceleration, or continue in uniform motion for ever. We represent this view of the Carnot cycle as the limit of all thermal transformations very simply and diagrammatically in Figure 24. The other extremity of this spectrum may be taken to be the simple conduction of heat without the generation of mechanical power:

what we have called a heat-engine of zero thermodynamic efficiency. The intermediate positions between these two extremes are occupied by such engines as those of Newcomen,

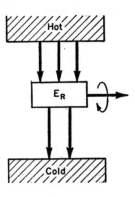

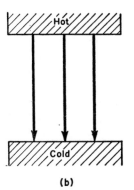

(a)

(b)

Figure 24. *Showing the two extremes of the spectrum of thermodynamic transformations.*

(a) *A heat-engine of maximum efficiency (Carnot).*
(b) *A heat-engine of zero efficiency (simple conduction).*

Watt, Trevithick and, approximating to the Carnot ideal, of Rudolf Diesel.

Clausius further recognised that in all cases when heat is converted into work in a continuous or cyclic fashion there is, unavoidably, a flow of heat from the hot body to the cold body. Indeed, it seems to have been the realisation of this that enabled Clausius to salve Carnot's ideas and reconcile them with the dynamical theory of heat; Kelvin on the other hand seems to have supposed that the doctrines of the *conversion* of heat into mechanical energy and of the *transmission* of heat *associated* with the production of mechanical energy were entirely incompatible with one another: if one is correct, the other must be wrong. This conjectured failure of insight on Kelvin's part provides us with another explanation, possibly the most convincing of all, of Kelvin's failure to grasp the significance of the dynamical theory of heat: he did not realise that there was a third option available; one which reconciled Carnot's doctrine of transmission with Joule's doctrine of conversion. Clausius did realise this.

If we now detach our minds from the accumulated prejudices of the past and think of heat and mechanical energy in the same context, and not as two radically different things, we can see that, in the Carnot cycle, two related *transformations* take place at the same time: a transformation of heat energy into mechanical energy and a transformation of high-temperature heat energy into low-temperature heat energy. The flow of heat energy from hot to cold is inevitable even in the limiting, ideal Carnot cycle. In the latter, in fact, the two transformations are mutually determined by the temperatures of the source and the sink, t_n and t_c, or in absolute temperatures, T_1 and T_2. For if this were not so, were one reversible engine to transform more energy into mechanical work and less into low-temperature heat energy than another engine, then it would only be necessary to use the first engine to drive the second one backwards to achieve the flow of heat from a cold body to a hot body without any other change taking place; and this would violate the second law of thermodynamics. Accordingly in a reversible cycle the relationship between the two transformations is fixed rigorously by the temperature differences only.

The elimination of this distinction between 'heat' conceived as a subtle fluid, or at least as conserved as a distinct entity, on the one hand, and mechanical power on the other, and the establishment of both as forms of 'energy', together with the general notion of the *transformation* of energy, invented by Clausius, call—indeed almost beg—for quantification in terms of a suitable mathematical law of nature.* The situation was

* The merging of 'heat' into 'energy' presents certain psychological but not intellectual or conceptual difficulties. Thermodynamic and in fact

rather like that in the seventeenth century when hitherto unrelated tracts of human experience were being reduced to
scientific law: planetary motion, projectiles, tides, pendulums.
In these cases the concepts used—mass, momentum, force,
gravitation—were related in the scheme of Newtonian laws of
motion and the principle of central forces. In the field of
thermodynamics it was now necessary to reduce the phenomena of the transformation of energy—a common feature of
the cosmos—to scientific law. Broadly speaking the three
important aspects of thermodynamics were:

(1) Heat is a form of energy, and the mechanical value of a
unit of heat has been determined by Joule (first law of
thermodynamics).
(2) Thermal and thermo-mechanical processes are essentially
transformations of energy.
(3) Nevertheless there seems to be a built-in bias in nature
whereby the energy transformed by such *irreversible* processes as conduction, friction, percussion, etc., is, as Kelvin
put it, dissipated, and cannot be transformed back, as the
unqualified first law of thermodynamics would seem to
imply should be possible.

The problem for Clausius was to derive from these established principles suitable laws—analogous for those of motion—
governing the transformation processes.
    Let us revert to Clausius' initial insight. We have two related
transformations which we can, in imagination, separate quite
distinctly into a flow of heat energy and a conversion of heat
energy into mechanical energy. But then we are struck by a
limitation which restricts the *generality* of our idea: the heat to
be transformed into mechanical energy is absorbed at the same
temperature as that to be transformed into low-temperature
heat energy.[19] *Generally* we suppose that this is not the case; why
should we assume that *all* transformations will be of this simple
and rather special sort? Here again Clausius reveals the

---

thermal processes are ones in which energy is transformed, not ones whereby
an elusive fluid, or some undefined entity, produces thermal and mechanical
effects. That the old conservationist attitude is still powerful is revealed by
the persistence of such regrettable expressions as 'latent heat', 'heat
capacity', etc.

penetration of his scientific insights: the significance of the
general case lying behind the special instance.

How, then, in our general process does the flow of heat
energy from a high to a low temperature determine the amount
of heat energy converted into the mechanical form at any other
temperature? To answer this question, which lies near the heart
of thermodynamics, we need only devise a reversible cycle
which has *three* heat reservoirs, hot, intermediate and cold, and
suppose that the working substance is taken through a compound cycle of three stages. Then if Q_1 is the heat absorbed

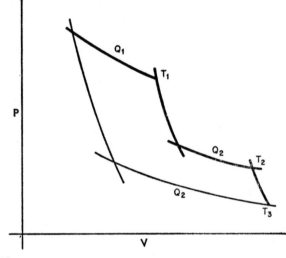

Figure 25

during the 'isothermal' expansion at temperature T_1, and Q_2
is the heat absorbed at T_2, we can easily arrange the cycle so
that during the 'isothermal' compression at T_3 (Figure 25) as
much heat, Q_2, is given up to the cold reservoir as was absorbed
from the intermediate one. We have now, in fact, clearly separated the two transformations: the passage of heat energy Q_2
from a body at temperature T_2 to one at T_3, on the one hand,
and the transformation of heat energy Q_1 entirely into mechanical energy. Furthermore, these two transformations can be
equated; they are, in fact, *equivalent*. Remembering that we are
dealing with perfectly reversible cyclic operations and that we
have separated the two related transformations we can say that

the flow of heat energy from a hot to a cold body has the same *equivalence value* as the transformation of a given amount of mechanical energy into heat energy at a third temperature, since the one can be substituted for the other without causing any change; or, a flow of heat can be negated by a conversion of mechanical energy into heat, and *vice versa*. In a few words, we are dealing with transformations that *compensate* one another.

Clausius' aim is to quantify these transformations so that mathematical equalities can be introduced. It is reasonable to assume that the value, the equivalence value, of a transformation of heat into mechanical energy must be the product of the quantity of heat produced and some function of the temperature; the sign—positive or negative—will depend upon the direction of the transformation, and Clausius proposes that the transformation of mechanical energy into heat be taken as positive, that of heat into mechanical energy as negative. Nature herself seems to favour the transformation of mechanical energy into heat energy—there is a bias in that direction—so it seems logical to denote this as the positive process.

Let us write $Q_1.f(T_1)$ as the equivalence value of the generation of heat energy Q_1 from mechanical energy at the temperature T_1, and $Q_2.f(T_2, T_3)$ as the equivalence value of the flow of heat energy Q_2 from T_2 to T_3. What is the nature of the function $f(T)$? At this point Kelvin's paper of 1854 on thermoelectric currents must have given Clausius some help, for he took $f(T)$ as being, in fact, $1/T$. He then showed that $f(T_2, T_3)$ was equivalent to $(1/T_3) - (1/T_2)$, so that his equation asserting the equivalence between a flow of heat energy and the conversion of heat into mechanical energy becomes:

$$-\frac{Q_1}{T} + \left(\frac{Q_2}{T_3} - \frac{Q_2}{T_2}\right) = 0$$

He pointed out that the flow of heat energy is equivalent to *two* transformations of the first sort—that is to say, it can be replaced by a transformation of heat energy into mechanical energy at the higher temperature (equivalence value: $-Q_2/T_2$) and this can in turn be transformed back into heat energy at the lower temperature (equivalence value: $+Q_2/T_3$). The total effect of these must be a positive transformation, for T_3 is smaller than T_2, and this accords with our convention that the flow of heat from hot to cold should be taken as *positive*, in terms of transformation value.

In this way Clausius quantified the two types of transformation that occur in reversible cyclic changes. He found, in the case of both types of transformation, a measure whereby the equivalence value of one can be compared with that of the other. In the case of the general reversible cycle the net equivalence value of all the transformations will be $\int \frac{dQ}{T} = 0$.

But what, now, of *irreversible* changes?

In such cases it is quite clear that, compared with reversible operations, relatively more heat flows from the hot body to the cold one than is converted into mechanical energy; such had been the experience of all engine-builders from the time of Newcomen onwards. Indeed the whole aim of engineers had been to reduce the former transformation and increase the latter as much as possible, and it was by attending to such things as thermal lagging, expansive operation and the pre-heating of boiler-feed water that the Cornish engineers had achieved such remarkable successes in the first half of the nineteenth century.

In these, the usual and commonly occurring cases, $\int \frac{dQ}{T} > 0$, for the positive and negative transformations do not compensate each other; the equivalence value of the negative transformation, heat energy into mechanical energy, is too small to compensate for the equivalence value of the positive transformation of high-temperature heat into low-temperature heat. There must, therefore, be a net positive change.

We can combine our expressions, quantifying all thermodynamic transformations in the form of a general expression:

$$\int \frac{dQ}{T} \geqslant 0.$$

The sum total of the equivalence values of all transformations is either positive or, in the case of a reversible cycle of operations, equal to zero. This expression can be taken as the mathematical statement of the second law of thermodynamics. To understand this, let us consider the ideal case of the reversible cycle. If the value of $\int \frac{dQ}{T}$ was negative, or less than zero, it would mean that (we remember the principle of the equivalence of transformations) the net effect must be equivalent to a flow of heat from a cold body to a hot body. This would violate the second law of thermodynamics. If, on the other hand, the expression is greater than zero, or positive, we have only to reverse the cycle to make the value negative, and thus again violate the second law. As the value of the expression must be positive for irreversible cycles we are left with the conclusion that $\int \frac{dQ}{T} \geqslant 0$ is the mathematical expression for the second law of thermodynamics.

Kelvin, as we saw, answered his own query about the work that could be done by heat when it flowed by simple conduction from a hot to a cold body. He observed that the energy was not destroyed, not annihilated, but made unavailable: transformed, in Clausius' terminology, into heat energy at low temperature. In the formula $\int \frac{dQ}{T} > 0$, Clausius has succeeded in expressing this process in a mathematical form.

However, the theorem of the equivalence of transformations is not restricted to cases of *external* transformations of energy. There is, Clausius remarks, the problem of the *internal* energy of a body, too. Once again Kelvin had led the way by recognising the concept of the internal energy of material bodies and, as we know, the notion of inter-molecular forces against which the expansive power of heat operated was a very old one, going back well into caloric times, to Dalton at the latest. We remember, too, that the difference between the two principal specific heats of a gas had long been ascribed to the effort supposedly required to separate molecules when the gas was allowed to expand on being heated. But in a cyclic operation, whether reversible or not, the net *internal* work that is done must be zero when the working substance is brought back to the original state at the end of the cycle. The only work that is done in such a case is external work.

At this point Clausius faced a difficulty that caused him to delay for some eight years the publication of his theory of the equivalence of transformations in its complete form: that is, extended to include the transformations of the internal energy of material bodies.[20] When a quantity of heat is imparted to a body some of it goes towards increasing the thermal content of the body, some of it is used to do work against external forces, and some of it to perform internal work against the forces of mutual attraction between constituent molecules. The problem for Clausius was: what, in these circumstances, becomes of the specific heat of the body? It appeared to him that differences in specific heats must be more apparent than real, for they are surely no more than the differences in the internal work performed when a body expands? In fact, the 'real' specific heat of water would be that measured when steam was at a very low density; that is, when it approximated to a perfect gas, and inter-molecular forces were eliminated. Otherwise, he argued, the specific heat of water is always the same, and the differences which appeared between the measured specific heats of ice, steam and water simply reflected the different amounts of internal work performed by the heat in the three cases. A consequence of this was that Clausius asserted that there was no difference between the amounts of heat in water and in ice at 0°C; the 'latent' heat of fusion being merely the energy required to change the state of the substance. This was, in fact, an echo of Southern's opinion, as expressed to Watt sixty years earlier.

When heat is imparted to a body it causes it to expand, or at any rate—we recall the behaviour of water between 0° and 4°C —to undergo some sort of change of molecular arrangement. It is by virtue of such changes that the body as a working substance does external work, while the change of arrangement itself implies internal work. If we impart a given amount of heat, $-dQ$, from a source to a material body we may write: $dQ + dH + AdL = 0$, where dH is the increment of the internal heat, dL is the increment of internal *and* external work, and A is equal to $1/J$.*

Clausius goes on to point out that when the heat in a body does work by overcoming resistances, the magnitude of the resistances that it can overcome is proportional to the absolute temperature. In the case of a gas, the pressure measures the separative force of heat, and the pressure must be proportional to the absolute temperature. He adds the words that we have

---

* Writing dU and dI for the increments in internal energy and internal work respectively, we have: $dU = dH + AdI$. dL is equal to $dI + dW$, where dW is the increment in external work.

already quoted in connection with Dalton's insight: 'The internal probability of the truth of this result is indeed so great that many physicists since Gay-Lussac and Dalton have without hesitation presupposed this proportionality and have employed it for calculating the absolute temperature.' But as it is easier to measure the work done than to measure an internal force, Clausius proposes that the mechanical work that can be done by heat in changing the arrangement of a body is taken as proportional to the absolute temperature.

At this point he introduces a new concept, a new magnitude: the *disgregation*—a term which is almost self-explanatory, for it indicates the degree of dispersion of a body, or the rearrangement of its particles. The disgregation of a liquid is greater than that of a solid, that of a gas is greater than that of a liquid. Generally speaking disgregation is the action by which heat performs work and it reflects the change of internal arrangement consequent upon heating. In short, the disgregation is the equivalence value of the dispersion of the body, or of the rearrangement of its particles. Thus if heat is transformed into work, as in the expansive phases of the operation of the Carnot engine, the equivalence value of the transformation is negative; but the accompanying disgregation, which has the same magnitude, is obviously positive. Conversely, in the compressive phases the equivalence values are positive and negative respectively.

If we take Z to be the disgregation of a body, dZ to be a vanishingly small increment in its value, and dL (expressed in appropriate units) to be the corresponding work done, we have: $AdL = TdZ$. So that the relationship becomes,

$$dQ + dH + TdZ = 0.$$

Or, dividing by T, $\dfrac{dQ}{T} + \dfrac{dH}{T} + dZ = 0.$

Or for a finite and reversible change of condition,

$$\int \frac{dQ + dH}{T} + \int dZ = 0.$$

In a complete cycle the disgregation must be the same at the end as it was at the beginning, so that $\int dZ = 0$ and hence

$\int \dfrac{dQ + dH}{T} = 0$. For such a cycle $\int \dfrac{dQ}{T} = 0$, so that $\int \dfrac{dH}{T} = 0$,

which, Clausius shows, implies that the specific heats of water, ice and steam are the same! This is consistent with his belief that water and ice at the same temperature contain the same amounts of heat.

We have therefore three types of transformations: mechanical energy into heat and vice versa, the flow of heat from a hot body to a cold one (which, as we have seen, can be reduced to two transformations of the first sort), and thirdly, a change in the arrangement of the constituent molecules. These transformations represent the total effects of heat, and the equivalence values are given by the expressions discussed above. The first integral, $\int \dfrac{dQ + dH}{T}$ comprises the equivalence value of all changes of the kind work-into-heat and vice versa; the second integral $\int dZ$ gives the total change of disintegration, and the sum of the two must, in the case of reversible changes, amount to zero.

If we now consider *irreversible* changes the above integral equation cannot hold. Let us take the simple, limiting case of Joule's experiment of allowing a gas to double its volume by expanding into an evacuated vessel equal in volume to its original container. In this case the gas does no work, and when it has doubled in volume it has failed to provide the mechanical energy which could be used to restore the original situation. The gas, which we can consider to be perfect, has the same internal energy at the end of the expansion as it had at the beginning; indeed, its state is exactly the same as if the expansion had been 'isothermal', as in the Carnot cycle. But the change has been irreversible and work would have to be done by some outside agency to restore the original condition. Evidently there has been a change in molecular arrangement; the dispersion has increased and so, accordingly, has its equivalence value, the disgregation. There has been no gain or loss of heat and the final temperature of the expanded gas is unchanged, as Joule had shown. The total thermal, or rather thermodynamic effect is, then, that there has been an increase in the disgregation so that the equation reduces to an inequality.

We conclude, on the basis of this and the other examples discussed by Clausius, that for all irreversible changes the net value of $\int \dfrac{dQ + dH}{T} + \int dZ$ must be greater than zero.

'The whole mechanical theory of heat rests on two fundamental theorems—that of the equivalence of heat and work and that of the equivalence of transformations.'[21] We can now muster the ideas on thermodynamics that were expounded by Clausius in his ninth memoir, published in 1865. When heat is imparted to a body it may increase either the thermal content or the internal work content; the sum of these being the energy of the body. A change in the thermal content has the equivalence value $\int \dfrac{dH}{T}$ and a change in the arrangement of the molecules (by virtue of which heat does internal and external work) has the equivalence value $\int dZ$. If we now take the heat imparted to the body as positive, writing dQ for $-dQ$ in the integral equation, we obtain $\int \dfrac{dQ}{T} = \int \dfrac{dH}{T} + \int dZ$, for a finite, reversible change. The two quantities on the right hand side of the equation are the equivalence value of the heat content and the equivalence value of the change of molecular arrangement, or the disgregation of the body. Taken together these constitute the transformational content; the total equivalence value of the energy changes that the body has undergone from its original condition. This magnitude, the transformational content, Clausius proposes to call the 'entropy', S, of the body.

We can see how the concept of entropy resolved itself in Clausius' mind into two separate components; even though today we have long forgotten them and retain only the comprehensive idea of entropy. In the case of the expansion of a perfect gas without the performance of work, there is no net thermal change—heat is not overtly involved; yet there must be a change of some kind: the energy condition or state of the body is altered, for otherwise heat could not be generated when the body is restored by compression to its original volume. What, then, is the nature of this thermodynamic change? The answer is simple: the disgregation is increased.

On the other hand when a body, say a gas, is heated at

constant volume the disgregation cannot possibly increase since there is no overt molecular change. But the heat content changes, and this means that $\int \dfrac{dH}{T}$ increases. Thus one or the other—the transformational value of the heat in the body, *or* the disgregation, *or* both—can increase to give a net increase in the entropy of the body.

We can now interpret the Carnot cycle in terms of the transformational content of the working substance. In the 'isothermal' expansion there would certainly be an increase in the disgregation, but no increase in the internal heat of the working substance since all the heat entering the gas is converted into work; the entropy therefore increases. In the 'adiabatic' expansion, the increase in disgregation is matched by a decrease in the thermal content, since the gas now performs work by drawing on its own internal heat energy; there is, therefore, no net change in entropy. Similar considerations apply to the two compressive phases, with the difference that in the 'isothermal' compressive phase the entropy is reduced so that the total, overall change of entropy for the cycle is zero.

The principle of entropy as developed by Clausius, the idea of the transformational content of a body, provides a new and fundamental measure of the thermodynamic state of bodies. Heat, or caloric, is no longer conserved; entropy does not replace caloric as has sometimes been implied, but it does provide the measure made necessary by the breakdown of a fundamental conservation axiom. The substitution of the axiom of the conservation of energy is not in itself sufficient, for as we have seen, there is a natural bias in the distribution of energy and in the direction which energy changes tend to take. Entropy gives us a measure of this bias in the case of material bodies or systems of bodies.

Let us conclude this section by quoting a few relevant observations made by Clausius:[22]

> We might call 'S' the *transformational content* of the body, just as we termed the magnitude 'U' the *thermal* and *ergonal* content. But as I hold it better to borrow terms for important magnitudes from the ancient languages so that they may be adopted unchanged in all modern languages, I propose to call the magnitude *S*, the *entropy* of the body, from the Greek word τϱοπή, *transformation*

He continues by referring to Kelvin's comments on the universal characteristic of thermal processes—dissipation. How are we to characterise the limiting conditions? 'This can be done by considering, as I have done, transformations as mathematical quantities whose equivalence values may be calculated and by algebraic addition united in one sum.' He goes on to point out that two magnitudes present themselves, the transformational value of the thermal content and the disgregation; the sum of the two being the entropy. The latter concept is, he argues, capable of development and of wide application. Finally, he adds:

> If for the entire universe we conceive the same magnitude to be determined consistently and with due regard to all the circumstances, which for a single body I have called the *entropy*, and if at the same time we introduce the other and simpler conception of energy, we may express in the following manner the fundamental laws of the universe which correspond to the two fundamental theorems of the mechanical theory of heat.
> (1) The energy of the universe is constant
> (2) The entropy of the universe tends to a maximum.

The cosmic role of heat, first discerned at the end of the eighteenth century and eloquently described by writers like Fourier and Carnot had thus, by way of Joule, Rankine and Kelvin, achieved its final definition by Clausius. This is not a balanced, symmetrical, self-perpetuating universe, as the development of rational mechanics, building on the foundations of Newton's *System of the world*, seemed so confidently to indicate. It is a universe tending inexorably to doom, to the atrophy of a 'heat death', in which no energy at all will be available although none will have been destroyed; and the complementary condition is that the entropy of the universe will be at its maximum.

The concept of entropy was the result of Clausius' attempt to quantify the processes of transformation of energy associated with heat phenomena. The transformational content, or the entropy of a body is the measure of the change in character of its energy: the sum of the equivalence values of the heat and mechanical energy changes it undergoes in passing from one specified state to another. Partly because Clausius' writings, apart from the first memoir, were very little known in England

κ

the nature of entropy has caused some confusion among later British textbook writers.

A very important property of the entropy of a body is that the change in its value—all entropy measurements are relative—when a body passes from a given state (pressure, volume, and temperature) to a different state (different pressure, volume, and temperature) is quite independent of the route by which the change takes place. In Figure 26, a gas at p_1, v_1 and T_1 can be brought to p_2, v_2 and T_2 either by heating at constant volume to C and then expanding isothermally, or by heating at constant pressure to D and expanding isothermally, or by some more complex method indicated by the path via E. In all three cases, indeed in all possible cases, the change of entropy will be

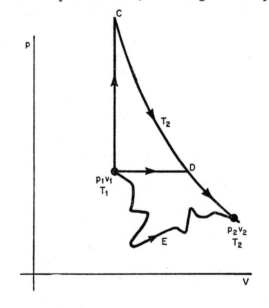

Figure 26

exactly the same. This it will be remembered was not the case with the change of heat, which in all cases is different, depending on the amount of work done. The entropy, on the other hand, depends solely on the condition of the body, and is quite independent of the route by which it reached that condition. This is indicated by the statement that for a reversible cycle the net change of entropy is zero, or $\int \dfrac{dQ}{T} = 0$.

It is very easy to show* that the change in entropy when a gas passes from the first to the second state shown on Figure 26 is,

$$S_2 - S_1 = C_v \log \frac{T_2}{T_1} + R \log \frac{v_2}{v_1},$$

or, writing $C_p = C_v + R$,

$$= C_p \log \frac{T_2}{T_1} + R \log \frac{p_1}{p_2}.$$

These very simple textbook examples suggest that we can look upon the entropy change as having two components, a thermal or temperature component and a pressure or volumetric component. These correspond to Clausius' idea of the equivalence value of the thermal content and the disgregation together constituting the transformational content, or entropy of the body. Clausius' idea has long been forgotten but it can still serve to aid one's understanding of the nature of entropy.

The concept of entropy was put forward as a necessary consequence of the mechanisation of the theory of heat and a desire to quantify—to express analytically—the natural tendency for all forms of energy to transform themselves into heat and for heat to flow from a hot body to a cold one. We can measure changes only in entropy (changes, that is, relative to the previous energy state of the body in question) and the net result of such changes is, in practice, always an increase in entropy. A hot and a cold body placed in contact soon reach a common intermediate temperature, and in doing this the entropy of the hot body is diminished while that of the cold

---

* $(1/T)$ is, mathematically, an integrating factor which converts dQ into the perfect differential dQ/T so that the equation $dQ = dU + pdv$ can be integrated; or, in physical terms, the path of the change specified. Thus, if we put dU equal to $C_v dT$ and $pv = RT$ we have:

$$\int_{T_1}^{T_2} \frac{dQ}{T} = \int_{T_1}^{T_2} \frac{C_v dT + RTdv/v}{T} = C_v \log \frac{T_2}{T_1} + R \log \frac{v_2}{v_1}$$

—assuming for the sake of simplicity that C_v is independent of the temperature. The second and identical equation results if we consider the gas heated at constant pressure, via D. It is sometimes said that as (dQ/T) is a perfect differential it was elevated to the status of separate concept, the entropy. This would, of course, make entropy solely a mathematical concept. But this is not satisfactory. Entropy was formulated as a physical concept and it is as such that its nature becomes apparent.

body is increased. But, as the expression $\int \dfrac{dQ}{T}$ indicates, the increase in entropy of the cold body must be greater than the diminution in entropy of the hot body, so that over all there is a net increase.

We may summarise our ideas by remarking that if, after a series of changes, the entropy of a body is found to have increased by so much, this means that the sum of the equivalence values of all the changes is the same as that of a specific increase in the thermal content of the body under known temperature conditions.

It is also important to remember that a change of entropy is not *necessarily* associated with heat entering or leaving a body. Maxwell himself was rather misleading on this point, for he wrote that when '. . . there is no communication of heat (the entropy) remains constant but when heat enters or leaves the body the quantity increases or diminishes' (*Theory of heat* [fifth edition, London 1877], p. 162). Now while this is true it is not the entire truth, for entropy can change without heat entering or leaving a body. It does so, for example, when a gas expands irreversibly without doing work, as in the Joule/Gay-Lussac experiment. The increase in entropy (entirely 'disgregation' in this case) is measured by the equivalence value of the heat generated when the gas is compressed to its original volume.

We conclude therefore that a change in entropy of a body is the measure of the change in the energy state of the body relative to its previous state. Since physical and chemical changes that take place in a system of bodies involve the transference of energy, the application of the principle of the increase of entropy, reflecting the bias of nature, enables us to find the laws that govern the course of such changes and the conditions of equilibrium. It is in fact easy to see that if a system of bodies is in equilibrium every possible infinitesimal change between the bodies is reversible, and the total entropy must remain constant. It is, in general terms, a necessary condition for equilibrium that $dS = 0$. But the further significance of this must be sought in the textbooks and in the application and use of the concept of entropy in actual practice.

SOME LOOSE ENDS

The establishment of thermodynamics during the middle years of the nineteenth century—one of the great creative epochs of science—was so rapid and extensive that a whole new branch of knowledge was suddenly created, where previously there had been the old science of heat plus the substantially empirical technology of heat-engines. It is curious that whereas meteorology had played an important and fruitful part in the development of the old science of heat it was almost untouched by, and unrelated to, the new thermodynamics. Why this should have been so is an interesting problem, for as the late Sir Napier Shaw remarked: '. . . the theory of heat . . . finds its application in the atmosphere if anywhere.'[23] Shaw, a distinguished meteorologist with a sensitive awareness of history, argued that the collateral development of electro-magnetic theory had diverted men's attentions from meteorology: 'There can be little doubt that if the applications of the principles of mass and energy to electricity and magnetism had not diverted the minds of natural philosophers they would have pursued the application of the great physical principles to the atmosphere as used to be their wont.' This may well be true, but another and hardly less convincing explanation can be propounded, which we shall discuss in a moment. In the meantime we observe that although meteorology did not play a significant part in the thermodynamic revolution, its contributions towards the development of the science of heat were such that we must at least attempt to round off the works of Halley and Erasmus Darwin, de Luc and Dalton with a brief resumé of the thermodynamic explanation of the main meteorological problem of atmospheric circulation.

In the first place we note that the theory of the trade winds, which had such a very long life, collapses on being submitted to the simple scientific test of continuity (as used, for example, by Joseph Black). The very moment that air starts to warm up it expands and therefore begins to ascend. As it rises its expansion cools it, adiabatically, and thus after a few feet the whole process stops. Halley's explanation is, therefore, inadequate.

In purely qualitative terms Halley is sound enough, for he

indicates how the heat energy from the sun provides the necessary power for the great atmospheric circulations. But an
additional component, unspecified by and indeed unknown to
Halley, is necessary in order to make the equatorial air rise
many thousands of feet into the atmosphere. This component
turns out to be the water-vapour mixed with the air, which by
virtue of its 'latent heat' constitutes an effective heat reservoir
for the rising air. As soon as adiabatic cooling begins the air
condenses some of its water-vapour and the 'latent heat' which
it gains thereby prevents its temperature falling markedly. The
air therefore can go on rising.

The height to which a given sample of equatorial air can rise
is evidently determined by its pressure, temperature and heat
sources available, compared with those of the surrounding air
through which it rises. Now the developments in the design of
instruments, of meteorological balloons, of radio and of aircraft
have greatly extended our knowledge of the upper atmosphere.
When accurate data on the temperature and pressure of the air
at different altitudes became available Napier Shaw pointed
out that the atmosphere appeared to be horizontally stratified
into layers of equal entropy—'isentropic' layers. The difference
between the entropy of the air at a given altitude and that of
the same air at a standard temperature and pressure can easily
be computed by the simple formulae given on page 275, and in
this way the layers of equal entropy shown on Figure 27 can be
drawn. The data show that the entropy increases progressively
with height above the earth's surface.

The heated moist air will have, at the beginning of its ascent,
a higher entropy than its surroundings. As it rises it draws on
its 'latent heat' reservoir so that its temperature does not fall
appreciably; $\log \dfrac{T_2}{T_1}$, although negative, remains small, so that
the entropy of the rising air continues to be greater than that
of the surrounding air. This process will continue until the air
rises to that altitude at which its entropy is the same as that of
the surroundings. Then, having reached its appropriate layer,
the air will be free to travel along the isentropic surface without
having to absorb or emit heat. If it is to rise further it will have
to be heated up so that its entropy is increased; if it is to descend
it will have to get rid of some heat, either by radiation or by
some other method, so that its entropy is reduced. Therefore,

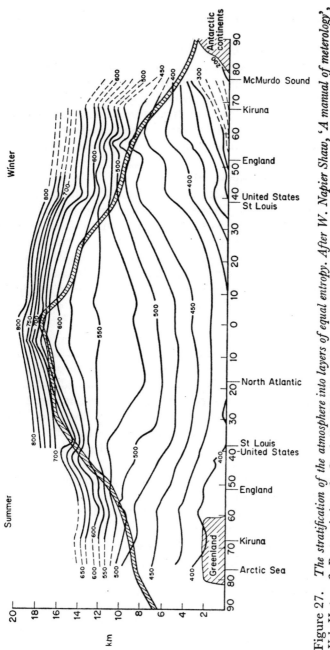

Figure 27. *The stratification of the atmosphere into layers of equal entropy. After W. Napier Shaw, 'A manual of meterology', Vol. II, p. 116. By permission, the Cambridge University Press.*

according to Shaw, *the entropy of the air tells us its proper place in the firmament.*[24]

In brief, to say that a sample of air will rise to a height determined by its pressure and temperature amounts to saying that it will rise to a height determined by its entropy (compare the simple equations set out on page 275). We can easily appreciate that unless the 'transformational content' of a sample of air is the same as that of its surroundings it cannot be in equilibrium, and must either rise or fall until equilibrium is attained.

The rise of equatorial air, the descent of cooled polar air, indeed the whole process of circulation, is equivalent to the operation of a great heat-engine, the indicator diagram of which will denote the work done. The source is, of course, the energy of solar radiant heat, and the work is manifest in the lifting of great volumes of air against the force of gravity. At long last, over a hundred years later, the bold speculations of Hachette were fulfilled in an intellectual context of which he cannot have dreamed: the atmosphere regarded as a great heat-engine.

Many of the old disputes which we described in the earlier parts of this work were terminated when the authoritative experiments of Victor Regnault were published. One point, however, must be mentioned: the problem of the specific heats of gases. The kinetic theory of gases was finally established through the efforts of Joule and Krönig, Clausius and Maxwell. And according to the doctrine of the equipartition of energy, enunciated by Maxwell, Willard Gibbs and Boltzmann, the energy of a molecule is equally divided among different modes of motion of which it is capable. A monatomic molecule, such as that of helium gas for example, is capable only of projectile motion in three dimensions, so that it has three degrees of freedom, and this leads to the conclusion that its specific heat per mole must be $(3/2).R_m$, while γ, the ratio of the principal specific heats must be 5/3, or about 1·667. Measurements of the specific heats of the (monatomic) inert gases have confirmed this. For the diatomic gases the theory predicts that the (common) specific heat should be $(5/2).R_m$ per mole while γ should be 7/5, or 1·400. The experimental values for such diatomic gases as oxygen, nitrogen, hydrogen and air generally agree with this prediction. In the case of the polyatomic gases the correspondence between theory and experimental data is much less satisfactory, being hardly more than qualitative. In fact for a fully satisfactory explanation of the specific heats of gases it is necessary to take account of the internal vibrations of the atoms in the gaseous molecules, and for this one must use quantum dynamics.

We are, however, concerned only with the simpler gases, and it is enough for our purposes to note that the prediction of the theory of the equipartition of energy that all gases having the same number of atoms per molecule have the same specific heats per mole is exactly the same thing as asserting that the *volumetric* specific heats of such gases are the same.* Hence the conclusion that Dulong reached in 1827 is given a foundation in the kinetic theory of gases and the doctrine of energy.

In terms of technology the consequences of the thermo-dynamic revolution were almost equally important. The development of the internal combustion engine began in the 1860s and reached its most interesting stage, at least from our point of view, with the appearance of the high-efficiency diesel engine at the end of the century. This engine uses air as the working substance and approximates closely in operation to the Carnot cycle. The development and the manufacture of such engines necessitated more advanced metallurgy, fuel technology and engineering standards than were available in Carnot's time.

From another point of view what is significant about the development of the internal-combustion engine is that while the steam-engine had been the work of Englishmen and Scotsmen, almost exclusively, in the case of its successor the main contributions came from Germans and Frenchmen. The reasons for this seem to have been partly intellectual and educational, and partly economic and social. It is to a consideration of one aspect of this very general problem that we must now turn.

---

* According to Avogadro's hypothesis the weight of a given volume of gas must be proportional to its molecular weight. Since the specific heat per mole is the heat capacity of a weight of gas numerically equal to its molecular weight (C_v.M.W.) it follows that the heat capacity of a given volume of gas—which is the *volumetric* specific heat must be the same for all such gases.

SCIENCE AND MODERN CULTURE

The establishment of the doctrine of energy and Rankine's identification of its two forms as 'actual' and 'potential'—an Aristotelean dichotomy—mark an important stage in the mid-nineteenth century revolution in physics. At much the same time the energy concept was applied in thermodynamics and, a few years later, in electro-magnetic field theory. But we cannot concern ourselves with the latter, fundamental as its consequences were,[25] for the scale of our ambitions limits us to mechanical energy and the theories of heat.

The doctrine of energy in mechanics was not, historically, a deduction from Newton's system. The approach to mechanics through 'energy' was, in embryonic form, adopted by Huygens and Leibnitz, as an alternative to the Newtonian system. Its generalisation and fruitful application were delayed, not so much by the success of Newtonianism as by the backwardness of those branches of physical science, to which it was later found to be as well adapted as the Newtonian system had been to astronomy and classical mechanics. When, in time, the science of heat had developed sufficiently the doctrine of energy did indeed come into its own. But by then it had been systematically improved and refined through the labours of men like Parent, Smeaton, Lazare Carnot and Coriolis, whose main concern had been the study of machines. The conceptual tools they elaborated—work, 'reformed' *vis viva* ($\frac{1}{2}mv^2$), and power— were the ones needed for the establishment of the mechanical theory of heat.

In 1877 however P. G. Tait (1831–1901), Professor of Natural Philosophy at Edinburgh University, claimed that Newton must be regarded as the true originator of the doctrine of energy. If, he remarked, you study the great Scholium to the third law of motion and Propositions 40 and 41 of the *Principia* sufficiently closely you will come to see that this is so. No doubt; and if one's devotion is intense enough and if one's interpretation sufficiently flexible one can trace it back to the pre-Socratic fragments. What Tait was doing, in fact, was perpetuating the Robison–Playfair tradition of writing history to suit one's presuppositions of how things ought to have happened.*

* Tait's words imply that his interpretation of Newton was all his own

There was probably no great harm in all this—had not that ardent Smeatonian Peter Ewart been quite happy to sprinkle a pinch of incense on the altar of the great Newton? But Tait's activities turned out to have rather unfortunate consequences. It all began with an address given by John Tyndall before a Royal Institution audience in 1862. The subject was *force*, which was rather unsuitable as the actual topic happened to be the doctrine of the conservation of energy. Now Tyndall was an Irishman* and had a generous temperament; he sympathised with the unfortunate Mayer. Accordingly at the end of his lecture, when he came to mention the discovery of the mechanical equivalence of heat, he went on to suggest that it had all been the work of a comparatively unknown and quite unfairly neglected little doctor who was, even then, cultivating his vineyard in Heilbronn. Tyndall deserves credit for calling attention to Mayer's work, but the form of words he used could have been more happily chosen.

Joule, at any rate, felt that he had been unjustly treated and wrote that his particular claim to have established, independently and scientifically, the doctrine of energy conversion could not be denied.[26] With this Tyndall entirely agreed and the whole matter could—and certainly should—have ended amicably with Joule's credit unimpaired and justice at last done to the unhappy Mayer. Unfortunately Tait intervened: with Kelvin he wrote an article on the subject of energy for the contemporary journal *Good Words*, in the course of which the controversy was reopened, Mayer's claims rejected and Tyndall's scientific abilities called in question.

We are not concerned with the subsequent course of this unnecessary and harmful squabble.[27] The area of combat was widened, Kelvin's claims were urged by Tait as against those of Clausius, and a radical interpretation of the history of thermodynamics was begun; an interpretation in which the

---

work. But it was surely an astonishing coincidence that a very similar argument had been put forward by a predecessor in Edinburgh and published in the best known of all Scottish journals—the *Edinburgh Review*—then at the height of its fame.

* John Tyndall (1820–93) was Professor of Physics at the Royal Institution. He was one of the most colourful of Victorian scientists. A courageous mountaineer, he was also a typical nineteenth-century rationalist and a notable controversialist, in the style of T. H. Huxley.

contributions of British scientists were emphasised compared with those of foreigners. In 1868 Tait published his *Sketch of the history of thermodynamics*[28] in which he begins by confessing that he found it almost impossible to be strictly impartial, and then proceeds to make some extraordinary historical assertions. Rumford's and Davy's claims are accepted without question; indeed, he says that Davy's ice-rubbing experiment, 'rightly viewed', conclusively demonstrates the dynamical theory of heat.* At the beginning of the nineteenth century, Tait continues, a number of facts had been established, prominent among them being that heat is a form of motion (by which was *really* meant 'energy'), that a mechanical equivalent of heat exists, and that this equivalent should be capable of experimental determination. Tait, however, omits to tell us who demonstrated these extraordinary facts.

The tone having been set at the beginning, the rest of the work follows suit. Mayer is effectively denied consideration—the stress on Rumford and Davy implies that his work was supererogatory—and Clausius is, as far as possible, subordinated to Kelvin. The origin of the concept of entropy is traced back to Kelvin's brilliant insight in his paper on thermoelectricity (1854), and Clausius' long, detailed discussions of the equivalence of transformations, with their deep insights, are brutally truncated. His recognition that Carnot's insistence on considering the *flow* of heat in the thermo-mechanical process was still valid in the context of the dynamical theory of heat was quite ignored by Tait. Further, the concept of entropy is virtually stood on its head in an apparent attempt to subordinate it to Kelvin's doctrine of the dissipation of energy. The entropy of the universe, announces Tait, is continually decreasing. In the second edition of his *Sketch* Tait was moved to withdraw this startling and idiosyncratic interpretation of the entropy concept.

What are we to make of all this? In the first place the history of thermodynamics without Clausius may not be exactly the play of *Hamlet* without the Prince of Denmark but it must assuredly be a poor thing. In fairness, however, we add that it

---

* Bearing in mind that an increase in pressure lowers the melting point of ice it is quite clear that Davy's experiment, 'rightly viewed', demonstrates the caloric theory of heat no less conclusively than it demonstrates the dynamical theory.

would still be a poor thing with the roles reversed: with Clausius exalted at the expense of Kelvin. It cannot, we suggest, be fairly understood unless the contributions of both these men are adequately and reasonably represented. And at this point Tait's efforts can be seen to be fundamentally and deplorably obscurantist. Before we go on to justify this observation let us consider one or two further instances of Tait's standards of scholarship and fair play.

He was seemingly indefatigable in his vendetta against the unfortunate and quite inoffensive Mayer. In 1876 he published in the *Philosophical Magazine*, a translation of Mohr's paper of 1837 explicitly to denigrate Mayer's work. For, he says, while Mohr's paper contains practically everything that is worthwhile in Mayer, it avoids some of the worst mistakes made by the latter—notably his 'false analogy and *a priori* reasoning'. Tait however had either overlooked or did not know of Ampère's paper of 1835, which makes all the points put forward by Mohr. And when he rebukes Mohr for ascribing the discovery of the polarisation of radiant heat to Melloni and not to Forbes he overlooks the fact that Bérard has a claim in the matter too— and that Bérard's claim pre-dates those of both Melloni and Forbes.

Tait was fond of pointing out that Mayer's determination of the mechanical equivalent of heat, based on the difference between the two principal specific heats of a gas, was invalid since, when a gas expands isothermally there is no *a priori* ground for believing that *all* the heat imparted has been used solely for *external* work. The gas expands and some heat energy may, in theory at any rate, be consumed in loosening the bonds between the gas molecules. The only way one can be quite sure that all the heat energy imparted to a body has been converted into external work is to ensure that the body is brought back, after its expansion, to its original pressure, volume and temperature. This criticism of Mayer's determination was repeated by G. G. Stokes[29] and it is, of course, quite valid. Mayer *was* unjustified in assuming that all the heat imparted to a gas is converted into external work when it expands isothermally; he would have been justified only in the light of Joule's researches, in particular of the famous demonstration that when a gas expands without doing work (i.e. into a vacuum) it does not need to absorb any heat. But having admitted as much one must ask whether this criticism is of the sort that can fairly be made of pioneer work. How many scientists would be left with their reputations intact if only those of their hypotheses that were rigorously justified at the time in all respects could be accepted for the historical record? What, for instance, would be the present status of Copernicus, who advanced the theory that the earth spins on its axis and orbits round the sun long before there was a Galilean–Newtonian system of mechanics to harmonise the theory with the elementary facts of human experience: stones dropped from high towers are not left behind, we are never flung out into space, there is no gale of wind always blowing from the east? What, for that matter, about the status of Carnot, the founder of thermodynamics? Some of his leading ideas were, as we have shown, wildly *a priori*, and a good deal of his reasoning was based on faulty data and faulty theory. How, if we were strict Taitists, could we conceal our contempt for such inadequate scientists as Copernicus and Carnot?

Enough has been said to establish that Tait was a very poor historian and an unreliable interpreter of science. Why, then, have we spent so much time discussing the activities of a minor figure in the annals of science? What, in fact, did Tait do? Apart from some original work that had merit he wrote textbooks, and the latter activity was extremely important at the time in question and in the circumstances we have discussed. Tait's personal position was curious and perhaps even invidious. He was a Scottish professor of physics who numbered among his colleagues—friends and fellow countrymen—such talented men as Kelvin and Rankine, as well as Maxwell, who was a man of genius. In such company Tait must have cut a comparatively poor figure. Perhaps, then, he felt that his mission was to defend his great contemporaries from all attacks, real or imagined. He would be at once their defender and interpreter. Neither Joule nor Kelvin wrote a book on heat or thermodynamics; Rankine wrote only for engineers,[30] and then in rather an opaque style (Maxwell found his statement of the second law of thermodynamics unintelligible); even the enemy, Tyndall, confined his activities to popular books or learned papers. The textbook field was clear for Tait; he was left as the only professional interpreter of the latest advances in heat and thermodynamics. Even Maxwell's famous book on the theory of heat, published in 1877, shows some of Tait's influence, while Tait's own book of 1884 was a master-textbook and a model for the many that followed. Nevertheless it is significant that Napier Shaw, whose ideas and insights we have just discussed, acknowledged explicitly his indebtedness to Maxwell[31] and did not mention Tait.

In the event, then, Tait's position was a peculiarly influential one; just how influential we can appreciate when we recall that it was Tait and Kelvin who established Rankine's 'potential' and 'actual' energy in the current terminology of potential and *kinetic* energy to suit the arrangement of their textbook on natural philosophy. It was also Tait who persuaded us, uncritically and unjustifiably, to ascribe Gay-Lussac's law of the expansion of gases to J. A. C. Charles—and, in the English speaking world, it has remained 'Charles' law' to this day. Tait even succeeded in mystifying Maxwell about the nature of entropy, for in 1873 Maxwell wrote to Tait, saying:

> It is only lately under the conduct of Professor Willard Gibbs that I have been led to recant an error which I had imbibed from your $\theta\varDelta$cs [i.e. *Sketch of thermodynamics*], namely that the entropy of Clausius is *unavailable energy*, while that of T' [Tait] is *available energy*.* The entropy of Clausius is neither one nor the other. It is only Rankine's Thermodynamic Function . . .[32]

Plainly, Tait was a very influential man. If he was guilty of misleading Maxwell and, as we have suggested, many others subsequently, through the bias of his polemical writings and his textbooks, is there anything we can say on his behalf? We all know that science has its heroes and even its martyrs; does it also have its villains, and if so was Tait among them? We must deplore the evident bigotry of his attacks on the inoffensive Mayer, and he was parochial in a way in which his eminent contemporaries were not.† But loyalty to a colleague or to a

---

* As I have remarked, Tait tried to subordinate the entropy concept to Kelvin's doctrine of the dissipation of energy. Thus he asserted that the entropy, or the *available* energy of the universe must continually diminish. This misled Maxwell. Gibbs, who had studied under the German masters, was not subjected to this delusion. There is, *pace* Kelvin, no impenetrable level below which the energy in a body becomes unavailable. As Carnot had in effect pointed out, availability depends on the nearest cold body to hand. This can be at any temperature, down almost to absolute zero.

† Tait lacked the sophistication and cosmopolitanism of Tyndall. Golf on his native links and not mountaineering in the Alps was Tait's favourite hobby.

fellow countryman is not a defect, and if we find his history contemptible we must remember that in his day the serious study of the history of science had not begun (Whewell excepted). Men were unfamiliar with the fact that discoveries and inventions are frequently—indeed usually—made simultaneously but quite independently by different men in different countries; and that Mayer's work was not an instance of German duplicity but one more example of the common phenomenon of simultaneous discovery. Nor could Tait be as aware as we are of the subtleties and ambiguities which lie at the hearts of so many scientific advances. For him the whole process would seem to be a much simpler and more individualistic affair. Perhaps he even regarded scientific discovery as a species of personal property: one owned one's discovery in the same way as one owned one's house, one's stocks and shares, one's horses. . . . It may not be entirely fanciful to point out that Tait was, like his more distinguished contemporaries Kelvin, Rankine, Joule and Tyndall, a conservative in politics.*

If, then, the picture of Tait which emerges from the cold print of the controversies of long ago is an unsympathetic one, we should remember that there must have been another side to the man, a side which would have been known only to his contemporaries in their personal contacts. Perhaps we should leave the last word to Tyndall. In 1874 Tyndall wrote to Tait commenting on an attack on his (Tyndall's) work by a forgotten scientist named Zöllner: 'I would rather see you and Clausius friends than Zöllner and myself. Trust me, C. is through and through an honest, high minded man.'

The ultimate significance of Tait's writings on energy and thermodynamics only begins to emerge when we consider the social changes that were affecting science at that time. After about 1851 the sciences, in England at any rate, began to disintegrate into separate disciplines or specialisms and inevitably separate accounts of thermodynamics came to be written for engineers, for chemists and for physicists and mathematicians.

---

* Joule's political attitude is revealed in a letter which he wrote to a nephew in 1886: 'I do not know anyone who does not rejoice at the overthrow of Gladstone's mischievous legislation and the deep disgrace of having a minister like Dilke washed away.' The common view which associates scientists with radical or progressive political views would seem to have been founded on an illusion.

It had fallen to Tait to have the last opportunity to write a constructive, balanced and fair book on the foundation, subsequent development and nature of thermodynamics. He was an able writer and a competent scientist, and he knew all the British contributors, some of them intimately. Thereafter the constructive criticism and general appreciation of the energy revolution and thermodynamics would become increasingly more difficult as academic specialism tightened its grip on the world of learning.[33] For this development Tait was in no way to blame; the responsibility was that of the universities, the schools and examination boards, and beyond these of governments and peoples. But the biases and misinterpretations that Tait foisted on the doctrines of heat, thermodynamics and energy would almost inevitably be perpetuated. So it is possible to detect in modern textbooks the outlines of the form which Tait originally gave to thermodynamics in this country: the relatively minor role ascribed to Clausius, the mystification or frank defeatism over the concept of entropy—dismissed uneasily as a mathematical device or not explained at all—and the neglect of criticism of such anachronisms as 'latent heat'; which, of course, had been one of Clausius' initial insights.

It may be, however, that even if Tait had never lived it would have been necessary to have invented him. It is surely notable that the subsequent development of the theories of energy and thermodynamics, to which the British had contributed so much, thereafter left this country and was carried forward by men like Willard Gibbs, Helmholtz, Boltzmann, Stefan and Planck. Even in the technological field the country of Newcomen, Trevithick and Watt was outpaced and it was men like Rudolf Diesel who took up Carnot's splendid suggestion at the time when engineering techniques made the high-efficiency hot-air engine possible. One is left with the impression that by 1900 the social, economic and intellectual tides had turned.

The rationalisation of learning, its subdivision into various separate disciplines taught to an ever-increasing number of students, has inevitably increased the importance of the textbook. The textbook is, of course, designed to meet the examination requirements set out in the syllabus; but at the same time the availability of textbooks helps to determine the content of the syllabus: the relationship is a two-way one. It is, we suggest, by means of this curious social dialectic that most people's ideas of the nature and content of the sciences are ultimately formed.

A new textbook is one which includes the latest scientific work and never, or almost never, one which reconsiders the basic assumptions and procedures of the science; for to do that would be to challenge the social mechanism of education. In this way, we argue, the mis-interpretations of the past, if not the actual errors, are carried forward from one generation to the next. So it has come about that the great re-casting of thought that constituted the scientific revolution of the mid-nineteenth century has been flattened out and then buried by the successive generations of textbooks. The science of thermo-dynamics and the doctrine of energy have been forced into the same mould as the abstract Newtonian system of point-masses moving in space and subject to the laws of motion and the central force of gravitation. But how can we reconcile the notion of the Carnot cycle, performed by an ideal, reversible engine with the surely very different Newtonian procedures and concepts? What kind of thinking is it that can, without so much as a by-your-leave, jump from concepts like quantity of heat, conductivity and specific heat to the consideration of the performance of what at first sight appears to be a Platonic engine? Scientific thinking, so imparted, must surely seem quite inscrutable to the person who would wish to understand two of the most important developments in modern thought: the second law of thermodynamics and the concept of entropy.*

The answer to the enigma can, I believe, only lie in the recognition of the true nature of scientific progress, which con-sists not in a series of deductions from an axiomatic system, given once-for-all, but in a recasting of knowledge in response to the changing experience of succeeding generations. Thus the quantum theory and the theory of relativity grew, not out of reflections on Newton's system, as recent popular accounts have tended to suggest, but out of the developments of the very characteristic nineteenth-century concepts of energy and field. The awareness of the growing, developing nature of scientific

* A distinguished modern writer has tried to explain the second law of thermodynamics to the lay reader by likening its implications to the empirically established principle that one cannot unscramble omelettes. Such an oversimplification must surely generate more confusion than edification.

knowledge is the vital component that is missing from the textbook, with its authoritarian and scholastic bias. Without an awareness of this it is hard to see how intelligent people can be brought to realise the significance of science as a basic cultural activity.

It is worth recalling here that during the first quarter of the nineteenth century a new cosmology, that of heat, was set up and that it was in fairly sharp contrast to the mechanical—clockwork—cosmology of the seventeenth century. Proclaimed by men as diverse as Rumford, Fourier and Carnot, the new cosmology was based on the science of heat, which had been formally established by Lavoisier and Laplace on the surely self-evident axiom of conservation. The experimental results achieved by the new science were impressive enough and formed the basis for elaborate theoretical developments by men like Fourier and Poisson. Hence we can understand Kelvin's distress when, confronted with Joule's arguments, he felt that the whole science was threatened with collapse. The simple truth is this: in the years after 1850 a developed, established science with extensive theoretical structures and satisfactory experimental verification had to be re-established on a new basis; that of the axiom of the conservation of *energy*. *This was the first time in history that such a thing had happened.** It pre-dates the famous re-establishments that followed the quantum and relativity theories by something like fifty years. Thus the assertion made by popular writers of a generation ago that the calm certainty of science was not disturbed between the times of Newton and the first papers of Planck and Einstein is seen to be, in historical terms, a dangerously misleading half-truth.

Our view of the development of science—a continual re-casting by generation after generation—inevitably raises the old problem of the relationship between science and technics. During the nineteen-thirties a number of very articulate scientists argued that science is, in effect, the spearhead of technology and industry. 'Science is a mode of behaviour whereby man gains mastery over the environment' was a brief

---

* Lavoisier's chemical revolution was not a similar case, for the scientific status of the phlogiston theory, which he overthrew, may be disputed. There was precious little theory about the phlogiston 'theory' and the experimental evidence in its support was scanty and unimpressive. It was accepted in its time, *faute de mieux*.

statement of this belief. And it carries with it the implication that science, having a utilitarian end, should be subject to appropriate political control and direction. This view of science, despite its laudable concern for human welfare, understandably provoked a counter-attack. There have been plenty to argue that science is justified for its own sake. It is the pursuit of knowledge, no more and no less, and its connections with technics have been only incidental in that technics can provide science with improved philosophical instruments: microscopes, telescopes, spectroscopes etc. Science is a highly philosophical activity concerned only with the structure of the outside world, so far as it can be known, *unmodified by technics*.

It is impossible to agree with this view. The development of European technics has resulted in a dramatic widening of man's horizons and an increased understanding of the world he inhabits. Just as the great geographical discoveries of the fifteenth century radically changed man's knowledge of his world, informing him of new continents and oceans undreamed of by the old writers, shifting his vision from the confines of his own back-yard to a rich and diverse world (and incidentally making things like the Copernican revolution acceptable and fruitful) so too the development of machines like the steam-engine in the eighteenth century almost forced man to recognise the enormous power, the *puissance*, of heat, the grand moving-agent of the universe. The sight of a primitive steam-engine tirelessly pumping ton after ton of water out of a mine, or of a crude early locomotive hauling a train of trucks along a rough, uneven railway-track, did more for science than all the speculations of the philosophers about the nature of heat since the world began. He who would deny this must either prove that Sadi Carnot was and is unimportant or that he did not mean what he said in the opening pages of the *Réflexions*.

Besides revealing in sharp and dramatic form the great physical power of heat, the work of the power technologists, from Smeaton and Watt onwards, has delineated the physical circumstances in which energy is transformed; that is, it has indicated what happens when heat is transformed or, following Clausius, transmitted to produce mechanical work. Thus a great scientific revolution was effected as a result of man's experiences of an enormously important technological development: the invention of the heat-engine. To suggest that the same advance could have been achieved by philosophical speculation and 'pure' scientific experiment unaided by technics is merely idle fancy. We have to take the world as it is, not as we think it ought to be. Man and his technics are part of the natural order; there is no independent, pre-existing order, no utopia where only 'pure' science need be studied.

On the other hand the utilitarian position seems to be no less vulnerable. Had technological, industrial or utilitarian considerations been allowed to determine the development of heat and thermodynamics it seems clear that the course indicated by men like Tredgold, Fairbairn, Farey and de Pambours would have been favoured against the highly speculative course charted by Ewart, Carnot and Clapeyron. The element of free speculation is therefore, as the philosophers have argued, absolutely indispensable. And this brings us to our last point.

Looking back, we see that we have been concerned, among other things, with the rise of *physics*. In the seventeenth and eighteenth centuries there were only two established sciences: Newtonian mechanics and planetary astronomy. All the rest were, as Whewell later put it, in a progressive, empirical state. The establishment of physics as an autonomous discipline was a major achievement of the nineteenth century[34]—that branch of it called heat and thermodynamics having had its roots in eighteenth-century chemistry, power technology, meteorology and medicine. Nineteenth-century physics was an extremely successful enterprise; so successful that it justifies the observation that if any history can be called Whig history it must be the history of science. One may regret the great reform bills of the nineteenth century and deplore the repeal of the Test Act, but one can hardly deny the validity of Faraday's discovery of electromagnetic induction or of Archimedes' discovery of the principle of displacement.

And yet the Whig view of the history of science is not without its dangers.[35] It may be difficult, and it would certainly be rash, to venture to judge recent scientific work on the basis of history. But it is possible to discover in the history of heat and thermodynamics certain outstanding features which seem to be lacking in the modern scientific scene, and whose absence may represent a defect in our arrangements for science. Three key figures in our narrative—Carnot, Joule and Mayer—were in strict terms amateurs, or devotees, of science. They stood outside the 'establishments' of their days and because their ideas were unorthodox they were ignored. Carnot died unrecognised and Mayer was driven to despair, while Joule succeeded only through good fortune and very favourable personal circumstances. These men and the independent lines they took were essential for the progress of science. If we consider the strong and complex pressures which modern society exerts on all its members, scientists included, it is reasonable to wonder whether such men could work effectively today, and if not, whether science can continue to progress satisfactorily without them. Has it been shown that scientific 'establishments' are today more intelligent, more sensitive and more tolerant of unorthodox opinions than they were in the past? The lives of Carnot and Mayer remind us that our narrative has indeed included the elements of tragedy; it may be a greater tragedy if our society makes such careers impossible today.

Abbreviations used below

A. de C.	*Annales de Chimie* (1789–1815).
A. de C. & P.	*Annales de Chimie et de Physique* (1816–).
A. R. de S.	*Académie Royale des Sciences.*
B.J.H.S.	*British Journal for the History of Science.*
B.R.L.	*City of Birmingham Central Reference Library.*
Manchester Memoirs	*Memoirs and Proceedings of the Manchester Literary and Philosophical Society.*
P.T.R.S.	*Philosophical Transactions of the Royal Society of London.*
Phil. Mag.	*The Philosophical Magazine.*
T.N.S.	*The Transactions of the Newcomen Society, London.*
T.R.S.E.	*Transactions of the Royal Society of Edinburgh.*

Notes and References

CHAPTER ONE

1. The most recent account of the invention and development of the thermometer is by W. E. Knowles Middleton, *A History of the Thermometer and its Use in Meteorology* (Johns Hopkins University Press, Baltimore 1966).
2. G. R. Talbot, *Origins and Solutions of some Problems in Heat in the Eighteenth Century* (Manchester University Ph.D. Thesis 1967), p. 4.10.
3. ibid.
4. For an account of this aspect of Boyle's works see Marie Boas, *Robert Boyle and Seventeenth Century Chemistry* (Cambridge University Press 1958).
5. Arthur Raistrick, *Dynasty of Ironfounders: the Darbys and Coalbrookdale* (Longmans Green & Co., London 1953).
6. This was first pointed out by H. E. Roscoe and A. Harden in *A New View of the Origin of Dalton's Atomic Theory* (London 1896). See also F. Greenaway, *John Dalton and the Atom* (Heinemann, London 1966).
7. Accounts of the development of these instruments are given by W. E. Knowles Middleton in his *History of the Barometer* (Johns Hopkins University Press, Baltimore 1964) and in op. cit. (1). A. K. Biswas, *A History of Hydrology* (Strathclyde University Ph.D. Thesis 1967) gives a very interesting account of the history of the rain gauge.
8. Martin Lister, 'The Origins of the Trade Winds', *Philosophical Transactions of the Royal Society* (London, 1684), p. 494.
9. George Garden, *P.T.R.S.* (Lowthorp abridgement), ii, p. 129.
10. Edmund Halley, 'An Historical Account of the Trade Winds', *P.T.R.S.* (1686–7), p. 153.
11. George Hadley, 'Concerning the Cause of the General Trade Winds', *P.T.R.S.* (1735–6), p. 58.
12. J. T. Desaguliers, 'On the Rise of Vapours, Formation of Clouds and Descent of Rain', *P.T.R.S.* (1729), p. 6.
13. W. E. Knowles Middleton, *A History of the Theories of Rainfall* (Oldbourne History of Science Library, London 1965), p. 48.
14. Gordon Manley, 'Dalton's Contribution to Meteorology' in *John*

Dalton and the Progress of Science, edited by D. S. L. Cardwell (Manchester University Press 1968), p. 140. Also A. K. Biswas, op. cit. (7).

15. Edmund Halley, 'A Discourse Concerning the Proportional Heat of the Sun', *P.T.R.S.* (1691–3), p. 878.

16. W. Napier Shaw, *A Manual of Meteorology*, three volumes (Cambridge University Press 1926, 1928, 1930), i, p. 123; ii, p. 398, and iii, pp. 255–6.

17. Lynn White, Jr., 'Eilmer of Malmesbury, an Eleventh Century Aviator', *Technology and Culture* (1961), ii, p. 97.

18. Otto von Guericke, *Experimenta nova Magdeburgica* (Amsterdam 1672; reprinted Otto Zeller, Aalen 1963), pp. 100–24.

19. For a short account of these early experiments see D. S. L. Cardwell, *Steam Power in the Eighteenth Century* (Sheed and Ward, London 1963), p. 7 ff.

20. Thomas Savery, *The Miner's Friend* (London 1702; reprinted 1827), p. 19.

21. The only biography of Thomas Newcomen is by L. T. C. Rolt, *Thomas Newcomen and the Early History of the Steam Engine* (David and Charles, Dawlish; MacDonald, London 1966). There are some interesting articles on the history of the Newcomen engine in the *Transactions of the Newcomen Society* (1962–3), xxxv.

22. Lynn White, Jr., argues (in *Galileo reappraised* edited by C. L. Golino [University of California Press 1966], pp. 107–8) that because we know so little about Newcomen we must assume that his invention was empirical and owed nothing to contemporary scientific knowledge of the atmosphere and the void. But it would have been an extraordinary coincidence if, immediately after the efforts of Papin, Savery, Huygens etc., all of whom knew of atmospheric pressure, Newcomen should have succeeded in harnessing it *at that particular time* and by methods similar to the ones they used *while ignorant of the basic principle*. We therefore interpret Newcomen's invention as further evidence of the widespread diffusion of interest in and knowledge of the new philosophy. Professor White also argues that the invention of the snifting valve confirms the empirical nature of Newcomen's achievement, for, he says, the scientists of the time did not know that air is dissolved in water. On the contrary, this had been well established long before Newcomen's time. See, for example, J. R. Partington, *A History of Chemistry* (Macmillans Ltd., London 1961), ii, pp. 525, 529, 607, 628–9. This detail is therefore evidence in the opposite direction to that which Professor White supposes.

23. William Lee's invention of the stocking-frame (1589) might be supposed to be an exception. But it was an isolated event and did not have any of the consequences that mark the major inventions

of history: the printing press, the weight-driven clock, the steam-engine, the water-frame.

24. I. B. Cohen, *Franklin and Newton* (American Philosophical Society, Philadelphia 1956) discusses at some length the influence of Newton's work on Benjamin Franklin's ideas and procedures.
25. Isaac Newton, *Opticks* (London 1730 edition; reprinted Dover Books, New York 1952), pp. 395–6.
26. ibid. p. 340.
27. A. R. and M. B. Hall, *Some Unpublished Scientific Papers of Isaac Newton* (Cambridge University Press 1962), p. 224.
28. Isaac Newton, *Principia mathematica*, Proposition 23, Book 2.
29. Isaac Newton, 'A Scale of the Degrees of Heat', *P.T.R.S.* (1701), p. 824.
30. Isaac Newton, op. cit. (28), Proposition 1, Theorem 1, Book 2.
31. G. Amontons, 'Method of Substituting the Force of Fire for Horse and Man Power to Move Machines', *Histoire et Mémoirs de l'Académie Royale des Sciences* (1699), p. 112.
32. This knowledge was not available until after the establishment of Lavoisier's system of chemistry, following the publication of the *Traité élémentaire de chimie* (Paris 1789).
33. G. Amontons, op. cit. (31). E. Halley, *P.T.R.S.* (1691–3), p. 650.
34. See the *Abridgments to Specifications of patents and inventions relating to air, gas and other Motive Power Engines*, 1835–66 (London 1873); *Part ii, 1867–1876* (London 1881) and the subsequent issues of *Abridgments of Specifications* (H.M.S.O., London).
35. Such devices were sometimes used to turn spits. The hotter the fire the more rapidly the spit turned, thus ensuring that the roast was evenly cooked. See Lynn White, Jr., *Medieval Technology and Social Change* (Oxford University Press 1964).
36. G. Amontons, op. cit. (31).
37. G. Martine, *Essays and Observations on the Constitution and Graduation of thermometers and on the Heating and Cooling of Bodies* (third edition Edinburgh 1780), p. 12.
38. D. G. Fahrenheit, 'Experiments on the Degree of Heat in Boiling Liquids', *P.T.R.S.* (1724), p. 1. Fahrenheit introduced mercury as the common thermometric liquid. A possible advantage, apart from the more obvious ones, was that it could be obtained in purer form than other thermometric liquids. Alcohol and linseed oil might well mean different things in different places, but 'quicksilver' would be more or less the same everywhere. It is, after all, an element!
39. Brook Taylor, 'An Experiment Proving the Expansion of the Liquid in the Thermometer to be as the Degree of Heat', *P.T.R.S.* (1723), p. 291. W. E. Knowles Middleton has pointed out that Brook Taylor used linseed oil thermometers, and the agreement

he obtained was therefore almost certainly the result of a fortunate accident!

40. J. T. Desaguliers, *A Course of Experimental Philosophy* third
    edition (London 1763), ii, pp. 296–7.
41. ibid. p. 348. See also W. E. Knowles Middleton, op. cit.
    (13).
42. G. R. Talbot and A. J. Pacey, 'Some Early Kinetic Theories of
    Gases: Herapath and his Predecessors', *British Journal for the
    History of Science* (1966), iii, p. 133.
43. ibid. p. 138.
44. ibid. p. 139.
45. Daniel Bernoulli, *Hydrodynamica, sive de viribus et motibus fluidorum
    commentarii* (Basle 1738).
46. G. R. Talbot and A. J. Pacey, op. cit. (42), p. 141.
47. Isaac Newton, op. cit. (28). Preface to the first edition.
48. The three English editions of Boerhaave's *Elements of Chemistry*
    were translated by Peter Shaw (1727), T. Dallowe (1735), and
    finally, Peter Shaw again (1741). Dallowe's was the authorised
    edition.
49. ibid. (Dallowe), p. 153.
50. ibid.
51. I am much indebted to Dr G. R. Talbot for making this
    essential point clear to me. Boerhaave has, for far too long, been
    misunderstood in this respect.
52. Boerhaave (Dallowe), op. cit. (48), pp. 160–1.
53. ibid. p. 166.
54. For an account of the works of G. W. Krafft and G. W. Rich-
    mann see D. McKie and N. H. de V. Heathcote, *The Discovery of
    Specific and Latent Heats* (Edward Arnold, London 1935), p. 59 ff.

CHAPTER TWO

1. *Short Description of the Atmospheric Engine* (Stockholm 1734).
   Translated and published by the Newcomen Society. Extra Publication No. 1 (Cambridge University Press 1928).
2. The Schemnitz Bergswerksakademie was founded in 1733 at
   Joachimstal and transferred in 1763 to Prague. It was recon-
   stituted in 1770 at Schemnitz and soon became famous all over
   the world for the quality of its teaching. The syllabus influenced
   the founders of the École Polytechnique (1794). C. T. Delius'
   lectures were published as: *Anleitung zu der Bergbaukunst for die* . . .
   Bergswerksakademie (Vienna, two editions 1773, 1806).
3. Gabriel Jars, *Voyages métallurgiques* (Paris 1780), ii, p. 152 ff.
4. William Cullen, 'Of the Cold Produced by Evaporating Fluids,

and Some Other Means of Producing Cold', *Essays and Observations* (Edinburgh 1756), p. 11.

5. Joseph Black, *On Magnesia Alba, Quicklime and other Alcaline Substances*, Alembic Club Reprint No. 1 (1893).

6. Brougham was a great parliamentarian and had heard all the great orators of his time: Pitt, Fox, Sheridan etc. See D. S. L. Cardwell, *The Organisation of Science in England* (Heinemann, London 1957), p. 28.

7. For example, the custom of the Scottish universities whereby the professor was paid directly by the students who attended his lectures. See D. S. L. Cardwell, 'Reflections on Some Problems in the History of Science', *Memoirs and Proceedings of the Manchester Literary and Philosophical Society* (1963-4), cvi, p. 108.

8. Anon., *An Enquiry into the General Effects of Heat and Observations on the Theory of Mixtures* (London 1770). John Robison, *Lectures on the Elements of Chemistry by Joseph Black, M.D.* (Edinburgh 1803).

9. D. S. L. Cardwell, *Steam Power in the Eighteenth Century*, pp. 37-41.

10. G. R. Talbot, *Origins and Solutions of Some Problems in Heat in the Eighteenth Century*, pp. 13.27-13.30. See also the 'Richardson', 'Dobson' and 'Roscoe' MSS lecture notes. Dr Talbot points out that Black's use of the method of mixtures to measure the temperature of red-hot iron indicates his understanding of the principle of heat capacity and its utility. This is a most important observation.

11. See D. McKie and N. H. de V. Heathcote, *The Discovery of Specific and Latent Heats*, p. 15 ff., and G. R. Talbot, op. cit. (10), pp. 12.32-12.33.

12. For an excellent account of the basic principles involved in 'operational' measures see J. R. Ravetz, 'The Representation of Physical Quantities in Eighteenth Century Mathematical Physics', *Isis* (1961), lii, p. 7.

13. A complete account of Wilcke's work on specific and latent heats is given by D. McKie and N. H. de V. Heathcote, op. cit. (11), pp. 54-121. They are more sympathetic to Black and less sympathetic to Wilcke than I am and they lean rather too much on the authority of Robison. But generally speaking their monograph is a model study in the history of science.

14. D. McKie and N. H. de V. Heathcote, op. cit (11), p. 108 ff.

15. *Encyclopaedia Britannica*, third edition (1797), article 'Steam Engine'.

16. 'Robison, on historical points, was a very inaccurate writer.' See W. Vernon Harcourt, Letter to Lord Brougham, *Philosophical Magazine* (1856), p. 106.

17. See especially the authoritative letter that Watt wrote to David Brewster denying the legend. The letter is included in John

Robison's *A System of Mechanical Philosophy*, edited by Brewster (Edinburgh 1822), ii.

18. J. P. Muirhead, *The Origins and Progress of the Mechanical Inventions of James Watt*, three volumes (London 1854); and *The Life of James Watt with selections from his Correspondence* (London, 1859).

19. R. L. Galloway, *The Steam Engine and its Invention* (London 1881), p. 139 ff. T. M. Goodeve, *Text-book on the Steam Engine* (London 1878), pp. 19–22. R. H. Thurston, *A History of the Steam Engine* (London 1878) repeats the Robison account uncritically.

20. William Ramsay, *The Life and Letters of Joseph Black, M.D.* (London 1918).

21. Letter to Brewster. See Muirhead, *The Life of James Watt*, p. 75.

22. G. R. Talbot, op. cit (10), pp. 12.28–12.29.

23. J. P. Muirhead, op. cit. (21), p. 485.

24. From a footnote by Watt to Robison's article 'Steam' (from *Encyclopaedia Britannica*), reprinted in J. Robison, op. cit. (17), ii. This footnote and the others like it do not, of course, appear in the original 1797 version of the article.

25. See D. S. L. Cardwell, *Steam Power in the Eighteenth Century*, p. 53 ff., 60 ff.

26. 'I mentioned to you a method of still doubling the effect of the steam and that tolerably easy by using the power of steam rushing into a vacuum at present lost . . . shut the valve and the steam will continue to expand.' Letter from Watt to William Small, 28 May 1769, Assay Office Library, Birmingham.

27. Quoted by Robert Schofield, *The Lunar Society of Birmingham* (Oxford University Press 1963), p. 65. See also G. R. Talbot, op. cit. (10), p. 13.38.

28. J. R. Partington, *A History of Chemistry*, iii, p. 352.

29. J. A. de Luc, *Annales de Chimie* (1791), viii, p. 73. De Luc was controverting Monge's assertion that the process of evaporation was the same as that of solution: the water evaporated was actually *dissolved* in the air. In this case, according to Monge, less heat would be required the more abundant the solvent, air.

30. J. H. Magellan, *Essai, sur la nouvelle théorie du feu élémentaire* (London 1780). J. A. de Luc, *Idée sur la méteorologie* (London 1786), three volumes. P. D. Leslie, *A Philosophical Inquiry into the cause of Animal Heat* (London 1778).

31. Irvine's ideas were set out in *Essays, chiefly on Chemical Subjects*, edited and published posthumously by his son (London 1805). For a very clear account of Irvine's ideas see Robert Fox, 'Dalton's Caloric Theory', *John Dalton and the Progress of Science*, pp. 187–202.

32. Adair Crawford, *Experiments and Observations on Animal Heat* (London 1779, second edition 1788). Crawford's book was widely read and thus publicised Irvine's ideas as well as his own.

33. G. R. Talbot, op. cit. (10), p. 12.44.

34. P. J. Macquer, *Dictionnaire de chimie*, second edition (Paris 1778), ii.

35. See B. B. Kelham, 'Atomic Speculations in the Late Eighteenth Century' in *John Dalton and the Progress of Science*, pp. 109–22. It was an obvious extension of Newton's hypothesis (*Principia*, Proposition 23, Book 2) to suppose that heat increased the repulsive force between the atoms of an elastic fluid. See, for example, W. J. s'Gravesande, *Mathematical Elements of Natural Philosophy*, translated by J. T. Desaguliers (London 1721), p. 219; T. Rutherforth, *A System of Natural Philosophy: being a course of lectures in Mechanics, Optics, Hydrostatics and Astronomy* (London 1748), ii, p. 569. Bryan Higgins accounted for the repulsive force in terms of a material substance: fire.

36. Robert Fox, *The Caloric Theory of Gases from Lavoisier to Regnault* (Oxford University Press, forthcoming).

37. J. H. Lambert, *Pyrometrie, oder vom Maass d. Feurs und d. Wärme* (Berlin 1779), para. 492.

38. Erasmus Darwin, 'Frigorific Experiments on the Mechanical Expansion of Air', *P.T.R.S.* (1788), p. 43.

39. According to 'Watt's law', as the temperature of steam falls the latent heat increases, so that the 'total heat' is constant. Darwin asserted that the cooling of expanded steam in the Watt engine led to rapid condensation with consequent reduction in the amount of condensing water required. But this is inconsistent with 'Watt's law'. From this we must conclude either that Darwin knew nothing of the law, perhaps because Watt kept quiet about it, even among his friends, or that Watt himself was not convinced of its accuracy. Either or both these hypotheses are plausible. For an account of the Lunar Society and its members, see R. Schofield, op. cit. (27).

40. The essay 'On Air and Fire' is in the *Collected Papers of Carl Wilhelm Scheele*, translated by L. Dobbin (G. Bell & Sons, London 1931).

41. Robertson Buchanan, *Practical and Descriptive Essays on the Economy of Fuel and the Management of Heat* (Glasgow 1810), p. 43.

42. A. L. Lavoisier and P. S. Laplace, 'Memoir on Heat', *A. R. de S.* (1780), pp. 4, 355.

43. Robert Fox, op. cit. (36).

44. A. L. Lavoisier, *Traité élémentaire de chimie* (Paris 1789), i, pp. 1–21, 203.

45. J. Ingen Housz, *Nouvelles expériences et observations sur divers objets de physique* (Paris 1785–9), i, p. 380. See also *Observations sur la Physique* (1789), xxxiv, p. 68, and Sir B. Thompson (Rumford), 'New Experiments upon Heat', *P.T.R.S.* (1786), p. 273. Rumford's study, which was confined to liquids and gases, was gravely

defective in that he ignored both the specific heats of the different substances and the effects of convection.

46. A common error was to assume that conductivity and specific heat were in simple reciprocal relationship. Thus F. T. Mayer measured the conductivities of different woods by comparing the times that identical spheres made of these woods took to cool down; he then deduced the conductivities from the relationship $c = 1/\rho\ C$, where c is the specific heat, ρ the density and C the conductivity. But the relationship was incorrect and the method defective in that Mayer had not allowed for differing external conductivities of the different surfaces. See F. T. Mayer, 'Memoir on the Different Capacities of Woods to Conduct Caloric and on their Specific Heats', *A. de C.* (1799), xxx, p. 32.

47. Private communication.

48. As Professor A. R. Hall remarks, an ambition at that time was that '. . . the three related agents in nature, light, heat and electricity "shall be shown to arise from one and the same principle" (this was an idea later given more specific form as the "correlation of physical forces")'. A. R. Hall, 'Precursors to Dalton' in *John Dalton and the Progress of Science* (op. cit. [35]), pp. 40–55.

CHAPTER THREE

1. The rise of the textile industries has long fascinated the historians, and a number of authoritative works have been written about it. Among these are: W. C. Unwin, *Samuel Oldknow and the Arkwrights* (Longmans Green & Co., London 1924); A. P. Wadsworth and Julia Mann, *The Cotton Trade and Industrial Lancashire, 1600–1780* (Manchester University Press 1965); A. P. Wadsworth and R. S. Fitton, *The Strutts and the Arkwrights, 1758–1830* (Manchester University Press 1964); *The Leeds Woollen Industry, 1780–1821*, edited by W. B. Crump (Thoresby Society, Leeds 1931). Dr Fitton is now preparing a biography of Richard Arkwright. The *motivations* of eighteenth-century industrialists have been discussed by, for example, Max Weber and T. S. Ashton. But we know very little about their technological and scientific views. What prompted them to analyse textile processes into their basic elements, and what enabled them to mechanise these processes thereafter? We do not know, and this represents a serious gap in our knowledge.

2. On the subject of the central heating of industrial buildings at this time see M. C. Egerton, 'William Strutt and the Application of Convection to the Heating of Buildings', *Annals of Science* (1968), xxiv.

3. The definitive work on the problems of the application of power to the textile industries is R. L. Hills, *Studies in the History of Textile Technology* (Manchester University Ph.D. Thesis, 1968). See also Dr Hills' *Power in the Industrial Revolution* (Manchester University Press 1970).

4. Dean Swift authoritatively if quite unconsciously proved the point. The Laputan nobleman describes how he had had a very convenient mill, 'turned by the current from a large river and sufficient for his own family as well as a great number of his tenants'. But projectors suggested to him that the water should be made to go over the wheel instead of under it since water descending down a declivity would turn the mill with half the current of a river whose '. . . course is more upon the level'. The subsequent operations were as unsuccessful as they were expensive! (*Gulliver's Travels*). But the joke was, as Smeaton confirmed, on the gloomy Dean.

5. R. L. Hills, op. cit. (3), p. 6.1.

6. ibid.

7. ibid.

8. The Newcomen engine was applied to textile mills comparatively late in the day. A double-acting engine was developed and drive was effected by a rack-and-pinion method. See John Farey, *A Treatise on the Steam Engine* (London 1827).

9. For some details of early measurements of horse-power see D. S. L. Cardwell, 'Some Factors in the Early Development of the Concepts of Power, Work and Energy', *B.J.H.S.* (1967), iii, p. 209.

10. D. S. L. Cardwell, 'Power Technologies and the Advance of Science', *Technology and Culture* (1965), vi, p. 188.

11. A. E. Musson and E. Robinson, 'The Origins of Engineering in Lancashire', *Journal of Economic History* (1960), p. 209.

12. R. L. Hills, op. cit. (3), p. 8.8 ff.

13. For details of the Kier engine, see John Farey, op. cit. (8).

14. Professor T. S. Kuhn gives an interesting account of one of these buoyancy engines—the Cagnard Latour engine—in his article 'Sadi Carnot and the Cagnard Engine', *Isis* (1961), lii, p. 567. The engine is also described in *Nicholson's Journal* (1811), xxix, p. 175.

15. The expansion of solids and liquids may *seem* irresistible, but it is very slow and small compared with that of gases. To convert such slight expansions into useful mechanical work requires, therefore, very high gearing; so high, in fact, that any such engine would have to be capable of standing enormous pressures. Even if this were technically feasible, under the operating conditions the 'working substance' might no longer prove incompressible. An engineer named Pattu proposed such an engine

L

in 1818 (*Annales de Chimie et de Physique* [1818], ix, p. 91) but the fallacy was immediately exposed by A. T. Petit (ibid, p. 198).

16. See Thomas Ewbank, *A Descriptive and Historical Account of Hydraulic and other Machines for raising Water* (New York 1842). Discussing rotary pumps, Ewbank remarks: 'Such machines have often been patented, both as pumps and steam engines. In 1782 Mr Watt thus secured a 'rotative engine' of this kind and in 1797 Mr Cartwright . . . in 1818 Mr Routledge . . . but the principle or prominent feature in all these had been applied long before by French mehanicians. Nearly a hundred years before the date of Watt's patent, Amontons communicated to the French Academy a description of a rotary pump' (p. 287).

17. H. W. Dickinson, 'Some Unpublished Letters of James Watt', *Minutes of Proceedings of the Institution of Mechanical Engineers* (1915), p. 487. Letter from James Watt to Matthew Boulton, 11 May 1784. In Birmingham Reference Library, B. & W. Coll.

18. Watt to Boulton, 11 May 1784.

19. H. W. Dickinson, *James Watt, Craftsman and Engineer* (Cambridge University Press 1935), p. 133.

20. J. L. and Barbara Hammond, *The Town Labourer, 1760–1832* (Longmans Green & Co., London 1966) pp. 65–6, and *The Rise of Modern Industry* (Methuen, London 1966), p. 232. The Hammonds do not do justice to Gilbert's efforts on behalf of the scientific and technological community of Cornwall, to say nothing of his general services to science.

21. A. C. Todd, *T.N.S.*, xxxii (1959–60), p. 1.

22. Letter, James Watt to John Southern, 21 April 1792. B.R.L., B. & W. Coll. Also in Dickinson, op. cit. (17).

23. A. J. Pacey and S. J. Fisher, 'Daniel Bernoulli and the Vis-Viva of Compressed Air', *B.J.H.S.* (1967), iii, pp. 388–92.

24. Letter, George Lee to John Lawson (an employee of Boulton and Watt), 12 April 1796. B.R.L., B. & W. Coll.

25. H. W. Dickinson, *A Short History of the Steam Engine* (Cambridge University Press 1938), p. 85. Sadi Carnot was, *a fortiori*, ignorant of the indicator diagram when he wrote *Réflexions sur la puissance motrice du feu* (Paris 1824).

26. Letter, Robison to Watt, 22 July 1797. B.R.L., B. & W. Coll.

27. R. L. Hills, op. cit. (3).

28. ibid. p. 9.2.

29. ibid. p. 9.5 ff.

30. The speed of an engine driving textile machinery must not—for obvious reasons—vary appreciably. But as Ewart found, the application of the 'whirling regulator', or governor, to an expan-

sively operated engine was very difficult. R. L. Hills, op. cit. (3), p. 9.18.

31. D. S. L. Cardwell, *The Organisation of Science in England*, pp. 17–18.

32. Peter Ewart, 'On the Measure of Moving Force', *Manchester Memoirs* (1813), p. 105.

33. It is interesting to note that the Cornish engine-making firm, Harvey's of Hayle, outlasted the much-vaunted Boulton and Watt enterprise and indeed, survives to the present day. See Edmund Vale, *The Harveys of Hayle* (D. Bradford Barton, Truro 1965).

34. These engines were not low-pressure engines. For details see D. Bradford Barton, *The Cornish Engine* (D. Bradford Barton, Truro 1966) and Vale, op. cit. (33).

35. D. S. L. Cardwell, op. cit. (10), pp. 194–5.

36. ibid. Figures 1 and 2.

37. C. T. Delius, *Anleitung zu der Bergbaukunst* . . . The second volume of the 1806 edition contains accounts of the fire-engine, the air-engine (Hero's engine) and the column-of-water engine.

38. Gabriel Jars, *Voyages métallurgiques*, ii, p. 152.

39. John Smeaton, 'Description of a Statical Hydraulic Engine Invented and Made by the late Mr William Westgarth', *Transactions of the Society for the Encouragement of Arts Manufactures and Commerce* (1787), v, p. 179.

40. Thomas Ewbank, op. cit. (16), p. 356.

41. H. W. Dickinson and A. Titley, *Richard Trevithick, the Engineer and the Man* (Cambridge University Press, 1934), pp. 39–42. These engines were put up by Trevithick from about 1798 onwards. A very large machine was installed by him at the Alport lead mines in Derbyshire in 1803. See also *Nicholson's Journal* (1802), i, p. 161, and an interesting article by Nellie Kirkham, 'The Draining of the Alport Lead Mines Derbyshire', *T.N.S.* (1960–1), xxxiii, p. 67.

42. A Guenyveau, *Essai sur la science des machines* (Lyon 1810), pp. 192–3.

43. ibid. p. 194.

CHAPTER FOUR

1. J. H. Lambert, *Pyrometrie*, S. D. Poisson, *Théorie mathématique de la chaleur* (Paris 1835).

2. It has, however, been questioned by Douglas Freshfield in his biographical study, *The Life of Horace Benedict de Saussure* (London 1920), Chapter 1.

3. H. B. de Saussure, *Voyages dans les Alpes*, 8 vols. (Geneva and Neuchatel 1787–96), i, p. vi.
4. ibid. iv, pp. 119–22.
5. J. A. de Luc, *Recherches sur les modifications de l'atmosphère* (Geneva 1772), i, p. 274, and in *P.T.R.S.* (1778), p. 419.
6. The English translation (London 1791) was entitled *Essay on Fire*, and was published 'under the inspection of the author'.
7. According to Pictet, in de Luc, op. cit. (5).
8. M. A. Pictet, op. cit. (6), pp. 116–23. According to Pictet, the suggestion that cold might be transmitted and reflected was first suggested to him by a M. Bertrand, a disciple of the great Swiss mathematician and theoretical physicist Euler.
9. ibid. p. 301.
10. For an account of Prévost's theory of exchange, see his *Du calorique rayonnant* (Paris and Geneva 1809), p. 13 ff. 'De l'Equilibre du Feu', *Journal de Physique* (1791) and 'Réflexions sur la Chaleur Solaire', *Journal de Physique* (1793).
11. See, for example, R. B. Morrison, *Concise Physics for Ancillary Degree Students* (Edward Arnold & Co., London 1962), p. 122.
12. Graf von Rumford, 'An Enquiry into the Source of Heat which is Excited by Friction', *P.T.R.S.* (1798), p. 102.
13. Rumford, 'New Experiments upon Heat', *P.T.R.S.* (1786), p. 273.
14. Rumford, 'Experiments upon Heat', *P.T.R.S.* (1792), p. 48.
15. Rumford, 'On the Propagation of Heat in Fluids' (Seventh Essay), *Nicholson's Journal* (1797), i, pp. 289, 341.
16. He does not mention Dalton or the objection he raised. Rumford's position seems to be that if one destroys the mobility of a particle then liquids do become 'imperfect' conductors; they are non-conductors in so far as their particles are extremely mobile. But this is inconsistent with his first observation that in perfectly *still* liquids heat cannot be conducted downwards. Whether liquids whose constituent particles cannot move are still to be regarded as liquids is unresolved. See *Nicholson's Journal* (1806), p. 253 and *Bibliothèque Britannique* (1806), xxii, wherein he sets out the answer to Dalton without posing the objection.
17. John Leslie, *An Experimental Inquiry into the Nature and Propagation of Heat* (London 1804), note to p. 402.
18. Von Rumford's observations should be compared with those of Archdeacon Paley, who wrote: '. . . wool, in hot countries, degenerates as it is called, but in truth (most happily for the animal's sake) passes into hair; whilst on the contrary that hair in the dogs of the polar regions, is turned into wool, or something very like it'; and, '. . . bears, wolves, foxes, hares, which do not take to the water, have fur much thicker on the back than the belly, whereas

in the beaver it is the thickest upon the belly' (*Natural theology, or evidences of the existence and attributes of the Deity:* [London 1802], pp. 230–1).

19. Rumford, 'An Enquiry Concerning the Nature of Heat and the Mode of its Communication', *P.T.R.S.* (1804), p. 77.

20. D. S. L. Cardwell, *Manchester Memoirs* (1963–4), cvi, p. 108.

21. William Henry, 'A Review of Some Experiments which have been Supposed to Disprove the Materiality of Heat', *Manchester Memoirs* (1802), v, p. 603. Also *Nicholson's Journal* (1802), iii, p. 197.

22. One of the very few exceptions was John Robison who, in his article on 'Electricity' in the Supplement (1803) to the third edition of *Encyclopaedia Britannica*, adopts a cautious and open-minded view about the existence or otherwise of the 'electric fluid'.

23. In, for example, Thomas Thomson's article 'Caloric' in the Supplement ('Chemistry') to the third edition of *Encyclopaedia Britannica*, it is said: 'We are certainly not yet sufficiently acquainted with the laws of motion of caloric (allowing it to be a substance) to be able to affirm with certainty that friction could not cause it to accumulate in the bodies rubbed. This we know at least to be the case with electricity. Nobody has been hitherto able to demonstrate in what manner it is accumulated by friction; and yet this has not been thought a sufficient reason to deny its existence.'

24. J. J. Berzelius, 'Explanatory Statement of . . . Principles . . . adopted as the Basis . . . of Chemical Nomenclature', *Journal de Physique* (1811), lxxxiii, p. 253. *Nicholson's Journal* (1812), xxxiv, p. 142. Discussing the relationship between caloric and electricity Berzelius wrote: '. . . what may be true with regard to the materiality of one of them, must also be true with regard to that of the other.'

25. A. C. Becquerel, *Traité de physique* (Paris 1842), i, p. 461.

26. The fact that water expands on being cooled from 4° to 0°C made it *a priori* likely that *some* liquid(s) might be *cooled* by percussion or adiabatic compression, at any rate over a limited range of temperature. See *A. de C. & P.* (1827), xxxvi, p. 225; *Mémoires de la Société d'Arcueil*, ii, p. 441.

27. Dr Haldat, 'Inquiries Concerning the Heat Produced by Friction', *Journal de Physique* (1810), lxv, p. 213; *Nicholson's Journal* (1810), xxvi, p. 30. Haldat's critique of Rumford's 'experiment' was both penetrating and fair.

28. Rumford, 'An Enquiry Concerning the Weight Ascribed to Heat', *P.T.R.S.* (1799), p. 381.

29. Rumford, 'Historical Review of the Various Experiments of the

Author on the Subject of Heat', *The Complete Works of Count Rumford* (Boston, 1870), i, p. 222.

30. Thomas Young, *P.T.R.S.* (1800), p. 106; (1802), pp. 12, 387; Bakerian Lecture (1804), p. 1.

31. I am grateful to Dr B. B. Kelham for pointing this out.

32. Francis Trevithick, *The Life of Richard Trevithick* (London 1874), p. 112.

33. ibid. p. 113.

34. Were this the case there could be no historians of science, only philosophers of science. A successful scientist would appear to need four particular gifts: a minimum of technical competence, or ability to use the tools of the trade, whether laboratory equipment or pencil-and-paper; a flair for detecting possibilities not recognised by his colleagues—to recognise, that is, the 'growth points' before others do; the courage to continue working even though results are slow in coming and, lastly, the ability to convince others of the significance and validity of what he has done. Rumford possessed the last gift in abundance. It is not worth possessing the first unless one possesses the other three in some measure. However, modern educational and institutional arrangements seem to emphasise the importance of the first to the almost total exclusion of the others.

35. E. N. da C. Andrade, 'Two Historical Notes', *Nature* (1935), cxxxv, p. 359.

36. J. Leslie op. cit. (17), note to p. 136.

37. ibid. p. 140.

38. The study of heat phenomena was, at this stage, expedited by rapid advances in thermometry. Rumford and Leslie carried the technique to a much higher standard of instrumental accuracy. They were in turn surpassed by Nobili, with his invention of the thermopile (*vide infra*).

39. J. Leslie, op. cit. (17), p. 71.

40. ibid. p. 500.

41. ibid, p. 284.

42. J. R. Ravetz, *Isis* (1961), lii, p. 7.

43. Realisation of the importance of physical constants follows logically on the establishment of the theory of dimensions and the abandonment of the Galilean practice of 'proportions'. The significance of physical constants was obscured by the custom of expressing relationships in terms of *proportions*. It is arguable that physics, as an autonomous discipline, dates from the recognition of the importance of physical constants at the beginning of the nineteenth century. See Charles Babbage, 'On the Advantages of a Collection of the Constants of Nature and Art', *Edinburgh Journal of Science* (1832), p. 334.

44. Joseph Fourier, *The Analytical Theory of Heat* (Paris 1822; Dover Books, New York 1955). On a less exalted plane we wonder whether the facts that the familiar central-heating device and the most conspicuous part of a car's cooling system are both called —quite wrongly—'radiators', when both are plainly convectors, do not constitute an historical tribute to the prestige of the study of *radiant* heat at this time? Convection does not lend itself to rigorous analysis and was therefore comparatively neglected. A comprehensive history of domestic heating and cooling devices is long overdue.

45. William Herschel, 'Investigations of the Powers of the Prismatic Colours to Heat and to Illuminate Objects', *P.T.R.S.* (1800), p. 255. Herschel was not the first to study the distribution of heat in a solar spectrum. He had been anticipated by the Abbé Rochon, who published the results of his experiments in 1783. He had used a sensitive air-thermometer and claimed to have found that the heating effect at the red end of the spectrum was 8 to 1 compared with that at the violet end. But the order of his experimental accuracy was dismally low. See G. R. Talbot, *Origins and Solutions of some problems in Heat in the Eighteenth Century*, p. 15.24. The most recent work on Herschel's discovery is by D. J. Lovell, 'Herschel's Dilemma in the Interpretation of Thermal Radiation', *Isis* (1968), lix, p. 32.

46. J. Leslie, op. cit. (1), note to p. 458. See also Leslie's article in *Nicholson's Journal* (1800), iv, p. 344, 416, and the (posthumous) 'Fifth Dissertation on the History of Natural Philosophy', *Encyclopaedia Britannica*, eighth edition (1860), i.

47. Sir H. Englefield, *Journal of the Royal Institution* (1802), p. 202.

48. For contemporary accounts of this debate see J. B. Biot, *Traité de physique expérimentale et mathématique*, four volumes (Paris 1816), iv, Book 7; Thomas Thomson, *An Outline of the Sciences of Heat and Electricity* (London 1830); the Rev. Baden Powell, 'Report on the Present State of our Knowledge of the Science of Radiant Heat', *Reports of the first and second meetings of the British Association*, second edition (1835), p. 259.

49. F. Delaroche, 'Observations on Radiant Heat', *Journal de Physique* (1811), lxxv, p. 201. *Annals of Philosophy* (1813), ii, p. 100.

50. E. Malus, *Mem. Soc. d'Arcueil* (1810), ii, p. 143. The Society of Arcueil and its importance for science in Napoleonic France has been admirably discussed by M. P. Crosland, *The Society of Arcueil* (Heinemanns, London 1967).

51. See, for example, 'Report on a Memoir of M. Bérard . . . on the Different Rays . . . in Solar Light' by Berthollet, Chaptal, and Biot, *Nicholson's Journal* (1813), xxxv, p. 250.

52. T. J. Seebeck, *Abhandlungen der Kk. Academie Berlin* (1818–19), p. 305.
53. J. B. Biot, op. cit. (48), iv, p. 614.
54. ibid. p. 651.
55. Sir E. Whittaker, *A History of the Theories of the Aether and Electricity* (Thomas Nelson, London 1962), p. 107.
56. Henry Meikle, 'On Calorific Radiation', *Phil. Mag.* (1818), liii, p. 260.
57. Whittaker, op. cit. (55), p. 114.
58. ibid. pp. 115–217 (for an account of Fresnel's work).
59. Philip Kelland, *The Theories of Heat* (Cambridge 1837), p. 124.
60. The Rev. Baden Powell, *Edinburgh Journal of Science*, x, p. 207, and vi, p. 297.
61. J. B. Biot professed himself indifferent to questions of the ultimate nature of heat (op. cit. 48) but regarded light as particulate. For the development of French ideas about the nature of heat, see Robert Fox, *The Caloric Theory of Gases from Lavoisier to Regnault.*
62. W. C. Wells, *An Essay on Dew and several appearances Connected with it.* For a discussion of Wells' experiments as exemplars of scientific method see G. Burniston Brown, *Science, its Method and Philosophy* (Allen and Unwin, London 1950).
63. J. Fourier, 'Questions in the Physical Theory of Radiant Heat', *A. de C. & P.* (1816), iii, p. 350; (1817), iv, p. 128; (1817), v, p. 259.
64. *A. de C. & P.* (1817), vi, p. 263.
65. J. Fourier, op. cit. (44), p. 2.
66. ibid. pp. 28–9.
67. ibid. p. 52. It is interesting that Fourier's exact definition of interior conductivity (or just conductivity) was developed in the light of the exact experimental methods then being used in France. In this case experiment clarified a theoretical concept, which was therefore brought within range of experimental determination at a more advanced level of insight.
68. ibid. p. 23.
69. For some observations on dimensions and physical constants see *Analytical Theory*, pp. 126–30. This work is as important for its contributions to mathematical physics generally as it is for the theory of heat. Dr Herivel suggests, very plausibly, that Fourier's separation of heat and mechanics may well have retarded the acceptance of the dynamical theory of heat in France. It probably had other consequences as well; the influential Kelvin—a profound admirer of Fourier—rejected the evolutionary biologists' computations of the minimum age of the earth as he could not reconcile it with that obtained by means of the theory of heat.
70. *A. de C.* (1797), xix, is devoted to a detailed account of the con-

tributions which science, particularly chemistry, made to the national effort during the turbulent years of revolution.

CHAPTER FIVE

1. Joseph Reade, 'Remarkable Fact of an Increase in Temperature Produced in Water by Agitation', *Nicholson's Journal* (1808), xix, p. 113.
2. Marshall Hall, 'On the Nature of Heat', *Nicholson's Journal* (1811), xxix, pp. 213, 257.
3. Among them being Nuguet, la Hire, de Luc, Colonel Roy, de Saussure, Monge and Berthollet.
4. Guyton de Morveau and Duvernois, 'On the Expansion of Air and Gases by Heat', *A. de C.* (1789), i, p. 256.
5. F. Greenaway, *John Dalton and the Atom*, p. 71.
6. John Dalton, 'Experiments and Observations on the Heat and Cold Produced by the Mechanical Condensation and Rarefaction of Air', *Manchester Memoirs* (1798), v, p. 515.
7. John Dalton, *A New System of Chemical Philosophy* (Manchester 1808), p. 9.
8. John Dalton, 'On the Force of Steam or Vapour', *Nicholson's Journal* (1803), vi, p. 257. Also in *Manchester Memoirs*, v.
9. John Dalton, 'On the Expansion of Gases by Heat', *Nicholson's Journal* (1802), iii, p. 130. Also in *Manchester Memoirs*, v.
10. ibid.
11. John Dalton, op. cit. (7).
12. R. J. E. Clausius, *The Mechanical Theory of Heat, with its applications to the Steam Engine and to the Physical Properties of Bodies*, translated by T. Archer Hirst (London 1867), p. 221. Poggendorff's *Annalen* (1862), cxvi, p. 73. *Phil. Mag.* (18), 4, xxiv, pp. 81, 201.
13. Article 'Steam', *Encyclopaedia Britannica*, third edition (1797).
14. J. L. Gay-Lussac, 'Researches on the Expansion of Gases and Vapours', *A. de C.* (Year 10, 1802), xliii, p. 137.
15. ibid. pp. 156–7.
16. *Journal de Physique* (Year 7, 1799), xlviii, p. 166.
17. *A. de C.* (Year 11, 1803), xlv, p. 96. As we saw, de Luc had publicised this law in the same journal (1791), viii, p. 73.
18. By Dr Robert Fox. See *The Caloric Theory of Gases from Lavoisier to Regnault*.
19. J. B. Biot, *Journal de Physique* (Year 10, 1802), lv, p. 173.
20. J.-L. Gay-Lussac, *Mem. Soc. d'Arcueil* (1807), i, p. 3.
21. John Leslie, *An Experimental Inquiry*, notes.
22. J.-L. Gay-Lussac, 'On the Capacity of Elastic Fluids for Caloric', *A. de C.* (1812), lxxxi, p. 98.

314   *From Watt to Clausius*

23. John Dalton, *New system of Chemical Philosophy*, op. cit. (7), i, p. 74. Dalton does not mention Gay-Lussac's results, presumably because he did not know of them when the first volume of the *New system* went to print.
24. J.-L. Gay-Lussac, 'Notes on the Capacity of Gases for Caloric', *A. de C.* (1812), lxxxiii, p. 106.
25. F. Delaroche and J. E. Bérard, 'Memoir on the Determination of the Specific Heats of Different Gases', *A. de C.* (1813), lxxxv, pp. 72, 113. *Nicholson's Journal* (1813), xxxv, p. 281; (1813), xxxvi, pp. 140, 184. *Annals of Philosophy* (1813), ii.
26. Robert Fox, op. cit. (18).
27. See E. Mendoza, preface to S. Carnot, *Reflections on the Motive Power of Fire* (Dover Books, New York 1960).
28. P. J. Dulong and A. T. Petit, 'Researches on the Measure of Temperature and the Laws of Communication of Heat', *A. de C. & P.* (1816), ii, p. 240.
29. P. J. Dulong and A. T. Petit, 'Researches on the Measure of Temperature and the Laws of Communication of Heat', *A. de C. & P.* (1817), vii, pp. 113, 225, 337.
30. A. M. Ampère, *A. de C.* (1815), xciv, p. 145.
31. J. B. Biot, *Traité de physique*, i, Chapter 9.
32. If V is the rate of cooling, θ the temperature of the surroundings, t the excess temperature and m and a are constants, their formula becomes:

$$V = m.a^\theta \,(a^t - 1),$$

So that the only way to ensure a null rate of cooling is to make $m.a^\theta = 0$, which can only be true when $\theta = -\infty$.
33. P. J. Dulong and A. T. Petit, 'Researches on Several Points of Importance in the Theory of Heat', *A. de C. & P.* (1819), x, p. 395.
34. C. Despretz, 'Memoir on the cooling of Some Metals', *A. de C. & P.* (1817), vi, p. 95.
35. Dalton's pragmatic rule that when two elements combined to form one compound only, the compound must be assumed to be a binary one, when they formed two substances they must be a binary and a ternary, and so on, led him to postulate that water, for example, must be a binary compound of the form HO, and ammonia a binary NH, with consequent errors in computing the atomic weights of oxygen and nitrogen. His refusal to accept Avogadro's hypothesis compounded this uncertainty.
36. Robert Fox, 'The Background to the Discovery of Dulong and Petit's Law', *B.J.H.S.* (1968), iv, p. 1.
37. A. Fresnel, 'A Note on the Repulsion of Warm Bodies for one Another at Measurable Distances', *A. de C. & P.* (1825), xxix, p. 37.

38. A. Fresnel, 'Additional Observations on the Repulsion of Warm Bodies', *A. de C. & P.* (1825), xxix, p. 107.

39. Jacob Perkins used the idea of the 'repulsion of heat' to explain certain phenomena noticed in the operation of his ultra-high-pressure engine. See, 'Singular Phenomena of Steam in the "Generators" of M. Perkins' Machine', *A. de C. & P.* (1827), xxxvi, p. 435. The Rev. Baden Powell also wrote on the phenomenon of the repulsive power of heat; see *Phil. Mag.* (1838), xii.

40. Ironically enough it was Laplace, the caloricist, who asserted that the quantity of heat developed by 'adiabatic' compression is independent of the speed of compression. Herapath, the advocate of the dynamical theory of heat, argued that rapid compression is necessary for the generation of heat: '. . . in very slow compressions no rise in temperature has been observed.' John Herapath, 'Observations on M. Laplace's Communication . . .', *Phil. Mag.* (1824), lxii, p. 61.

41. P. S. Laplace, 'On the Speed of Sound in Air and in Water', *A. de C. & P.* (1816), iii, p. 230.

42. J.-L. Gay-Lussac, 'On the Cold Produced by the Expansion of a Gas', *A. de C. & P.* (1818), ix, p. 305.

43. N. Clément and C. B. Desormes, 'Experimental Determination of the Absolute Zero of the Specific Heats of Gases', *Journal de Physique* (1819), lxxxix, p. 321.

44. J.-L. Gay-Lussac, 'On the Caloric of a Vacuum', *A. de C. & P.* (1820), xiii, p. 304.

45. Note in *A. de C. & P.* (1822), xix, p. 436.

46. P. S. Laplace, *Mécanique céleste* (Paris 1823), v, pp. 119–23.

   In Gay-Lussac and Welter's modification of the Clément and Desormes experiment a large glass vessel is filled with the gas at a pressure, p_1, slightly above atmospheric pressure, p. The stopcock is opened and the pressure falls to p. The stopcock is now closed and the gas, which has been cooled 'adiabatically', warms up again until its pressure reaches p_2. These three pressure readings are sufficient to give the ratio of the two specific heats of the gas. For a long time this provided the only means of measuring C_v, the specific heat of the gas at constant volume.

47. S. D. Poisson, 'On the Heat of Gases and Vapours', *A. de C. & P.* (1823), p. 337.

48. Absolute temperature was not defined by Poisson so he did not write these equations in the modern form:

$$\frac{T}{T_0} = \left(\frac{V_0}{V}\right)^{\gamma-1} \text{ but as } (\theta_0 + 267) = (\theta + 267).\left(\frac{p_0}{p}\right)^{\gamma-1}$$

49. John Herapath, 'Observations on M. Laplace's Communication', *Phil. Mag.* (1824), lxii, p. 136.
50. G. R. Talbot and A. J. Pacey, *B.J.H.S.* (1966), iii, p. 145.
51. John Herapath, 'A Mathematical Inquiry into the Causes, Laws and Principle Phenomena of Heat, Gases, Gravitation, etc . . .', *Annals of Philosophy* (1821), i, p. 282.
52. G. R. Talbot and A. J. Pacey, op. cit. (50), pp. 140, 145.
53. If the pressure is proportional to (velocity)^2 and, by Gay-Lussac's law the pressure of a gas is proportional to its temperature, it follows that $t \, \alpha \, v^2$. As v measures the 'true' temperature, the Fahrenheit temperature must be proportional to the square of the 'true' temperature.
54. *Annals of Philosophy* (1812), ii, p. 303.
55. 'X', 'Remarks on Mr Herapath's Theory', *Annals of Philosophy* (1821), ii, p. 390.
56. G. R. Talbot and A. J. Pacey, op. cit. (50), pp. 145–6. The impression that Herapath leaves is that of a very able but basically undisciplined thinker. His highly idiosyncratic style of presentation suggests that he was also headstrong.

CHAPTER SIX

1. Robert Fox, 'The Fire Piston and its Origins in Europe', *Technology and Culture* (1969), x, p. 355.
2. *A. de C.* (Year 12, 1804), li, p. 157. *Nicholson's Journal* (1813), xxxv, p. 117.
3. John Leslie, 'A New Method of Producing and Maintaining Freezing', *A. de C.* (1811), lxxviii, p. 177.
4. The data has been obtained from the *Abridgements of specifications of patents and inventions relating to air, gas and other motive engines.*
5. ibid., see also A. F. Evans, *The History of the Oil Engine* (Sampson Low, Marston and Co., London n.d.), p. 5.
6. A. F. Evans, op. cit. (5), p. 7.
7. *Nicholson's Journal* (1807), xvii, p. 365.
8. *Nicholson's Journal* (1807), xviii, p. 260.
9. *Nicholson's Journal* (1811), xxix, p. 175.
10. *Nicholson's Journal* (1799), ii, p. 288.
11. The authority for this assertion is the economist W. Stanley Jevons, in *The Coal Question*, second edition (London 1866).
12. Arthur Raistrick, *Dynasty of Ironfounders: the Darbys and Coalbrookdale*, p. 163. H. W. Dickinson and A. Titley, *Richard Trevithick, the Engineer and the Man*, p. 43.
13. T. R. Harris, *Arthur Woolf, 1766-1837* (D. Bradford Barton, Truro 1966). Accounts of the Woolf engine and of its performance

are given in (Tilloch's) *Phil. Mag.* xix, p. 133; xxiii, pp. 123, 335; xlvi, p. 43. Also in *Nicholson's Journal* (1804), viii, p. 262.

14. At the end of the year 1813 *Nicholson's Journal* and Tilloch's *Philosophical Magazine* amalgamated under the title *Philosophical Magazine*.

15. *Phil. Mag.* (1815), xlvi, pp. 116, 295.

16. J. B. Biot, *Traité de physique*, iv, pp. 735-6.

17. *A. de C. & P.* (1816), i, p. 314.

18. *A. de C. & P.* (1816), iii, p. 329.

19. David Landes, *The Cambridge Economic History of Europe*, edited by M. M. Postan and H. J. Habbakuk (Cambridge University Press 1965), vi, Part I, p. 410.

20. W. Stanley Jevons, op. cit. (11).

21. Thomas Lean and Brother, *Historical statement of the improvements made in the duty performed by the steam engines in Cornwall* (London 1839), pp. 97–101.

22. F. Greenway, *John Dalton and the Atom*, p. 112.

23. John Sharpe, 'An Account of Some Experiments to Ascertain whether the Force of Steam be in Proportion to the Generating Heat', *Manchester Memoirs* (1813), p. 1.

24. John Robison, *A system of Mechanical Philosophy*, ii. See also John Farey, 'John Southern's Experiments on the Density, Latent Heat and Elasticity of Steam', *Phil. Mag.* (1847), xxx, p. 113.

25. For example, the Comte de Pambours, in his influential *Theory of the Steam Engine* (London 1839), p. 84 ff, contrasted Southern's observations unfavourably with those of Sharpe and came down firmly on the side of the latter (as he supposed) and of Watt's Law.

26. W. H. Wollaston, 'On the Force of Percussion', Bakerian Lecture, *P.T.R.S.* (1806), p. 13.

27. [John Playfair], *Edinburgh Review* (1808), xii, p. 120.

28. Peter Ewart, *Manchester Memoirs* (1812), p. 108.

29. J. N. P. Hachette, *Traité élémentaire des machines* (Paris 1811), p. xviii.

30. ibid. p. xv.

31. The characteristic in question was the size of the caloric atmospheres surrounding the atoms. This train of thought was one of the more important leads that took him to the concept of atomic weights and thus to the atomic theory in general.

32. W. E. Knowles Middleton, *A History of the Theories of Rain*, pp. 132–46.

33. John Dalton, 'Experiments and Observations to Determine whether the Quantity of Rain and Dew is Equal to the Quantity of Water carried off by Rivers . . .', *Manchester Memoirs* (1802), p. 346.

34. D. S. L. Cardwell, *B.J.H.S.* (1967), p. 209.

35. A. T. Petit, 'On the Principle of Living Forces in the Calculation of the Power of Machines', *A. de C. & P.* (1818), viii, p. 287.
36. The yield of steam from the combustion of a bushel of coal
varied with the quality of the coal and the efficiencies of the
furnace and boiler. However, in the case of Petit's computation it
is permissible to take the best possible yield. According to Davies
Gilbert this was 14 cubic feet for a bushel of coal burned. See
Davies Gilbert, *P.T.R.S.* (1828), p. 25.
37. C. L. M. H. Navier, 'Note on the Mechanical Action of Combustibles', *A. de C. & P.* (1821), xvii, p. 357. See also his paper,
'On the Variations in Temperature which Accompany the
Volume Changes of a Gas', ibid. p. 372.'
38. This enabled him to apply Boyle's law to the expansion and to
derive the familiar expression: log p_1/log p_2 Navier biased his
computation in the opposite sense to Petit. While he assumed that
the specific heat of air increased with volume, which therefore
made the air engine less efficient as it had to absorb more heat
than Petit had allowed for, he postulated that the total heat
absorbed in the steam engine was determined by the sum of the
latent and sensible heats.

A necessary consequence of the expansive operation of steam
engines was that, as the equations showed, the efficiency increased with the pressure. The work/heat ratio being of the form
$V.\log p_1/p_2.t$ divided by a function of t, it followed that the
higher the pressure, 'p_1', the greater the efficiency of the engine.

Navier's figures for the performance of the best-working steam-
engines under normal conditions indicated a duty of about
31 million ft. lb. This, of course, made nonsense of Petit's figures.
It should be remembered that French performances—like all
others—were markedly inferior to those achieved in Cornwall.
39. P. S. Laplace, 'On the Attraction of Spherical Bodies and on
the Repulsion of Elastic Fluids', *A. de C. & P.* (1821), xviii,
p. 181. See also John Herapath, *Phil. Mag.*, lxii, p. 136.
40. Charles Sylvester, 'On the Specific Gravity of Gases', *Annals of
Philosophy* (1822), iv, p. 29.
41. John Prideaux, 'Advantages of High Pressure Steam Engines',
Annals of Philosophy (1825), x, p. 432. Prideaux's method was to
calculate the pressure of steam at 300°F, assuming that the density
was constant. According to Gay-Lussac's law this pressure should
be 35·6 in. (30 in. at 212°F). The *observed* pressure of steam at
300°F was 140 in. so Prideaux reasoned that its density must, by
Boyle's law, be nearly four times that of steam at the same temperature but with a pressure of only 35·6 in.—and therefore also
of steam at 212°F. But if steam at 30 in. is increased in density by
nearly four times (3·9), its pressure must rise to 117 in. This is

23 in. less than the observed pressure, and the difference must be the 'profit' from working at the higher temperature.

42. *Bulletin des Sciences Technologiques* (1826), v, p. 360.

43. 'On the Motive Power of Steam', *Bull. des Sciences Tech.* (1826), vi, p. 42.

44. N. Clément and C. B. Desormes, 'Memoir on Several Important Points in the Theory of Heat', *Bulletin de la Société Philomatique* (1819), p. 103; and 'Memoir on the Theory of the Fire Engine', ibid. p. 115.

45. Thomas Thomson, 'On the influence of Humidity in Modifying the Specific Gravity of Gases', *Annals of Philosophy* (1822), iii, p. 302.

46. See, for example, the works published by Riche de Prony, Hachette, Borgnis Christian, Lanz and Bétancourt, Guenyveau, and Poncelet as well as those (on a less distinguished plane) by Banks, Emerson, Gregory and Tredgold. The tradition continued, so far as English-speaking engineers were concerned, up to the publication of Rankine's great treatise *On the Steam Engine and other Prime Movers* (London 1859).

47. Chevalier de Borda, 'On Water-wheels', *A. R. des S.* (1767), p. 270. For a discussion of Déparcieux's work, see D. S. L. Cardwell, *Technology and Culture* (1965), vi, p. 188.

48. A. R. Bouvier, 'Note on the Steam Engine', *A. de C. & P.* (1816), iii, p. 177.

49. Gengembre, *A. de C. & P.* (1817), iv, p. 190.

50. 'Report on a New Heat Engine', *A. de C. & P.* (1821), xviii, p. 133.

51. Davies Gilbert, 'On the Expediency of Assigning Specific Names (to) Physical Properties . . . and Some Observations on the Steam Engine', *P.T.R.S.* (1828), p. 25.

52. Peter Barlow and Charles Babbage, *Manufactures and Machinery of Great Britain* (London 1836), p. 154. Barlow and Babbage accepted Davies Gilbert's explanation without further comment.

53. M. I. Brunel, *Minutes of Proceedings of the Institution of Civil Engineers* (1849–50), p. 199.

54. Andrew Ure, 'New Experimental Researches on Some of the Leading Doctrines of Caloric . . .' *P.T.R.S.* (1818). Also in *Phil. Mag.*, liii, pp. 38, 87, 182.

55. *Phil. Mag.* (1822), lvii, p. 93. *Annals of Philosophy* (1822), iv, p. 12.

56. *Report from the select committee on steam boats . . . with minutes of evidence*, 24 June 1817 (House of Commons Papers, 1813–17).

57. See, for example, *A. de C. & P.* (1823), xxi, p. 306. There were several accounts of boiler explosions at this time.

58. Baron Dupin, *A. de C. & P.* (1823), xxiii, p. 320.

59. Girard had suggested that boilers should be tested up to 15 or 20 times their normal working pressures. *A. de C. & P.* (1823), xxi, p. 429.

60. Perkins' steam engine was extremely small, the stroke being only twelve inches, and it operated at the pressure of about 35 atmospheres with a steam temperature of about 200°C. It was, for its size, extremely powerful—developing about 10 h.p.—and was said to have been only one-fifth the size of a Watt engine of comparable power. It was also reported to be more economical. Dupin was interested in it, and an account of it may be found in *A. de C. & P.* (1823), xxii, p. 429.

61. John Farey, *A Treatise on the Steam Engine*, ii, pp. 209, 278. The second volume of Farey's monumental *Treatise* was never actually published. It exists only in the form of an uncorrected proof copy now at the Patents Office Library, London. This excellent work was devoted entirely to the high-pressure engine and is a mine of information on Woolf's machine.

62. Article 'Steam', *Encyclopaedia Britannica* (third edition). We leave aside the point that Robison's account is, in a serious sense, defective.

63. S. D. Poisson, *A. de C. & P.* (1823), xxiii, p. 337.

64. Thomas Lean and Brother, *Historical Statement*, p. 152. Woolf was, as we have remarked, well thought-of by Alexander Tilloch and by John Farey.

65. Prony and Girard, 'Report on a Memoir of M. Burdin entitled "Hydraulic Turbines, or High Speed Rotative Machines" ', *A. de C. & P.* (1824), xxvi, p. 207.

66. J. V. Poncelet, 'Memoir on Vertical Water-wheels with Curved Blades', *A. de C. & P.* (1825), xxx, p. 136.

67. J. A. Borgnis, *Théorie de la mécanique usuelle* (Paris 1821).

68. G. J. Christian, *Traité de la mécanique industrielle*, three volumes (Paris 1822–5).

69. Lanz and Bétancourt, *Essai sur la composition des machines*, second edition (Paris 1819).

70. A. M. Héron de Villefosse, *De la richesse minérale*, three volumes (Paris 1819).

71. Navier, Poncelet and Arago, 'Report on M. Juncker's Memoir on the Column-of-Water Engine at Huelgoat (Finistère)', *A. de C. & P.* (1835), lx, p. 202.

72. W. G. Armstrong was a leader in the application of the column-of-water principle. See D. S. L. Cardwell, op. cit. (47).

73. Navier, Poncelet and Arago, op. cit. (71).

74. J. V. Poncelet, *Mécanique industrielle* (Liége 1844), iii, pp. 167–8.

CHAPTER SEVEN

1. The diplomat's name was Spiker. See Trevor Turner, *A History of Fenton, Murray and Wood* (Manchester University M.Sc Thesis, 1965), p. 6.9.
2. J. Fourier, *The Analytical Theory of Heat*, pp. 14, 26.
3. J. Herivel, *Endeavour* (1968), xxvii, No. 101, p. 65.
4. See note (1), p. 304.
5. Quoted by W. Stanley Jevons, *The Coal Question*, second edition (London 1866), p. 125.
6. See S. J. Watson, *Carnot, 1753–1823* (The Bodley Head, London 1954), pp. 34–5. To the historian of science Carnot's competence in mathematics, physics and engineering is self-evident; his literary and artistic insights are perhaps less apparent. But, while serving on garrison duty in Artois in the late 1780s he had been a prominent member of a poetry society—anagrammatically the *Rosati* Society—and as a result of the taste for Persian poetry he acquired he named his son Sadi after a Persian poet of that name.
7. E. Mendoza, introduction to S. Carnot, *Reflection on the Motive Power of Fire*, p. xi. See op. cit. (9) *infra*.
8. ibid. p. xii.
9. S. Carnot, *Réflexions sur la puissance motrice du feu et sur les machines propres à développer cette puissance* (Paris 1824). The book has been through a number of French editions and was translated into English by R. H. Thurston in 1890. Recently Professor Mendoza has re-issued the Thurston translation together with Clapeyron's paper and Clausius' First Memoir of 1850 (translated by W. F. Magie). Professor Mendoza has added a useful introduction (Dover Books, New York 1960).
10. Joseph Larmor, 'On the Nature of Heat', *P.T.R.S.* (1917–18), p. 326.
11. C. F. Partington, *An Historical and Descriptive Account of the Steam Engine* (London 1822); Robert Stuart, *Historical and Descriptive Anecdotes of Steam Engines*, two volumes (London 1829); Dionysus Lardner, *The Steam Engine* (London 1830). Lardner's book was immensely popular and went through a number of editions.
12. Sadi Carnot, op. cit. (9), p. 28.
13. One very distinguished 'reader' who failed to appreciate the importance of this point was Professor J. S. Haldane, father of the equally distinguished Professor J. B. S. Haldane and brother of the politician Lord Haldane. See J. S. Haldane's article on the maximum efficiency of heat engines, *Transactions of the Institution of Mining Engineers* (1924–5), lxix. The 'Haldane cycle' would have consisted of adiabatic expansion followed by isothermal compression, completed by direct heating at constant volume

to restore the original pressure. The form of the indicator diagram would therefore have been that of a triangle and not a parallelogram.

14. D. S. L. Cardwell, *Technology and Culture* (1965), vi, p. 203.
15. ibid. p. 193.
16. We have used the word 'reversible' as synonymous with 'recoverable'. In this sense it seems particularly applicable to the hydraulic analogy and therefore to the material theory of heat. But it is also applicable to the modern, energetic theory of heat; and, indeed, 'recoverable' might be a more suitable word than 'reversible'. In asserting this I am merely following Bridgman's advice: 'It is not the reversibility of the process that is of primary importance; the importance of reversibility arises because when we have reversibility we have also *recoverability*. It is recoverability of the original situation that is important, not the detailed reversal of the steps which led to the original departure from the initial situation'—P. W. Bridgman, *The Nature of Thermodynamics* (Harvard University Press, Cambridge, Mass. 1941), p. 122.
17. Compressing the gas 'adiabatically' by 1/116th of its volume will, according to Poisson, raise its temperature by 1°C. If the gas is heated up 1°C at constant pressure its volume will, by Gay-

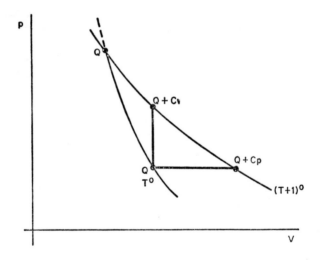

Lussac's law, increase by 1/267th. In both of these cases the temperature of the gas will be 1° higher but, by the axiom of conservation, the amount of heat present in the gas in the first case will be the same after the compression as it was before; the gas will be warmer only because the specific heat has been reduced.

It follows then that the difference in heat in the gas in the two cases will be the amount required to expand it 'isothermally' from $V(1 - 1/116)$ to $V(1 + 1/267)$; which is the same as the amount needed to raise it 1°C at constant pressure, or Cp. If we heat the gas directly, at constant volume, V, the amount of heat required to raise it 1° will be equal to that released by 'adiabatic' compression. The volume changes are relatively small and we may assume, says Carnot, that the heat absorbed in the two cases is simply proportional to the volume changes. In this case the heat absorbed during the change at constant pressure will be proportional to $(1/116 + 1/267)$ while that at constant volume will be proportional to $(1/116)$. The ratio of the two will therefore be $(1/116 + 1/267)/(1/116)$; or, $Cp/Cv = (267 + 116)/267$, which is equal to about 1·44. If we take Cp as equal to 1 we find that Cv is about 0·7, so that the difference between the two is 0·3, and this is the amount of heat required to account for the increase in *volume* when the gas is heated at constant pressure.

18. The proof is not rigorous as Carnot admits, for he does not know the law of variation (if any) of specific heat with temperature. After calling attention to the relationship between motive power and thermometric degree Carnot, in an analytical footnote, briefly explores the consequences that would follow the (hypothetical) variation of specific heat with temperature and, following the common assumption, with volume. If, conceivably, the specific heat did not vary with the volume the relationship would be very simple: the work done by a given amount of heat would be exactly proportional to the temperature difference.

19. S. Carnot, op. cit. (9), pp. 97–8.

20. The posthumous notes are included in the 1953 facsimile edition of the *Réflexions*, pp. 125–51. See also Mendoza, op. cit. (7), pp. 60–9.

The most important of the recently discovered manuscripts is presented and discussed by Drs Herivel and Gabbey, 'An Unpublished Manuscript by Sadi Carnot', *Revue d'Histoire des Sciences* (1966), xix, p. 151. The paper represents an attempt to calculate the power (expressed in Hachett's dynamic units) of an expansively operated steam-engine. It is undated but Dr Fox argues convincingly that it must have been written about 1826–7 and therefore after the publication of *Réflexions*; see below, Robert Fox (25). Carnot's manuscripts were burned after his death from cholera. This accounts for the paucity of documentation.

21. Among other factors are the frictional and percussive generation of heat (Rumford's arguments) and the relationship between light (undulatory) and radiant heat.

22. Sadi Carnot, op. cit. (9), p. 134.

23. ibid. p. 135. Professor Mendoza points out, op. cit. (7), that the modern value of Carnot's computation is 3·70 joules/calorie. The accepted figure today is 4·186 joules/calorie.

24. Nicholas Clément was a friend of Carnot. See Robert Fox, op. cit. (25), and B. S. Finn, *Isis* (1964), lv. Clément, who married Desormes' daughter, adopted his father-in-law's name and became known as Clément-Desormes.

25. J. Payen, 'A Source of Carnot's Ideas', *Archives Internationales d'Histoire des Sciences* (1968), p. 15. Robert Fox, 'Watt's Expansive Principle and its Place in the Work of Sadi Carnot', *Notes and Records of the Royal Society* (1969), xxiv.

26. (N. Clément), 'Industrial Applications of Heat', *Bulletin des Sciences Technologiques* (1826), v, p. 228.

27. W. T. Haycraft, 'On the Specific Heats of Gases', *Transactions of Royal Society of Edinburgh*; also *Phil. Mag.* (1823), lxiv, p. 200, and *A. de C. & P.* (1824), xxvi, p. 298.

28. A. de la Rive and P. Marcet, 'Researches on the Specific Heats of Gases', *A. de C. & P.* (1827), xxxv, p. 5. Also, ibid. (1829), xli, p. 78.

29. Prévost attempted to explain de la Rive and Marcet's results in terms of a caloric repulsion theory. See *A. de C. & P.* (1828), xxxix, p. 194 and (1828), xxxviii, p. 41. Prévost's theory was similar to the one Dalton put forward in 1802, but he concluded by questioning whether caloric was the unique cause of the elasticity of gases. It could, he thought, be explained in terms of the 'agitation' of molecules moved in diverse senses.

30. J. Joly, 'The Specific Heats of Gases at Constant Volume', *P.T.R.S.* (1891) and (1894). *Proceedings of the Royal Society* (1890), p. 440.

31. P. L. Dulong, 'Researches on the Specific Heats of Gases' (Acad. des Sciences), *A. de C. & P.* (1829), xli, p. 113.

32. A few of Dulong's results indicate the backing he had for his assertions:

Gas	Cp/Cv	Cv	Temp. rise for 1/267 comp.
Air	1·421	1·0	0·421°.
Oxygen	1·417	1·0	,,
Hydrogen	1·409	1·0	,,
CO_2	1·337	1·241.	0·337°

33. M. Melloni and L. Nobili, 'Researches into Calorific Phenomena by Means of the Thermo-Multiplier', *A. de C. & P.* (1831), xlviii, p. 198. M. Melloni, 'A New Property of Solar Heat', ibid. p. 385. M. Melloni, 'The Free Transmission of Radiant Heat by Different Solids and Liquids', ibid. (1833), liii, p. 5. M. Melloni, 'New Researches on the Instantaneous Transmission of Radiant

Heat by Different Solids and Liquids', ibid. (1833), lv, p. 337. M. Melloni, 'Notes on the Reflection of Radiant Heat', ibid. (1835), lx, p. 402.

34. J. D. Forbes, *Transactions of the R.S.E.* (1834), xiii, p. 153. See also *Reports of the first and second meetings of the British Association*. Forbes' papers on radiant heat were reprinted in the *Phil. Mag.*

35. M. Melloni, 'Memoir on the Polarisation of Heat', *A. de C. & P.* (1836), lxi and lxv, p. 5.

36. Sir E. Whittaker, *A History of the Theories of Aether and Electricity: the Classical Theories*, p. 122.

37. A. M. Ampère, *A. de C. & P.* (1835), lviii, p. 432.

38. J. von Wrede, 'On the Velocity of Propagation of Radiant Heat', Poggendorff's *Annalen* (1842), liii, pt. 4. See also *Phil. Mag.* (1842), xx, p. 379.

39. Thomas Tredgold, *The Steam Engine* (London 1827).

40. John Farey, *A Treatise on the Steam Engine*. As we have pointed out, a second volume of this work, devoted to the high-pressure engine, was prepared but never went beyond the proof stage.

41. Farey, op. cit. (40), p. 19. It is possible that Peter Ewart tried to resolve this problem when he envisaged that the atoms of steam were in motion. See his paper, 'Experiments on Compressed Elastic Fluids', *Phil. Mag.* (1829), v, p. 247.

42. Professor R. J. Forbes has remarked, in *A History of Technology* (Oxford University Press 1958), iv, p. 165, 'The idea of energy and the recognition of its laws form the real division between the old and the new engineering, and this will be clear from the perusal of early hand books on the steam engine by Farey and Tredgold.'

43. E. Clapeyron, 'On the Motive Power of Heat', *Journal de l'Ecole Polytechnique* (1834), xiv, p. 153. See also Richard Taylor's *Scientific Memoirs selected from the Transactions of Foreign Academies of Science and Learned Societies and from Foreign Journals* (London 1837), i, p. 347.

44. De Pambours' first work, *Treatise on the Steam Locomotive*, was published in 1835; his second, *The Theory of the Steam Engine*, appeared in an English edition in 1839.

45. He does however accept Watt's law and quotes Sharpe, surprisingly enough, in support of his criticisms of Southern. He also mentions Clément and Désormes' confirmation of Watt's law. See page 319, note 44.

46. C. Holtzmann, 'On the Heat and Elasticity of Gases and Vapours and on the Principles of the Theory of Steam Engines', from Richard Taylor's *Scientific Memoirs Selected . . .* (London 1846), iv, p. 189.

47. H. von Helmholtz, 'On the Conservation of Force a Physical

Memoir', from J. Tyndall and W. Francis, *Scientific Memoirs Selected* . . . (London 1853), p. 114.

48. F. Mohr, 'Views of the Nature of Heat', Liebig's *Annalen der Chimie* (1837), xxiv. This paper was translated by P. G. Tait and printed in *Phil. Mag.* (1876), ii, p. 110.

49. T. S. Kuhn, 'Energy Conservation as an Example of Simultaneous Discovery', in *Critical Problems in the History of Science*, edited by M. Claggett (Wisconsin University Press 1959), p. 321.

50. J. R. Mayer, 'Remarks on the Forces of Inorganic Nature', Liebig's *Annalen der Chimie* (1842).

51. Cf. G. G. Stokes' remarks, quoted by C. G. Knott in *The Life and Scientific Work of Peter Guthrie Tait* (Cambridge University Press 1911), p. 213. Carnot had observed that the only way to ensure that all the heat supplied to a 'working substance' is used to do work and not to change the state of the body is to take it through a closed cycle, ending up with the working substance in its original condition.

52. J. P. Joule later proposed that the incandescence of meteors was due to intense frictional heating on their entry into the atmosphere. It is probable that he did not know of Mayer's meteoritic hypothesis. See the *Manchester Courier* report of lecture, 12 May 1847. The meteoritic hypothesis later became popular with astronomers like Sir Norman Lockyer.

53. John Tyndall, 'The Copley Medallist of 1871', *Fragments of Science* (London 1872), i, p. 481.

54. J. P. Joule, 'On the Heat Evolved by Metallic Conductors of Electricity', *Phil. Mag.* (1841), xix, p. 260. Also in *The Scientific Papers of James Prescott Joule* (Dawson's, London 1963), p. 65.

55. J. P. Joule, *Phil. Mag.* (1843), xxiii, pp. 263, 347, 435. *Scientific Papers*, p. 123.

56. J. P. Joule, 'On the Rarefaction and Condensation of Air', *Phil. Mag.* (1845). *Scientific Papers*, p. 172.

57. Joule's paddle-wheel apparatus was very similar to that used by John Rennie's son, George Rennie, V.P.R.S., in his experiments to determine the viscosity of liquids. See *P.T.R.S.* (1831), p. 423.

58. J. P. Joule, 'On the Mechanical Equivalent of Heat', *P.T.R.S.* (1850), p. 61.

59. In the possession of the Department of the History of Science and Technology, University of Manchester Institute of Science and Technology. See also Professor Mendoza's paper in *Manchester Memoirs* (1962-3), cv, p. 15.

60. Silvanus P. Thompson, *Life of Lord Kelvin* (London 1910), i, p. 260.

61. Whewell wrote a number of treatises on mathematics, some of

them being specifically for the use of engineers. Among them were *An Elementary Treatise on Mechanics*, which went through seven editions by 1847, *First Principles of Mechanics*, and *Mechanics of Engineering*. Hachett's suggestion is discussed in *An Elementary Treatise on Mechanics* (London, second edition, 1824), Chapter 11.

62. J. P. Joule, op. cit. (58). *Scientific Papers*, p. 299.
63. In one of Joule's laboratory notebooks in the possession of the Department of the History of Science and Technology, University of Manchester Institute of Science and Technology.

CHAPTER EIGHT

1. S. P. Thompson, *Life of Lord Kelvin*, i, p. 113 ff.
2. W. Thomson (Kelvin), 'On an Absolute Temperature Scale', *Mathematical and Physical Papers* (hereafter referred to as *M.P.P.*) (Cambridge University Press 1882), i, p. 100. Also in *Phil. Mag.* (1848).
3. Kelvin, *M.P.P.* p. 113.
4. For an account of Victor Regnault's ideas see Robert Fox, *The Caloric Theory of Gases from Lavoisier to Regnault*.
5. Kelvin, *M.P.P.* p. 156.
6. R. J. E. Clausius, 'On the Moving Force of Heat and the Laws of Heat that may be Deduced Therefrom', Poggendorff's *Annalen* (1850), lxxix, pp. 368, 500. Also in *Phil. Mag.* (1851). This was the first of Clausius' great memoirs on the dynamical theory of heat. It and eight of the others that followed were collected into one volume and published as: *The Mechanical Theory of Heat with its Applications to the Steam Engine and to the Physical Properties of Bodies*, edited by T. Archer Hirst (London 1887). These memoirs have useful mathematical appendices which are all reproduced in Archer Hirst's little known volume. Hereafter references will be to *M.T.H.* as well as to Poggendorff's *Annalen*. The volume translated by W. R. Browne, *The Mechanical Theory of Heat* (London 1879) lacks the useful appendices.
7. Clausius, 'First Memoir', *M.T.H.* p. 20 ff.
8. Max Planck, *Treatise on Thermodynamics* (Dover Books, New York n.d.), p. 57, note.
9. See D. McKie and N. H. de V. Heathcote, *The Discovery of Specific and Latent Heats*, pp. 92–3.
10. Clausius, 'Seventh Memoir', *M.T.H.* p. 280 ff. Poggendorff's *Annalen* (1863), cxx, p. 426.
11. Kelvin, *M.P.P.* p. 174. Also in *Phil. Mag.* (1852).
12. W. J. M. Rankine, *Transactions of the Royal Society of Edinburgh* (1850), xx, p. 147. Also in *Phil. Mag.* (1854).

13. S. P. Thompson, op. cit. (1), p. 280.
14. Kelvin, *M.P.P.* p. 189.
15. Kelvin, *M.P.P.* pp. 222–3. Also in *T.R.S.E.*, xx.
16. Kelvin, *M.P.P.* p. 511. Also in *Phil. Mag.* (1852).
17. W. J. M. Rankine, 'On the Reconcentration of the Mechanical Energy of the Universe', *Phil. Mag.* (1852), p. 358.
18. Kelvin, *M.P.P.* p. 232. Also in *T.R.S.E.*, xxi.
19. Clausius, 'Fourth Memoir', *M.T.H.* p. 116 ff. Poggendorff's *Annalen* (1854), xciii, p. 481.
20. Clausius, 'Sixth Memoir'. 'On the Application of the Theorem of the Equivalence of Transformations to Interior Work', *M.T.H.* p. 215 ff. Poggendorff's *Annalen* (1862), cxvi, p. 73.
21. Clausius, 'Ninth Memoir', *M.T.H.* p. 327. Poggendorff's *Annalen* (1865), cxxv, p. 353.
22. ibid. pp. 357, 364, 365.
23. W. Napier Shaw, *Manual of Meteorology*, iii, p. 207.
24. ibid. p. 255 (Shaw's italics).
25. For an informative discussion of this fundamental topic see D. W. Theobald, *The Concept of Energy* (E. & F. N. Spon, London 1965), p. 89 ff.
26. J. P. Joule, 'Note on the History of the Dynamical Theory of Heat', *Phil. Mag.* (1862), p. 57.
27. It is instructive to compare S. P. Thompson's reticent mention of this dispute (op. cit. (1), Chapter Six) with the fuller and more convincing account by A. S. Eve and C. H. Creasey, *Life and work of John Tyndall* (Macmillans, London 1945), Chapter Nine.
28. P. G. Tait, *Sketch of Thermodynamics*, two editions (Edinburgh 1868 and 1877).
29. C. G. Knott, *Life and Scientific Work of Peter Guthrie Tait*, p. 213.
30. See his great treatise *On the Steam Engine and other Prime Movers*. The two statements of the second law of thermodynamics given on pages 306 and 307 are particularly obscure.
31. This is quite apparent from the *Manual of Meteorology*, op. cit. (23), but see in particular, iii, p. 207, where Shaw remarks: 'We would therefore ask the reader who wishes to explore the real sources of the ideas of the thermodynamics of the atmosphere which are contained in this book to keep by his side a copy of [Maxwell's] *Theory of Heat*.' See also i, p. 322.
32. Letter from J. C. Maxwell to P. G. Tait, 1 December 1873. Quoted by C. G. Knott, op. cit. (29), p. 115.
33. D. S. L. Cardwell, *The Organization of Science in England*, p. 175 ff.
34. The use of the word 'physics' in its modern connotation was unfamiliar even as late as the 1860s. This is proved by Walter Bagehot, whose *Physics and Politics* must mystify the unsuspecting modern reader when he finds that the 'physics' in question is

concerned with Darwin's theory of evolution by natural selection. The well-informed first editor of the *Economist* was, in fact, using the word 'physics' in the original Aristotelean sense of the *Physica*. See also p. 310, note 43.

35. D. S. L. Cardwell, *Manchester Memoirs* (1963–4), p. 108.

Index

Aether, 3, 4, 103, 106–7
Air-engines, 20, 152–3, 168, 207, 289
Amontons, G., 18–21, 25, 31, 37, 73–4, 77, 122, 132, 299
Ampère, A. M., 116, 139, 145, 178, 217–18, 228–9, 285, 314, 325
Anderson, J., 43
Andrade, E. N. da C., 106, 310
Arago, D. J. F., 113, 185, 217, 320
Archimedes, 293
Arkwright, R., 67, 71
Armstrong, W. G., 320
Ashton, T. S., 304
Avogadro, A., 102, 141, 148, 214, 218, 281 (n)

Babbage, C., 310, 319
Bacon, F., 1–2, 6–7, 27, 102, 189, 237
Bacon, R., 11
Baden Powell, Rev., 114, 217, 311, 312, 315
Bagehot, W., 328
Baillet, 131, 151, 184
Baines, E., 41
Barber, J., 152
Barker's mill, 77, 85
Barlow, P., 319
Barnes, T., 123
Barton, D. B., 307
Baudot, J. M., 211 (n)
Beckmann, J., 111
Becquerel, A. C., 101, 309
Belidor, B. F. de, 42, 183
Bérard, J. E., 112, 114, 217, 244, 285, 311. See also Delaroche, F.
Bergmann, T. O., 58
Bernoulli, D., 25, 79, 95, 100, 103, 147, 166, 300
Bernoulli, J., 24
Berthollet, C-L., 112, 129, 132, 135, 157 (n)
Berzelius, J. J., 101, 186, 309
Biot, J. B., 66, 93, 112–13, 131–2, 139, 157, 239, 311–14, 317
Birmingham, Lunar Society, 60, 303
Biswas, A. K., 297

Black, J., 31, 34, 35–55, 57, 59, 61, 65–6, 89, 92, 94, 99, 108, 110, 118, 131, 277, 301
Blagden, C., 157
Boas, M., 297. See also Hall, M. B.
Boerhaave, H., 26–36, 43, 57, 60, 93, 133, 300
Boltzmann, L., 280, 289
Borda, C. de, 70, 78, 173, 182, 185, 319
Borgnis, J. A., 183, 320
Bouguer, P., 91
Boulton, M., 49, 73, 306
Boulton, M., and Watt, J., 79–80, 82–3, 155, 162, 220, 306
Bouvier, A. R., 169, 173–4, 180, 200, 319
Boyle, R., 4–5, 27
Boyle, law of, 17–18, 23–5, 81, 91, 139, 148, 155, 161, 168, 170, 221–2, 318
Brahé, T., 9 (n), 116, 240–1
Bramah press, 157
Branca, G., 11, 73–4
Brewster, D., 44 (n), 301
Bridgewater Treatises, 99
Bridgman, P. W., 322
Brindley, J., 50
British Association, 233, 235
Brougham, Lord, 34, 301
Browne, W. R., 327
Brunel, I. K., 319
Brunel, M. I., 176
Buchanan, R., 61, 303
Burdin, C., 182
Burniston Brown, G., 312

Cagnard Latour engine, 152, 173, 305
caloric, 5, 61, 64–5, 83, 94–7, 101–6, 114, 120, 124, 126–7, 142, 163, 192, 193 et seq., 200, 203, 209, 211–13, 216, 226, 241, 245, 247, 263, 272, 309
Cannizzaro, S., 214
Cardwell, D. S. L., 199 (n), 298, 301–2, 305, 307, 309, 317, 322, 328–9

Carnot, Lazare, 70, 87, 152–3, 164–5, 182, 185, 191, 193, 212, 282, 321
Carnot, Sadi, 20, 52, 55, 181, 186 et seq., 231, 235, 240–7, 249–51, 254, 259–62, 269, 273, 276, 281, 284, 286, 289, 291–4, 306, 321–4, 326
Carnot, cycle, xiii, 194–201, 220 et seq., 260, 262, 270, 272, 281, 290
Carnot, function, 223–6, 239–40, 242, 250–1, 255–6, 260
Carnot, *Réflexions*, 191–212, 225, 323
Cartwright's engine, 177
Cauchy, A., 239
Cavendish, H., 32, 157 (n)
Cayley, G., 152, 197
Cecil, Rev. W., 177
Charles, J. A. C., 130, 153, 287
Charles' law. See Gay-Lussac, J.-L.
Christian, G. J., 104, 183, 212, 320
Clapeyron, E., 210, 220–7, 235, 239, 242, 244, 246, 250–1, 256 (n), 260, 293, 325
Clapeyron, equation, 223–4, 240. See also Clausius, R. J. E.
Clausius, R. J. E., 100, 128, 146, 201, 210, 231, 242 (n), 244, 246–54, 256, 259–75, 280, 283–5, 287–9, 292, 313, 327, 328
  and Clapeyron equation, 250 (n), 256, 260
Cleghorn, W., 39, 58
Clément, N., 211, 213, 324
  and Desormes, C. B., 135, 143–5, 167–9, 171–3, 180, 200, 211, 215, 319, 325
Cohen, I. B., 17, 299
Colding, L. A., 219
column-of-water engine, 85–8, 183–5, 208, 248
compound engine, 75
conduction, thermal, 36–7, 65–6, 98–100, 107–10, 117–18, 141, 200, 214, 241, 255, 267
conservation principle, 27, 35, 62–3, 65, 98, 102, 142, 196, 203–6, 209, 219–221, 226, 244–5, 247, 255, 272
convection, thermal, 8, 61, 108, 140
Copernicus, 286
Coriolis, G. G., 189, 282
Crawford, A., 55–7, 65, 108, 110, 127, 134 (n), 231, 302
Creasey, C. H., 328
Crompton, S., 190
Crosland, M. P., 311

Crump, W. B., 304
Cullen, W., 34, 46, 58, 300

Dallowe, T., 300
Dalton, J., 7, 9, 26, 61, 83, 99, 102, 122–131, 134 (n), 137–41, 146–7, 151, 160–1, 163–6, 171 (n), 186, 231–2, 236–7, 239 (n), 267, 269, 277, 302, 308, 313–14, 317, 324
Dancer, J. B., 236 (n)
Dannemora mine, 32
Darby, A., 7
Darwin, Charles, 99, 329
Darwin, Erasmus, 58, 60, 124, 131, 277, 303
Davy, H., 79, 106, 146–8, 177, 237, 284
De Caus, S., 11
De la Rive, A. and Marcet, P., 213–14, 324
Delaroche, F., 111, 140, 216, 244, 311
  and Bérard, J. E., 135–6, 143, 153, 167–8, 203, 213, 215, 223, 226, 314
Delius, C. T., 86, 300, 307
De Luc, J. A., 53, 55, 92–4, 277, 302, 308, 313
De Moura, 33
De Pambours, Comte, 224–5, 227, 247, 293, 317, 325
Déparcieux, Chev. A., 69, 173, 198, 319
Derham, W., 21
Desaguliers, J. T., 9, 22–3, 25, 27, 30–1, 35, 42, 129, 297, 300, 303
De Saussure, H. B., 92–4, 99, 166, 308
Descartes, R., 2–5, 23, 133, 187–8
Desormes, C. B., 324. See also Clément, N.
Despretz, C., 141, 314
Dickinson, H. W., 78 (n), 306–7, 316
Diesel, R., 151, 207, 261, 281, 289
disgregation, 269–72, 275–6
Dulong, P. L., 214–16, 223, 244, 281, 324
  and Petit, A. T., 137–9, 140–1, 214, 314
Dupin, C., 157, 178, 319–20
duty, 20, 33
Duvernois, 122
dynamode, 165–6, 213

Eddington, A., 2 (n)
Edwards, H., 157, 171, 210
Egerton, M. C., 304

Einstein, A., 291
energy, 5, 27, 67, 104, 108–9, 112, 167, 198, 219, 230–2, 236, 241, 248, 255, 259–60, 262–5, 282, 290
  conservation of, 209, 220, 235, 249, 273, 283
  dissipation of, 255, 257, 260, 263, 273, 287 (n)
Englefield, H., 111, 311
enthalpy, 122
entropy, xiii, 10, 260, 271–3, 278–9, 284, 289–90
equivalence values, 265–73
Evans, A. F., 316
Eve, A. S., 328
Ewart, P., 82–3, 123, 162–5, 189, 232, 237, 283, 293, 306–7, 317
Ewbank, T., 86, 306–7
expansive operation, 52, 78, 82, 87, 168, 170–4, 185, 196, 212

Fahrenheit, D. G., 21, 36, 299
Fairbairn, W., 182, 293
Faraday, M., 116, 124 (n), 293
Farey, J., 50 (n), 80, 175, 179, 219–20, 293, 305, 320, 325
Finn, B. S., 324
fire pistons, 151
Fitton, R. S., 304
Forbes, J. D., 31, 93, 114, 217, 285, 325
Forbes, R. J., 325
Foucault, L., 239
Fourier, J., xiv, 66, 110, 115–19, 140, 145, 150, 190, 231, 239, 273, 291, 311–12, 321
  Analytical theory of heat, 29
Fowey Consols engine, 159, 179, 180–1, 208, 243
Fox, R., xv, 127 (n), 131, 141, 165 (n), 211, 302–3, 312–14, 316, 323–4, 237
Freshfield, D., 307
Fresnel, A., 113–14, 141–2, 217, 314–15

Gabbey, A., 323
Gadolin, J., 40, 108
Galileo Galilei, xi, 2, 5, 7, 29, 102, 187, 199, 201, 251, 260
Galloway, R. L., 41, 302
gas-engine, 177
gases, expansion of, 137–9
Gay-Lussac, J.-L., 58 (n), 122, 128–37, 143, 146, 174, 178, 230, 234, 269, 276, 313–15

and Humboldt, A. von, law of, 214, 218
  and Welter, 144–5, 152, 203, 315
  law of, 19, 130–1, 139, 144, 147–8, 167, 170, 222, 236, 287, 316, 318, 322
Gengembre, 174, 319
Gensanne, 33
Gibbs, J. W., 212, 280, 287, 289
Giddy, D., See Gilbert, D.
Gilbert, D., 79–83, 148, 168, 173–5, 177–9, 210, 306, 318–19
Girard, 178–9, 182, 320
Goodeve, T. M., 41, 302
Gough, J., 124
Greenaway, F., 124, 313, 317
Grose, S., 159, 179
Guenyveau, A., 87, 307
Guericke, O. von, 12–13, 298
Guerlac, H., 23 (n)
Guyton de Morveau, 122, 126, 129, 313

Hachette, J. N. P., 104, 164–7, 169, 171, 180, 183, 185, 189, 213, 237, 280, 317, 327
Hadley, G., 8, 297
Haldane, J. S., 321
Haldat, Dr C.N.A., 101–2, 227, 309
Hall, A. R., 304
  and Hall, M. B., 299
Hall, Marshall, 121, 313
Halley, E., 8–10, 17, 19, 23, 25, 38, 59, 108, 166, 187, 277–8, 297–9
Hamilton, W. R., 217
Hammond, J. L. and B., 306
Harcourt, Rev. V., 41, 301
Harris, T. R., 316
Harvey's of Hayle, 307
Hautefeuille, Abbé, 13
Haycraft, W. T., 213, 324
heat,
  animal, 56–7
  latent, 37–40, 45, 51, 53–5, 92, 95, 132, 151, 161, 227–8, 240, 245, 251, 268, 278, 289
  radiant, 6, 30, 36, 60–1, 65, 94, 106–120, 217–18
  specific, xiii, 3, 28, 30–1, 36–7, 40, 57, 63–4, 110, 117, 124, 141, 144, 268
  specific, of gases, 56–7, 132–8, 143–5, 150, 161, 171, 203–4, 213–15, 226, 228, 230, 247, 267, 280

total, 52-4, 83, 122, 140, 164, 197, 221, 245-6
transfer, 90, 94, 96, 98, 119, 150
Helmholtz, H. von, xiv, 231, 235, 249, 289, 325
Henry, Thomas, 123, 232
Henry, William, 101-2, 123, 163, 232, 309
Herapath, J., 142, 145-9, 170, 236, 315, 316, 318
Herivel, J. W., 190, 312, 321, 323
Hermann, J., 24-5
Hero of Alexandria, 11, 73
Héron de Villefosse, A. M., 183, 320
Herschel, W., 110-12, 115, 216, 228, 311
Higgins, B., 57-8, 303
Hirn, G. A., 252, 254
Hirst, T. A., 313, 327
Hodgkinson, E., 232, 237
Höll, J., 86, 227
Holtzmann, C., 225-8, 230, 325
Hooke, R., xii
Hornblower, Jonathan, 78-9, 81-4, 155-6, 180, 207
Huelgoat mine, column-of-water engine, 184-5
Humboldt, A. von, 132, 228. See also Gay-Lussac, J.-L.
Hutton, J., 35, 99
Huxley, T. H., 99, 240, 283
Huygens, C., xii, 13, 70, 282
hydrologic cycle, 9-10, 187

indicator diagram, 53, 80-3, 195 (n), 220, 225, 280, 306
Ingen Housz, J., 66, 118, 303
Irvine, W., 55-7, 63-4, 96, 106, 108, 127, 133, 136, 141, 144, 302

Jars, G., 33, 86, 183, 300, 307
Jevons, W. S., 123, 316, 321
Joly, J., 214, 324
Joule, J. P., 40, 58 (n), 93, 100, 106, 123, 149, 163-4, 210, 216, 227, 229, 231-8, 241-55, 258-9, 262-3, 270, 273, 276, 280, 283, 285-6, 288, 291, 293-4, 326-8
Juncker, 184

Kelham, B. B., 303, 310
Kelland, P., 114, 219, 239 (n), 312
Kelvin, Lord, 31, 50, 210, 212, 223 (n), 235, 238-43, 249, 251, 254-60,

262-3, 265, 267, 273, 283-6, 288, 291, 312, 327, 328
Kempelen's engine, 77-8
Kepler, J., 116, 240, 242
Kier's engine, 305
Kirkham, N., 307
Kirwan, R., 131
Knott, C. G., 328
Krafft, G. W., 31, 300
Krönig, A., 100, 280
Kuhn, T. S., 210 (n), 305, 326

Lambert, J. H., 58, 90, 93, 133, 143, 227, 303, 307
Landes, D., 157, 317
Lanz and Bétancourt, 183, 320
Laplace, P. S., 17 (n), 59, 62, 113, 132, 142, 144-5, 169-70, 178, 315, 318.
See also Lavoisier, A. L.
Lardner, D., 191, 321
Larmor, J., 51, 191, 222 (n), 321
Lavoisier, A. L., 17 (n), 58-9, 62, 93, 95-7, 122, 124, 187, 291, 299
Traité élémentaire de chimie, 65
and Laplace, P. S., 62-5, 96, 101, 116 (n), 127, 136, 141, 209-10, 291 (n), 303
Law, R. J., 43, 80 (n)
Lawson, J., 306
Lean, J., 156-7, 159
Lean, T., 178 (n)
and brother, 180-1, 208, 317, 320
Lebon, Phillipe, 152, 197
Lee, George, 306
Lee, William, 298
Le Sage, 95
Leslie, J., 10, 31, 61, 93, 99, 102, 107-12, 115, 119, 132-3, 140, 151, 190, 308, 310, 313, 316
Leslie, P. D., 302
Liebig, J. von, 190, 228 (n)
Lister, M., 8, 297
Lloyd, H., 217
Locke, J., 237
Lockyer, N., 326
London, Royal Institution, 103, 146, 283
Lovell, D. J., 311

McKie, D. and Heathcote, N. H. de V., *The discovery of specific and latent heats*, 41-2, 251 (n), 300-1, 327
McNaught, W., 159
Macquer, P. J., 57, 303

Magellan, J. H., 55, 302
Malus, E., 112, 311
Manchester, Dissenting Academy, 123
Manchester, Literary and Philosophical
   Society, 82–3, 123, 232
Manley, G., 297
Manoury d' Ectot, 174–5
Marcet, 213–14
Mariotte, E., 4 (n), 6, 9, 166
Martine, G., 21, 30–1, 34–5, 122, 139,
   299
Mayer, F. T., 227, 304
Mayer, J. R., xiv, 40, 210, 226–7, 229–
   231, 233–4, 237–8, 249–50, 283–5,
   287–8, 293–4, 326
Maxwell, J. C., 31, 100, 116, 218 (n),
   276, 280, 286–7, 328
Meikle, H., 312
Melloni, M., 114, 216–17, 285
Mendoza, E., 191, 222 (n), 321, 323–4,
   326
meteorology, 6–10, 23, 25, 90, 119, 131,
   187, 212, 277
Middleton, W. E. K., 297, 299, 317
Mohr, F., 228–9, 245, 285, 326
Mollet, J., 151
Monge, G., 165 (n)
Montgolfier, J., 153, 176
Moricini, 112
Muirhead, J. P., 41, 302
Musschenbroek, P., 21, 30

naturphilosophie, xiv
N.S.U.-Wankel engine, 76
Navier, C. L. M. H., 167–9, 173, 180,
   183, 185, 200, 318, 320
Newcomen, T., 15, 266, 289
   engine, 15–17, 24–5, 32–3, 42, 49–51,
   54, 72–3, 80, 84–5, 155 (n), 157,
   175, 185 (n), 188–9, 208, 261, 305
Newcomen Society, 300
Newton, Isaac, xi, xii, 7, 17–18, 22–4,
   26–7, 35, 58, 100, 124 (n), 132,
   187–90, 237, 263, 273, 282–3, 290–
   291, 293, 299, 300, 303
   law of cooling, 18–19, 22, 117, 125–
   126, 140
   Opticks, 17, 62, 129
   Principia mathematica, 17, 23, 58, 129,
   282
Newtonianism, xii, xiv, 58, 70, 82, 114,
   146–8
Niepse cousins, 152
Nobili, L., 216, 310, 324

Oersted, H. C., 116
Ohm, G. S., 66
Ostwald, W., 249

Pacey, A. J., 25, 54, 89, 199, 300, 306,
   316
Paley, W., 99, 308
Papin, D., 13, 42
Parent, A., 68, 78, 85, 89, 168, 198,
   282
Paris, École Polytechnique, 119, 157,
   164, 166, 191, 220
Parliament, Committee on high-
   pressure steam-engines, 178
Partington, C. F., 191, 321
Partington, J. R., 53, 302
Pattu's engine, 206, 305
Payen, J., 211, 324
Penydarren railway, 154
Perkin, Jacob, 179, 219, 315, 320
Petit, A. T., 167–9, 173, 177, 200, 206,
   244, 306, 318. See also Dulong,
   P. L.
phlogiston theory, 291 (n)
Pictet, M.-A., 93–5, 103, 308
Placidus à Spescha, Father, 92
Planck, Max, 289, 291, 327
Playfair, J., 162, 282, 317
Pledge, H. T., 42
Poisson, S. D., 110, 113, 142, 145, 180,
   204, 208, 221, 244, 291, 315, 320,
   322
Pole, C., 178
Poncelet, J. V., 183, 185, 212, 320
Potter, I., 33 (n)
Prévost, P., 93, 95, 111
   theory of exchange, 95, 104, 115, 128,
   254, 308, 324
Prideaux, J., 170, 176, 318
Priestley, J., 32, 77, 122
Prony, R. de, 153, 174, 320
Prout, W., 8

quantum theory, ii, 290
Quetelet, A., 217

Raistrick, A., 297, 316
Ramsay, W., 41, 302
Rankine, W. J. M., 31, 50, 210, 244,
   254, 257, 259, 273, 282, 286–8,
   319, 327–8
Ravetz, J. R., 66, 301, 310
Reade, J., 121, 236, 313
Réaumur, R. A. F. de, 21 (n), 91

recoverability, 85, 89, 198, 322. See also reversibility
Regnault, V., 136, 239–40, 242, 244, 247, 260, 280, 327
Reichenbach, G. von, 184, 227
relativity, theory of, ii, 290
Rennie, G., 326
reversibility, xiii, 198, 227, 259, 322. See also recoverability
Reynolds, O., 11, 123, 186
Richmann, G. W., 31, 300
Ritter, J. W., 111
Robinson, E., 305
Robison, J., 34, 37, 41–2, 54–5, 81–3, 111, 129, 161, 168, 174, 179, 282, 301–2, 306, 309, 317
Roebuck, J., 49
Roscoe, H. E., 123
and Harden, A., *A new view of . . . Dalton's atomic theory*, 297
Rousseau, J. J., 91
Rowning, J., 9
Rumford, Graf von, (B. Thomson), 36, 93, 96–109, 113, 115, 118–19, 124, 151, 167, 189–90, 209, 218 (n), 231, 237–8, 252, 284, 291, 303, 308–10, 323
Rutherford, Lord, 123

Sadler's engine, 155 (n)
Savery, T., 13–15
engine, 33, 42, 75–6, 175
Scheele, C. W., 60–2, 65, 107, 111, 119, 303
Schemnitz, (Bánska Štiavnica), 32–3, 59, 67, 85–6, 131, 143, 151, 183
Bergwerksakademie, 300
Schofield, R., 303
Schorlemmer, C., 123
Schuster, A., 123
Scott, J. F., 3 (n)
Seebeck, T. J., 112, 216, 227, 312
Séguin, A., 101
Séguin, M., 229
s'Gravesande, W. J., 303
Sharpe, J., 160–1, 163, 171 (n), 174, 317, 325
Shaw, P., 300
Shaw, W. Napier, 212, 277–80, 287, 298, 328
Sims, W., 159
Singer, C., 42
Slatcher, W. N., 86 (n)
Small, W., 302

Smeaton, J., 32–3, 51, 69–70, 78, 86, 164–5, 175, 182, 198, 208, 282, 292, 305, 307
technique of evolutionary improvement, 33, 50
Smiles, S., 50 (n)
Smith, Adam, 35
Smith, R., 23 (n)
Somerville, Mary, 112
Sorocold, G., 185 (n)
Southern, J., 53, 79, 80, 82–3, 161–2, 268, 325
Stefan, J., 289
Stirling's engine, 176
Stokes, G. G., 285, 326
Strutt, W., 304
Stuart, R., 191, 321
Sturgeon, W., 232
Sylvester, C., 170, 318

Tait, P. G., 130, 276, 282–5, 286–9, 328
Talbot, G. R., 3, 25, 44, 127 (n), 297, 300–3, 310, 316
Taylor, Brook, 22–3, 31, 35, 299
temperature, absolute scales of, 126–7, 147, 226, 239–40, 243, 258–60, 268–9
absolute zero of, xiv, 27, 31, 39, 55–6, 64, 104, 108, 126–7, 140, 143, 147–148, 226, 228, 236, 276
Terzi, Father Lana, 12
textile industries, 67, 70–2
Theobald, D. W., xv
thermodynamics, first law of, 232, 235, 240, 259, 263
second law of, 40, 103, 247, 251 (n), 253, 255, 258–60, 290, 328
thermometers, 6
fixed points of, 6, 19–22
thermo-multiplier, 216
Thompson, S. P., 237, 244, 254, 326–8
Thomson, B. See Rumford, Graf von
Thomson, James, 243
Thomson, Joseph J., 51
Thomson, Thomas, 101–2, 171 (n), 309, 319
Thomson, William. See Kelvin, Lord
Thurston, R. H., 302
Tilloch, A., 156
Titley, A., 307, 316
Todd, A. C., 306
Towneley, R., 4, 91
Tregold, G., 219, 224, 293, 325

Treloar, L. R. G., 124 (n)
Trevithick, F., 105, 310
Trevithick, R., 67, 79, 83–4, 86, 89, 105, 154–6, 159, 180–1, 198–9, 289
Triewald, M., 32
turbine, steam, 11
Turner, T., 321
Tyndall, J., 93, 99, 231, 283, 286, 288

Unwin, W. C., 304
Ure, Andrew, 177, 319

Vale, E., 307
Vienna medical school, 27

Wadsworth, A. P., and Fitton, R. S., 304
    and Mann, Julia, 304
Wallace, A. R., 99
water-wheel, efficiency of, 68–70
Watson, S. J., 321
Watt, James, 23, 29, 31, 33–4, 40–55, 60, 72–3, 75–9, 82–3, 89, 122, 126, 156–7, 159, 161–2, 174, 177, 181, 183, 188, 196, 199, 254, 261, 268, 289, 292, 301, 303, 306
Watt's law, 52, 54, 83, 131, 161–2, 164, 168, 170–1, 174, 227, 247, 303, 312, 325
Weber, Max, 304
Wells, W. C., 115, 119, 312
West, W., 159
Wheal Abraham mine, 155, 159, 179
Wheal Towan mine, 169, 179
Whewell, W., 109, 237, 288, 293, 326–327
White, Lynn, 298–9
Whitehead, A. N., 41
Whittaker, E., 113, 312, 325
Wilcke, J. C., 40, 62, 65, 94, 97, 108, 110, 251, 301
Winds, trade, 7–10, 38, 118, 277
wire-drawing, 179
Wollaston, W. H., 104, 111, 162, 165
Woolf, A., 155–9, 170, 173, 175, 179–81, 207–8, 316, 320
Worcester, Marquis of, 11
Wrede, J. von, 218–19, 325

'X', 148, 316

Young, T., 103–4, 113–14, 213, 310

Ziegler, J. H., 151
Zöllner, J. C. F., 208